SPACE SHUTTLE ENVIRONMENT

Proceedings of the Engineering Foundation Conference,
Space Shuttle Experiment and Environment Workshop
held at New England College, Henniker, New Hampshire, U.S.A.
August 6-10, 1984

EDITORS
Thomas D. Wilkerson
Michael Lauriente
Gerald W. Sharp

SPONSORED BY:
Engineering Foundation

ORGANIZED BY:
National Aeronautics and Space Administration
European Space Agency

Additional financial support for the conduct of this conference was provided by the National Aeronautics and Space Administration.

Proceedings of this conference were published by the Engineering Foundation. Any findings, opinions, recommendations or conclusions contained herein are those of the individual authors and do not represent the opinion of nor imply endorsement by the sponsors, organizers, or financial contributors.

Cover: A fixed camera on Astronaut Bruce McCandless's helmet recorded this rare scene of the Space Shuttle Challenger some 50-60 meters away during a history-making extra-vehicular activity (EVA), February 7, 1984. Visible in the cargo bay are the support stations for the two MMU back-packs, the sunshields for the Palapa B and Westar VI satellites, Mu-band antenna and a number of getaway special (GAS) canisters. Photo courtesy of NASA.

Copyright 1985 United Engineering Trustees, Inc.
Library of Congress Catalog Card No.: 85-81606
ISBN 0-939204-28-2

ACKNOWLEDGEMENT

We are grateful to the participants whose enthusiastic discussions and contributions made this conference a success. In the completion of the <u>Proceedings</u>, we particularly thank Bonnie Bernstein for her tireless and thorough work on all aspects of the document preparation. Also we thank Debra Friedrick and Marilyn Spell for their contribution to the process of turning a wide variety of notes into the manuscript which is this book. The patience and close cooperation of the authors made a vital difference in our work.

Gerald Sharp is commended for doing a yeoman performance of planning and organizing the technical sessions. Violeta Vera arranged the pre-meeting details with the College, and additionally, in her capacity as senior staff of the Information Management Panel, attended to all of the details of tracking and systematically packaging the voluminous material presented so that it could be resurrected for insertion in the <u>Proceedings</u>. Bob Theis and John Cavanaugh arranged the logistics, installations, and demonstrations of the computers that were transported from NASA Goddard Space Flight Center.

Thanks are due the Engineering Foundation for its help in handling arrangements at New England College, Henniker, New Hampshire and financial support of some attendees. The Director, Harold Comerer, and Donna Romano at the New York office were especially understanding and cooperative in making it possible to organize this meeting within the constraints of an incredibly short advance schedule. Ted Von Wettberg, as the Engineering Foundation staff member for the meeting, was most cooperative in resolving the problems that cropped up at the most inconvenient times. Nate Promisel and Alan Grobecker, as senior technical consultants provided by the Engineering Foundation, were most helpful in defusing chaotic situations that were always threatening due to the pace of activities.

The Director of Programs for the College, Kim Berry, and the Conference Coordinator, David Malloy, were very patient with our unique requests. Sara Mowry insured success of the recreation sessions. Walter Kahn was appreciated for his support as audio and projection assistant. Last but not least were the excellent meals and service enjoyed at the meeting.

The support of the European Space Agency and the National Aeronautics and Space Agency made the conference possible.

The Editors

TABLE OF CONTENTS

Acknowledgement..iii
Table of Contents..v
Foreword..ix
Preface...xi
Executive Summary...xiii

CHAPTER 1: Plenary Sessions

Shuttle Role in the Space Program
    Michael Sander 3
Role of ESA in the Space Transportation System
    Dai Shapland 5
Shuttle and the Future
    Isaac T. Gillam 13
Spacelab Science in the Next Decade
    Robert Chapman 55
People in the Loop: Space Operations of the Future
    Byron Lichtenberg 65
STS Environment Measurements and Present Plans
    Robert Blount 69
Spacelab-Natural and Induced Environments Applicable to Experimenters
    Alan Thirkettle 77
The Impact of the Shuttle Environment on Spacelab-1
    Karl Knott 99
Online Scientific Data Bases and Data Banks Available From ESA-IRS
    George Proca 113
Overview of Environment Handbook
    Lyle Bareiss 127
1985 AIAA Meeting: Shuttle Environment and Operations II
    Billy McCormac 139

The Natural Environment
    Paul Robinson 143
User Concerns: Environmental Factors Affecting Scientific Users of
the Space Shuttle
    Laurence R. Young 155

CHAPTER 2: Subpanel Reports

Electromagnetic Interference
    William Cutler 161
Orbiter Motion
    Roger Chassay 165
The Particulate Environment
    Jack Barengoltz 171
Loads and Low Frequency Dynamics
    John Garba 175
Thermal and Humidity Environment
    James Clawson 187
Surface Interactions
    Henry Garrett 197
Vibrations and Acoustics
    Don Wong 215
The Gaseous Background
    Jack Barengoltz 223
Microbial and Toxic Contaminants
    Bonnie Dalton 227

CHAPTER 3: Future Directions

Loads and Low Frequency Dynamics
    John Garba 249
Vibrations and Acoustics
    Don Wong 253
Electromagnetic Interference
    William Cutler 255

Particulate Environment
    Jack Barengoltz 257
Molecular Contamination Environmment
    Jack Barengoltz 259
Microbial and Toxic Contaminants
    Bonnie Dalton 261
Summary Comments by INEP Chairman
    Gerald Sharp...263
Summary of ESA Reactions to thé Workshop
    Dai Shapland...265
Closing Remarks by General Chairman
    Michael Lauriente..271

APPENDICES

Appendix A--Surface Interactions...................................A-1
Appendix B--Electromagnetic Interference...........................B-1
Appendix C--Thermal and Humidity Environment.......................C-1
Appendix D--Orbiter Motion...D-1
Appendix E--Abbreviations and Acronyms.............................E-1
Appendix F--Conversion Units.......................................F-1
Appendix G--The Shuttle Environment Data Base (ENVIRONET)..........G-1
Appendix H--List of Attendees......................................H-1
Appendix I--Index..I-1

FOREWORD

Over the past decade, a wealth of knowledge has been reaped in the design, construction, and operational usage of instrumentation for remote sensing from space. Prior to this era, the activity of astronomers and space physicists was simpler, consisting essentially of mathematical predictions alternating with observations through optical and radio telescopes and cosmic ray detectors.

With man's step into space, the field of astrophysics has exploded into new vistas of knowledge by exploiting the entire electromagnetic spectrum. This growth of opportunities for research and information has been accomplished at the expense of the simpler and more tranquil environment that formerly existed for the astronomer. The present instrumentation and observational programs are conducted by teams of scientists and engineers organized to develop the ideas fundamental to the investigator's concepts, and programmed to meet the exigencies of fixed launch dates.

One of the biggest problems facing this multidisciplinary operation is communication in all forms. It is in the best interests of all concerned to have available a centralized depository for information with an access capability, and a system that can integrate recent flight-verified information as it becomes available.

The design of spaceborne astrophysical observing systems is going through a metamorphosis. Astronomers are learning the new principles of instrument design from the ground up. There is an ongoing effort to characterize and improve the operating environment which is quite complex.

Contamination has often been discovered to be the source of problems in the development and operation of space instruments. The risk of this is great when the effects of the operational environment are discounted or not even considered. Permanent, large space facilities such as the Space Station will require a more persistent consideration of this issue on contamination hazards.

The response to the 1982 Workshop on the Shuttle Environment verified the concern held by experimenters, users, and other individuals who needed to know what operational environments may be expected for payloads. The results of the 1982 Workshop revealed several areas of concern that were considered in the planning of this present meeting. The technical subpanels were based on these same areas of concern and others subsequently disclosed. Outlines were prepared ahead of time to help the subpanel leaders.

The conference brought together more than 120 people and included scientists and engineers from Europe and the United States. A unique opportunity to communicate was provided by the environment, free from the distractions of an urban area. Dr. Byron Lichtenberg, Payload Specialist on Spacelab I, was on hand to personally describe his experience with over 70 different experiments from five scientific disciplines.

We believe that this meeting has made significant progress to meeting the objectives needed to improve performance on future missions, including the long duration missions.

These objectives are:

a) Characterization of the operational environment.

b) A data base that will be on-line, easily updated, and available nationwide.

c) A medium for transferring knowledge from mission to mission.

General Chairman
Michael Lauriente

PREFACE

The Space Shuttle Experiment and Environment Workshop held at New England College, Henniker, New Hampshire, from August 6-10, 1984, provided an opportunity for the space, academic, government, and industry communities to carry on in-depth discussions and exchange of information. The conference was sponsored by the Engineering Foundation with the National Aeronautics and Space Administration (NASA) and the European Space Agency (ESA) as co-sponsors.

An evaluation was made of the concerns expressed at the NASA Shuttle Environment Workshop of October 1982, and of continuing and new concerns since that time. Formal presentations were made at the plenary sessions during the first day to set a base line for the working sessions to follow.

Ad hoc panels were organized which met for half a day to give the discussion leaders an opportunity to present discussion outlines on their respective disciplines. Numerous informal discussions filled the afternoons. In the evening there was a return to plenary sessions where the group was filled in on the progess made by all the sub-panels. The final day was devoted to discussing needs for the future and information needed for the planned data base.

The Workshop results will provide the initial formulation of a new Shuttle Environment Data Base (ENVIRONET), an on-line system that will be made available in the future to users of the Space Shuttle. A description of ENVIRONET is given in Appendix G.

Most participants attended all of the presentations, resulting in lively discussions and a totally successful meeting.

The organizing committee for the Workshop was composed of M. Lauriente, M. Dubin, R. Chapman, R. Blount, R. Rhodes, R. Kruger, E.

Miller, E. J. Marian, R. Theis, L. Young, G. Sharp, D. Shapland, and K. Knott. H. A. Comerer of the Engineering Foundation was responsible for all the administrative details.

                                                    General Chairman
                                                    Michael Lauriente

EXECUTIVE SUMMARY

Michael Lauriente
Goddard Space Flight Center
National Aeronautics and Space Administration

INTRODUCTION

After many weeks of planning and organizing, a Workshop was held August 5-10, 1984 in the bucolic atmosphere of New England to provide a forum for in-depth discussions on issues and problems encountered with the Space Shuttle. More than 125 engineers and scientists attended to this event, which was sponsored by the Engineering Foundation, the National Aeronautics and Space Administration (NASA) and the European Space Agency (ESA). Dr. Dai Shapland of ESA shared the Chairman duties with Dr. Michael Lauriente of Goddard Space Flight Center (GSFC), and Dr. Robert Chapman of GSFC was the Technical Chairman.

The objectives of the Workshop were to carry on in-depth discussions of the types of information needed, and issues and problems indigenous to the Shuttle, as perceived by users and those who have actually measured the environments. This meeting provided a forum for evaluating information from previous flights, and for reviewing the requirements and concerns of the scientific community including future experimenters. An evaluation was made of the concerns voiced at the NASA Shuttle Environment Workshop of October 1982 and of the continuing and new concerns.

The participants were involved in the initial phases of an activity that will be established as a focal point for an information/database pertinent to the environment surrounding the Shuttle and on the Shuttle itself. This will be a modern operating procedure using computer technology to input and process the information, and to output the product on remote terminals. A copy of the meeting agenda is included in the summary. As the Executive Summary of the Proceedings of the Workshop, this presentation summarizes the highlights of the meeting.

BACKGROUND

Some of you may recall the 1982 Workshop held October 5-7 at the Ramada Inn in Calverton, Maryland. The meeting was organized by Jules Lehmann of the NASA Office of Space Science and Applications. There was some trepidation about the turnout for this meeting, but the interest proved to be overwhelming with over 350 participants.

The 1982 Workshop was organized to present data, collected during flights STS-1 through STS-4, relating to the definition of the Shuttle environment. Although environment was generally considered to include all aspects, i.e., vibroacoustics, loads, thermal, electromagnetic and contamination in the form of light, emission, particles and gases, only a limited number of these could be covered.

The primary emphasis was placed on payload data that included results from the Induced Environment Contamination Monitor (IECM). The "vehicle glow" and surface charging of payload bay material were identified as candidates for the surprise element of Shuttle flight experience. Following these data presentations, three panels were convened with users interested in plasma measurements, infrared measurements, or ultraviolet measurements. The resultant output from these panels was a set of "User Concerns" about the Shuttle environment which also recommended certain actions to be taken to mitigate these concerns.

WORKSHOP ORGANIZATION

For the 1984 Workshop, a Working Group and three panels, Users, Environment, and Information Management were established beforehand to represent the constituents of the Shuttle environment organization, as shown in Fig. 1. In keeping with the philosophy of the meeting, the

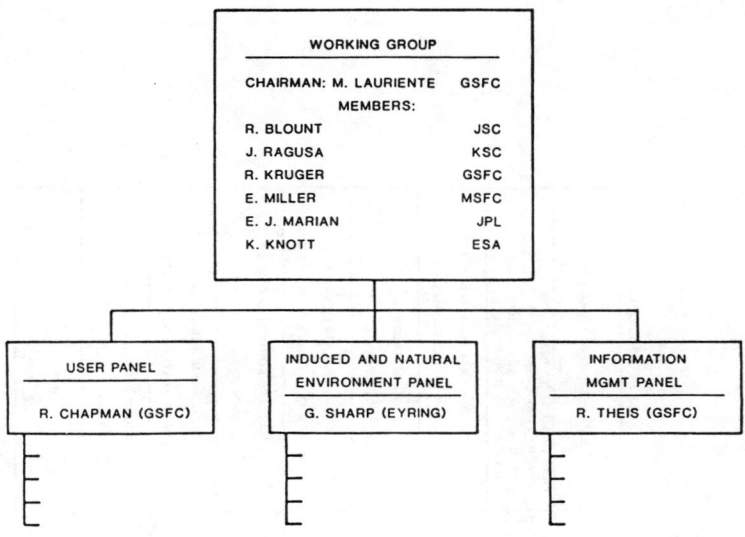

Fig. 1. Shuttle Environment Organization

agenda contained a minimum of formal papers, which in the main were presented at the plenary sessions during the first day to help set the stage for the working sessions to follow.

In contrast to the earlier conference, all aspects of the environment were considered. Table 1 shows the ten areas of concern identified in 1982 and how they relate to the technical subpanels organized for the 1984 Workshop. The third column lists the recommendations made in 1982. It is noted that the User Panel was concerned with the Shuttle management interface. The Natural and Induced panel was concerned with only the science and engineering disciplines.

A chairman was selected for each of these panels, as shown in Fig. 1, and appropriate subpanels were identified and invited to participate in the Workshop. At the same time members were selected

TABLE 1

AREAS OF CONCERN

1984 SUBPANELS	1982 AREAS OF CONCERN	1982 RECOMENDATIONS
LOADS & LOW FREQUENCY VIBRATION & ACOUSTICS ORBITAL MOTION	• VEHICLE INFLUENCE BY OPERATORS	DATA MORE EASILY AVAILABLE
EXPERIMENTER/USER	• USER/OPERATORS INTERFACE: MISMATCH OF ENVIRONMENT & EXPERIMENT REQUIREMENTS • OPERATIONAL MONITORING: FLIGHT COMPARISONS NEEDED FOR PLANNING	• MANAGEMENT TO REEXAMINE & IMPROVE USER INTERFACES • DEVELOP STANDARD MONITORING MODULE WITH OTHER USERS
NATURAL ENVIRONMENT	• ENVIRONMENTAL QUALITY: FEEDBACK TO FUTURE OPS	REVIEW PROCEDURES BASED ON MEASUREMENTS
THERMAL & HUMIDITY	• DAMAGE TO INSTRUMENTS COMPROMISE DATA	MORE THERMAL MEASUREMENT/OPTIONS
ELECTROMAGNETIC INTERFERENCE	• INDUCED FIELDS: UNCERTAIN VEHICLE & μWAVE TRANS. EFFECTS	REVIEW EMI TEST PLAN & INCLUDE ALL FREQ. & EXPER. COND.
PARTICULATE ENVIRONMENT	• DAMAGE TO OPTICAL SURFACES	ELIMINATE SOURCE, CLEAN-UP GROUND ENVIRONMENT
GASEOUS CONTAMINATION	• ROLE OF PAYLOAD, THRUSTER, & ATMOS.	DETERMINE PARAMETERS UNDER VARYING CONDITIONS
SURFACE INTERACTIONS	• GLOW: OPTICAL CONTAMINATION • EROSION OF MATLS. COMPONENT DAMAGE	STUDY GLOW & COORDINATE WITH OTHER AGENCIES
MICROBIOLOGY & TOXICITY	NOT CONSIDERED	

for the User, and the Information Management Panels, and they too were also invited to participate in the Workshop. Table 2 shows the fields of interest covered by the respective panels.

Table 2
Panel's Field of Interest

USER PANEL	INDUCED AND NATURAL ENVIRONMENTAL PANEL	INFORMATION MANAGEMENT PANEL
SUB-PANELS:	FLIGHT DYNAMICS	SUB-PANELS:
UV/OPTICAL	VIBRATION AND ACOUSTICS	PROTOCOL (RULES FOR DATA)
HIGH ENERGY	ORBITER MOTION	
ATMOSPHERIC SCIENCE	EMI	CONTENT
IR/RADIO/RADAR	THERMAL AND HUMIDITY	INFORMATION RETRIEVAL
LIFE SCIENCES	GASEOUS BACKGROUND	DOCUMENTATION
PLASMA PHYSICS	PARTICULATE ENVIRONMENT	SOFTWARE
MATERIALS SCIENCE	SURFACE INTERACTIONS	INTERNATIONAL EXCHANGE
	NATURAL ENVIRONMENT	SCIENCE LIAISON
	BIOLOGICAL	

The agenda of the formal meetings, held during the four and a half days of the Workshop, is given in at the end of this summary. Each of the subpanels was given one formal meeting time (a half day) during which they would organize what was known about the Shuttle environment in their area and present or discuss it. These formal meetings were supplemented with a number of ad hoc meetings of the subpanels, generally taking place in the afternoons of each day. Members of the Users Panel and of the Information and Management Panel were encouraged to attend these formal and ad hoc meetings to help determine if the subpanels were indeed collecting data helpful to their discipline interests. In the evening the Workshop returned to a plenary session where the subpanel leaders were given an opportunity to summarize for the group what they had accomplished in their formal

TABLE 3

SUB-PANEL SUMMARIES AND CONCLUSIONS

SUB-PANELS	DATA SUMMARY	GAPS/NEEDED
LOADS AND LOW FREQUENCIES	• DFI DATA FROM STS-1 THROUGH STS-5 FOR LIGHT-WEIGHT CARGOES • GOOD DESIGN DATA IN SPACELAB PAYLOAD ACCOMMODATION HANDBOOK	• MORE 0-50Hz DATA NEEDED • DESIGN DATA FOR ALL STS CARRIERS NEEDED • MORE DATA FOR STATISTICAL SAMPLE NEEDED
VIBRATIONS AND ACOUSTICS	• LIMITED MEASUREMENTS AVAILABLE IN P/L BAY • LIFTOFF ACOUSTICS GRADIENTS VARY FROM PREDICTIONS • VIBROACOUSTIC PAYLOAD ENVIRONMENT PREDICTION SYSTEM DEVELOPED AND AVAILABLE FOR USE	• MEASUREMENTS NEEDED IN FORWARD 1/3 OF P/L BAY • DATA NEEDED TO DEFINE TRANSONIC NOISE • MEASUREMENTS NEEDED AT P/L ATTACH AND RESPONSE POINTS
THERMAL AND HUMIDITY	• GOOD DATA AVAILABLE FOR KSC P/L FACILITIES AND PRELAUNCH ENVIRONS • ON-ORBIT P/L BAY TEMPERATURES AVAILABLE FOR LOW BETA ANGLES • TWO GOOD THERMAL MODELS AVAILABLE – NOT FOR NOVICES	• THERMAL DATA NEEDED FOR ALL STS CARRIERS • SAFEGUARDS ON DATABASE NEEDED TO PROTECT FROM MISUSE • SIMPLIFIED "EQUIVALENT SINK" PROGRAM IS NEEDED
ELECTROMAGNETIC INTERFERENCE	• MOST EMI DATA AVAILABLE TAKEN IN SHUTTLE AVIONICS INTEGRATION LAB • MEASUREMENTS LIMITED TO S- AND Ku- BAND RADIATED LEVELS • IN-FLIGHT MEASUREMENTS TAKEN ON STS-3 WITH PLASMA DIAGNOSTICS PACKAGE	• NEED DATA BETWEEN 1Hz AND 14Hz • NEED DATA ABOVE 20Hz • NEED S-BAND RADIATED LEVELS OUTSIDE P/L BAY
ORBITER MOTION	• LOW-G DATA TAKEN ON A LIMITED NUMBER OF STS FLIGHTS • VERY LIMITED ANALYSIS OF ACCELEROMETER DATA, SO LITTLE CORRELATION WITH POTENTIAL SOURCES AVAILABLE	• NEED ANALYSIS OF EXISTING DATA • NEED WIDE BAND (UP TO 50Hz) DATA • NEED TO MAP TYPICAL ORBITER ACCELERATIONS
PARTICULATE ENVIRONMENT	• ELEMENTAL COMPOSITION OF PARTICLES COLLECTED IN LAUNCH PHASE AVAILABLE • IECM DATA TAKEN ON STS-9 AND SOME EARLIER FLIGHTS • MODELS DEVELOPED BUT UNTESTED FOR SMALL AND VERY FAST PARTICLES	• NEED CARGO BAY AND P/L SURFACE MEASUREMENTS • NEED TIME AND MISSION PHASE DEPENDENCIES IN DATA • NEED IMPROVED MODELS TESTED AGAINST REAL DATA
MOLECULAR CONTAMINATIONS	• NON-VOLATILE RESIDUE MEASUREMENTS AT KSC SHOW REQUIREMENTS ARE MET • IECM MEASUREMENTS ON STS-9 SHOW WATER COLUMN DENSITIES OF 1 TO 2x10(12) per SQ. CM. • THRUSTER FIRINGS CAUSED PRESSURE PULSES WHICH DECAYED WITHIN A FEW SECONDS OF CUTOFF	• TIME HISTORY OF MOLECULAR CONTAMINANTS NEEDED TO HELP PLAN PROTECTIVE SHUTTLE OPS. • DEVELOPMENT OF MODELS WHICH REALISTICALLY DESCRIBE STS WITH P/L • MORE DATA NEEDED IN P/L BAY
SURFACE INTERACTIONS	• THE STS GENERATES TURBULENCE AND HEATS ELECTRONS IN ITS VICINITY • GLOW HAS BEEN OBSERVED IN THE VISIBLE AND IN THE RED WAVELENGTHS • GLOW IS ASSOCIATED WITH STS RAM EFFECTS AND THRUSTER FIRINGS, ITS BRIGHTNESS DEPENDS UPON SURFACE MATERIAL, IT EXTENDS UP TO 15 CM. FROM THE STS SURFACE • SURFACE EROSION HAS BEEN OBSERVED TO DEPEND UPON THE SURFACE AND THE OXYGEN CONCENTRATION • SURFACE EROSION DATA EXISTS FOR A LARGE NUMBER OF SAMPLES	• MEASUREMENTS ARE NEEDED TO CHARACTERIZE GLOW FULLY • STS CHARGING MEASUREMENTS ARE NEEDED
MICROBIAL AND TOXIC CONTAMINANTS	• TOXIC SUBSTANCES HAVE BEEN MEASURED IN THE ORBITER CABIN ON EVERY FLIGHT TO DATE • MICROBIAL MEASUREMENTS ARE MADE DURING EXPERIMENT PREPARATION AND INTEGRATION AS WELL AS DURING AND AFTER THE ACTUAL FLIGHT	• HUMAN TO HUMAN AND ANIMAL TO HUMAN PROTECTION IS NEEDED • DATA FOR NEW CONFIGURATIONS AND EXPERIMENTS IS NEEDED

sessions that day (Table 3). Other talks of general interest to the Workshop attendees were also given in these evening sessions.

The final day of the Workshop was devoted largely to looking to the future and assessing the measurements yet needed that would make the planned data base more useful to potential Shuttle experimenters.

WORKSHOP HIGHLIGHTS AND RESULTS

PLENARY SESSIONS

The purpose of the Plenary Sessions was to provide background and set the stage for the actual work of the group assembled. In addition, the international character of the workshop was demonstrated by presentations from participants representing the European Space Agency (ESA).

Michael Sander, Director of the Shuttle Payload Engineering Division at NASA Headquarters, spoke on the "Shuttle Role in the Space Program". He described in some detail the various ways in which the Space Shuttle is being and will be used by experimenters. As a sponsor of the Workshop, Mr. Sander laid a charge and responsibility upon the Workshop participants to produce an ongoing and detailed data base of Shuttle environment data which can be made available to the widest possible set of potential Shuttle users. He pointed out that if this is done it will be possible to reduce the costs of performing experiments in and on the Shuttle.

Dr. Dai Shapland, representing the ESA management, spoke on the "Role of ESA in the Space Transportation System". He described the European Space Agency and its member state participation in the development of the Spacelab, a prime Shuttle payload carrier. ESA and individual European countries have active plans and programs for use of the Shuttle. He showed how Spacelab elements and EURECA, ESA's Shuttle launched and retrieved payload carrier, could be used in the development of the U.S. Space Station. He suggested that the data base being planned by this Workshop could someday contribute to a data base needed for Space Station users.

Isaac T. Gillam, IV, Assistant Associate Administrator, Office of Space Flight, NASA Headquarters, talked on "Shuttle and the Future". He described the accomplishments of the Shuttle to date along with the short and long range goals for its use and operation. The Shuttle carriers and upper stages that will become available for the use of experimenters in the near future were discussed. He stressed that emphasis to date has been on getting the Shuttle operational but that future STS efforts will be to operate the Shuttle so that it will provide the support that experimenters and commercial payloads need for their success.

In "Spacelab Science in the Next Decade," Dr. Robert Chapman stated that the current trend in NASA's Office of Space Science and Applications is toward the use of "discipline" Spacelab laboratories which will reduce the costs of integration and optimize the use of Shuttle resources and services (pointing, etc.). Professor Young identified system inflexibility and excessive lead time in the planning and selection process as two of the greatest concerns of experimenters today. Inadequacy of exact knowledge of the environment in which experiments are expected to operate is another big problem to experimenters.

Dr. Byron Lichtenberg narrated a film taken during the launch and orbital operations of the Spacelab 1 Mission in which he pointed out the various ways a "Person in the Loop", in orbit, can be helpful to experimenters on the ground.

Robert Blount stated that the STS program released, in early 1984, new thermal models which can predict temperatures in the payload bay to within 5 to 10 degrees of those measured. He also indicated that the STS plans to install a few accelerometers on some existing and future orbiters to help better characterize the Shuttle vibroacoustic environment.

Alan Thirkettle described the contents of the Spacelab Payload Accommodations Handbook, particularly as it relates to the environment that experiments would experience in flying on or in the Spacelab, and also discussed the fact that the Verification Flight Instrumentation flown on Spacelab 1 should, when analyzed, provide a great deal of information on the Spacelab environment for payloads.

Dr. Karl Knott reported that Spacelab 1 experiment data showed that Spacelab was a good platform for research in atmospheric physics, plasma physics, solar physics, earth observations, and life sciences. It is yet to be proven that the Spacelab is a good platform for either ultraviolet or infrared astronomy. The Spacelab microgravity environment may not be satisfactory for material science research.

NATURAL AND INDUCED ENVIRONMENT PANEL

The conclusions of the subpanels of the Natural and Induced Environment Panel are given in Table 3. "Vehicle Glow" is still a very lively topic of interest to experimenters. Although still not understood, it has been found that glow activity is dependent upon the spacecraft surface material. It can be up to 15 cm thick, its intensity does not follow the N_2 concentration with altitude, and the orbiter thruster firings can cause glow to appear. On Spacelab 1 the glow spectrum included the N_2 1PG bands. It is now thought that the vehicle glow phenomenon is more a complex chemical (rather than a single) plasma reaction, because it is difficult to relate present observations with any single, reasonable mechanism. Clearly more measurements and computations are needed.

A detailed report was presented on the efforts at the Kennedy Space Center to clean up the payload bay environment during orbiter and payload processing. A considerable improvement was shown; it now appears that payload bay particulate contamination may be more dependent upon the payloads themselves, rather than the orbiter processing activity. Also reported was the need for a detailed bookkeeping of the particles present on orbit, with particular emphasis upon very large particles and the very small, fast ones. It was recognized that the Infrared Telescope, planned to be flown on Spacelab 2, will add considerable knowledge on the size distribution, number, and spatial distribution of the small particles.

There was a concern that the attrition of certain surfaces in the Shuttle environment needed to be studied in relation to: thermal control (both for vehicle and payload), contributions to gaseous species, sources for observed particulates, effects on optical reflectors and coatings, and the possibility of material substitutions in both vehicle and payload. Surface changes are seen only in the ram facing surfaces, and these changes are apparently due to atomic oxygen interacting with certain organic materials. It was also shown that a rather substantial list of organic and metallic materials have been tested for material loss, and the results were made available.

Of great concern was the definition of the gaseous environment of the Orbiter, including thruster firings, vehicle outgassing/venting, payload outgassing/venting, atmospheric "ram" heating , cabin pressure leaks, and various chemical interactions which might take place. It was learned that the Space Transportation System now maintains a rigorous materials control program, and they feel there is little more they can do for this problem. The STS-4 measurements showed that water vapor was the largest observed gaseous contaminant and its concentration could be correlated with Orbiter surface temperature, except during thruster firings and water dumps. Helium was also found to be a major contaminant. During thruster firings, the gaseous pressure rose from 1 to 2×10^{-6} Torr; the major products were molecules of nitrogen, water and hydrogen. On STS-9 a water vapor column density of 1 to 2×10^{12} cm^{-2} was measured.

The STS program office has responded in part to a request for a standardized monitoring module by including limited monitoring on all flights, depending on the practicality of installing the appropriate instrumentation. Ten accelerometers have been installed on OV-99, and a similar installation is planned for OV-103.

The concern related to the measurement and control of induced electric fields calls for (1) better understanding of the relationship of the field to vehicle operations, and (2) assurance that possible EMI in the payload bay during Ku band operations will be mitigated. It is now known that the induced electric field measurements in the payload bay match closely to the model used to specify electric fields for payloads. It is planned to make S-band and Ku-band radiated field measurements outside the payload bay on Spacelab 2. In addition, the STS has developed software to provide a "mask" to protect payloads from Ku-band radiation. The Ku-band transmission will be deenergized whenever the antenna gets too close to the payload.

The final concern is with temperature measurements relating to payloads in the Orbiter. Although existing thermal models seemed to adequately describe the temperature in the bay, it was felt that better documentation is needed. It was reported that the STS has released (Spring of 1984) new thermal models which make possible more accurate on-orbit prediction of the temperatures that payloads will experience. With these models it has been possible to correlate, to within 5 to 10 degrees, predicted and measured temperatures. The STS does not plan to make further measurements of payload bay temperatures.

EXPERIMENTERS' PANEL

The Experimenters Panel raised the issue of the difficulty of an experimenter receiving help before being manifested on a flight, but there was some relief over the prospect of a database such as has been proposed. There was also a concern raised over standards for paper processing, decision processing, and Center policies. The last issue was related to post-flight data. Future plans proposed were:
- o Complete recruitment of representatives for all disciplines
- o Review inputs to the data base
- o Define requirements of data base
- o Compile user directory
- o Support new measurements and data

INFORMATION MANAGEMENT PANEL

The Information Management Panel met with each of the Panels to describe the operation of the newly conceived Shuttle Environment database. The information flow plan to be established and maintained under NASA is shown in Figure 2.

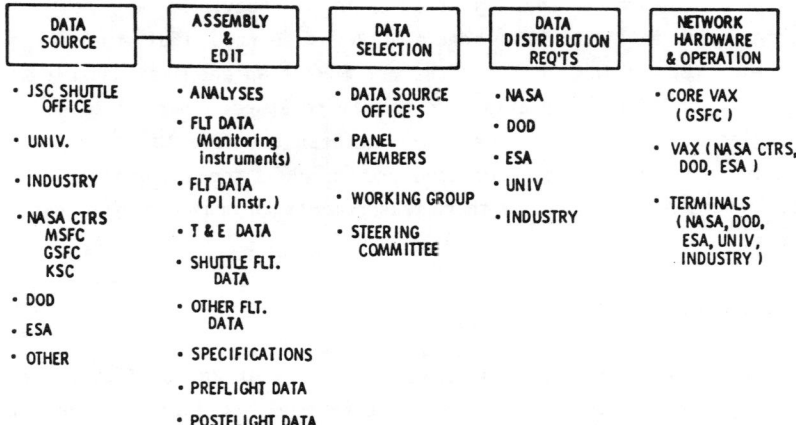

FIGURE 2. DATA MANAGEMENT AND FLOW PLAN

The data will first be collected, verified and reviewed by panels that are composed of national experts in the respective disciplines. The participation in these panels will be by NASA, other government agencies, industry, universities, and research institutes. The Working Group, responsible for coordinating with the Shuttle operations, will then review this information for content and for consistency with operational requirements and policy. The data are then sent to NASA Headquarters, for final review in accordance with a protocol established for this purpose. The data are then formatted and entered into the computer. The Information Management Panel Chairman is responsible for issuing accounts for users of the system. A schematic of the system is shown in Fig. 3.

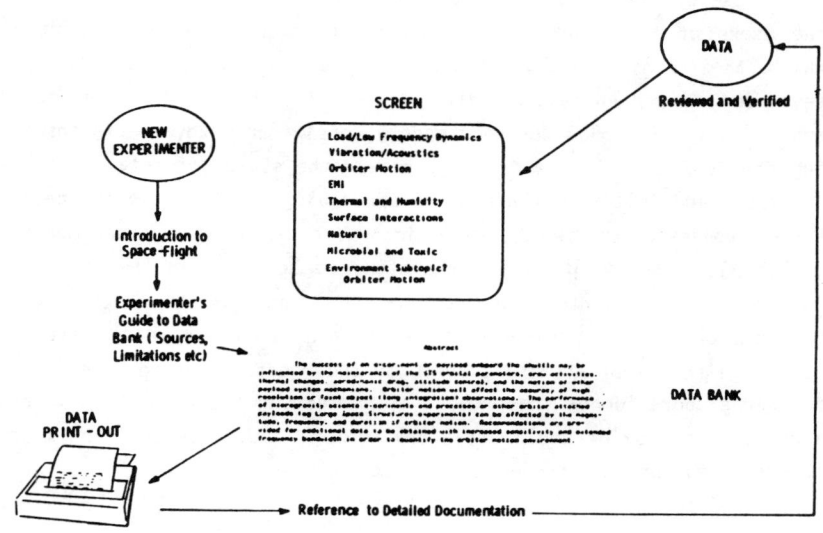

Fig. 3. Schematic of Shuttle Environment Information System

CONCLUSION

Concluding remarks by ESA Co-chairman Dai Shapland were:

"This Workshop is an important event for the use of Shuttle and Spacelab...The known environment must be communicated to future experimenters. ESA recognizes the importance of these factors and wishes to be a part of the overall Shuttle Environment Plan. It is ESA's hope that its sound investment in Spacelab will pave the way for continuing ESA-NASA cooperation."

Michael Lauriente, the NASA co-chairman of the Workshop, closed the conference by expressing appreciation for the hard work by participants in the five days of activity. He noted that the Workshop had uncovered considerations not dealt with in the earlier conference.

These considerations in 1982 had represented a "call for help" by the users of the Shuttle in trying to understand the environment in which their instruments are expected to operate (Table 1). It developed few quantitative data useful for the designs of later experiments. The 1984 Workshop made a considerable advance in answering the 1982 call for help: it reviewed the status of data currently available and initiated the process of cataloguing the data on the ten topics believed to be the most important to Shuttle experimenters (Table 3). Also it projected the steps to be taken in the immediate future. Dr Lauriente observed that NASA expects to have another Workshop on similar environmental topics in about two years. It may be that a review of accomplishments on the needs expressed here, and a further elaboration of the process of perfecting the comprehensiveness and accessibility of the Shuttle Environment data base could be major agenda items for this projected 1986 conference.

WORKSHOP AGENDA

Monday, August 6, 1984

9:00 a.m. - 12:30 p.m.   SESSION I

"Shuttle Role in the Space Program" Michael Sander, Shuttle Payload Engineering Division, NASA

"Role of ESA in the Space Transportation System" Dai Shapland, ESA

"Shuttle and the Future" Isaac T. Gillam, IV, NASA

2:00 p.m. - 5:00 p.m.   AD HOC SESSIONS

Loads and Low Frequencies, Thermal and Humidity, Orbiter Motion, Surface Interactions.

7:00 p.m. - 9:00 p.m.   SESSION II

"Spacelab Science in the Next Decade" Robert Chapman, NASA/Goddard Space Flight Center; and
Laurence Young, Professor of Aeronautics and Astronautics, Massachusetts Institute of Technology

"Person in the Loop" Byron Lichtenberg, Payload Specialist-Spacelab 1, NASA Johnson Space Center.

"STS Environment Measurements and Present Plans" Robert Blount, NASA Johnson Space Center.

Tuesday, August 7, 1984

9:00 a.m. - 12:30 a.m.   CONCURRENT SESSIONS:

Electromagnetic Interference, William Cutler, Aerospace Corp.
Orbiter Motion, Roger Chassay, NASA/Marshall Space Flight Center
Particulate Environment, Jack Barengoltz, Jet Propulsion Lab

2:00 p.m. - 5:00 p.m.   AD HOC SESSIONS

    Loads and Low Frequencies, Surface Interactions,
    Orbiter Motion, Microbial and Toxic Contaminants,
    Thermal and Humidity, Molecular Contamination.

7:00 p.m. - 9:00 p.m.   PLENARY SESSION
  Chairman: Laurance Young, MIT
  Subpanel Reports:
    Particulate Environment, Jack Barengoltz
    Electromagnetic Interference, William Cutler
    Orbiter Motion, Roger Chassay

  "Spacelab - Natural and Induced Environment Applicable to
  Experimenters", Alan Thirkettle, ESA

  "A Summary of the Experiment Results of Spacelab I"
  Karl Knott, ESA

Wednesday, August 8, 1984

9:00 a.m. - 12:30 p.m.   CONCURRENT SESSIONS:

  Loads and Low Frequencies, John Garba, Jet Propulsion Lab.
  Thermal and Humidity, James Clawson, Rockwell Int.
  Surface Interactions, Henry Garrett, Jet Propulsion Lab.

2:00 p.m. - 5:00 p.m.   AD HOC MEETINGS:

  Loads and Low Frequencies, Surface Interactions,
  Particulate Environment, Thermal and Humidity, Orbiter Motion.
  Microbial and Toxic Contaminants, Information Management Panel.

7:00 p.m. - 9:00 p.m.   PLENARY SESSION
    Chairman, Byron Lichtenberg, MIT

Subpanel Reports:
 Loads and Low Frequencies, John Garba
 Thermal and Humidity, James Clawson
 Surface Interactions, Henry Garrett

"Outline of Scientific Data Bases and Data Banks Available from ESA-IRS", George Proca, ESA.

"An Overview of the Shuttle/Spacelab Contamination Environment and Effects Handbook", Lyle Bareiss, Martin Marietta.

Thursday, August 9, 1984

9:00 a.m. - 12:30 p.m.   CONCURRENT SESSIONS:

 Vibrations and Acoustics, Don Wong, Aerospace Corp.
 Molecular Contamination, Jack Barengoltz, Jet Propulsion Lab.
 Microbial and Toxic Contaminants, Duane Pierson, Johnson Space Center

2:00 p.m. - 5:00 p.m.   <u>AD</u> <u>HOC</u> SESSIONS

 "KSC Processing of Payloads"

 Vibrations and Acoustics, Microbial and Toxic Contaminants, Loads and Low Frequencies, Thermal and Humidity.

7:00 p.m. - 9:00 p.m.   PLENARY SESSION
      Chairman, Robert Chapman, NASA/GSFC

 Subpanel Reports:
 Vibrations and Acoustics, Don Wong
 Molecular Contamination, Jack Barengoltz
 Microbial   Toxic Contaminants, Bonnie Dalton, NASA/ARC

"1985 AIAA Meeting on Shuttle Environment and Operations",

Billy M. McCormac, Lockheed MSC.
"The Natural Environment", Paul Robinson, Jet Propulsion Lab.

"Spacelab I Development" - a film, Dai Shapland, ESA.

Friday, August 10, 1984
9:00 a.m. - 12:00 p.m. FINAL PLENARY SESSION, FUTURE PLANS:

Natural   Induced Environments Panel:
 Loads and Low Frequencies, John Garba
 Vibrations and Acoustics, Don Wong
 Thermal and Humidity, James Clawson
 Orbiter Motion, John Winter
 Electromagnetic Interference, William Cutler
 Particulate Environment, Jack Barengoltz
 Molecular Contamination, Jack Barengoltz
 Surface Interactions, Henry Garrett
 Microbial   Toxic Contaminants, Bonnie Dalton
 Summary, Gerald W. Sharp, Chairman

Information Management Panel, Robert Theis, Chairman

Experimenters Panel, Robert Chapman and Karl Knott

Summary and Closing Remarks, Michael Lauriente   Dai Shapland

CHAPTER 1

PLENARY SESSIONS

SHUTTLE ROLE IN THE SPACE PROGRAM

Michael Sander*
Director, Shuttle Payload Engineering Division
National Aeronautics and Space Administration

Early in the Shuttle program, 160 experiment proposals were submitted. Thirty-seven were acceptable to NASA. Of these, because of fund limitations, only three were funded. As the Space Shuttle program advances, and we fly increasing numbers of experiments, we will require a data base that describes the environment faced by the Shuttle instruments. We have classified payloads into three categories: (1) attached payloads, (2) momentarily detached payloads, and (3) major observatories. Data derived from such payloads are available now and will form the first echelon of a data base that will describe the environment to which experiments are exposed when they are transported in the Shuttle bay.

Speaking on behalf of Burt Edelson, we hope this meeting will produce a good, ongoing and detailed data base that defines the environment. Contributions to that data base will be forthcoming from all the experiments conducted by the Office of Space Science and Applications. We hope that as DoD, ESA and other Shuttle users acquire data, they too will share them with us and that we can put together one fairly uniform, standard data base. I certainly expect this program to be ongoing and long-term in nature. There will always be new data sets. Rationalizing and mapping them into older data sets is going to be a complicated and important job. It is going to require innovation.

---

* Now with the Jet Propulsion Laboratory, Pasadena, California.

The Spacelab program can save costs by utilizing NASA flight-qualified hardware to accomplish many objectives of the Attached Payloads Program. JSC did a very nice job on an in-orbit refueling experiment. They found in NASA's backyard a couple of tanks fully qualified by the Planetary Program to carry hydrazine and incorporated two of those tanks into their on-orbit refueling, bringing the experiment in at an amazingly low cost. This is the kind of thing we have the opportunity to do in Spacelab, and by using ingenuity, we can conduct many experiments to measure and sample that environment at a relatively low cost.

I hope, in summary, that this week will be challenging. I hope I have shown that getting this data set--to which you will all be contributing--is important to OSSA, especially in context of what will be happening at NASA during the next ten years.

THE ROLE OF ESA IN THE SPACE TRANSPORTATION SYSTEM (STS)

D. J. Shapland
Directorate of Space Transportation Systems
European Space Agency

The European Space Agency (ESA) was formed in 1975 from two existing bodies--ELDO (European Launcher Development Organization) responsible for launchers and ESRO (European Space Research Organization) concerned with satellites for scientific research (Figure 1). Integration of these two bodies brought all European space activities

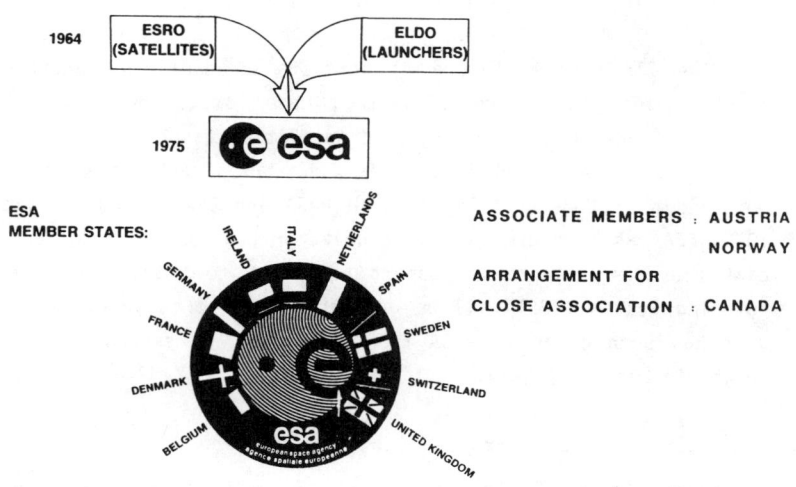

Fig. 1. Organization of ESA

under the umbrella of a single Agency. ESA's chartered purpose is to provide for and to promote cooperation in space activities among its Member States* for exclusively peaceful purposes (Figure 2). The total budget for 1984 amounts to about 955 MAU (1AU is currently just less than $1 but varies from year to year), of which some 13% is devoted to Spacelab-related programmes.

- **PROVIDE FOR AND PROMOTE, FOR EXCLUSIVELY PEACEFUL PURPOSES, COOPERATION BETWEEN EUROPEAN STATES IN THE FIELDS OF:**

    SPACE RESEARCH
    SPACE TECHNOLOGY
    SPACE APPLICATIONS

- **ELABORATE AND IMPLEMENT A LONG-TERM EUROPEAN SPACE POLICY**
- **ELABORATE AND IMPLEMENT A EUROPEAN SPACE PROGRAMME**
- **PROGRESSIVELY 'EUROPEANISE' NATIONAL SPACE PROGRAMMES**
- **ELABORATE AND IMPLEMENT AN INDUSTRIAL POLICY**

Fig. 2. Main Aims of ESA

The STS family (Figure 3) consists of the Space Shuttle, the Tracking and Data Relay satellites, upper stages for use with the Orbiter, Spacelab and those ground facilities associated with the STS operation. ESA's contribution is Spacelab. Under the terms of a Memorandum of Understanding between NASA and ESA signed in September 1973, ESA has delivered a Flight Unit, an Engineering Model and associated spares and ground support equipment to NASA. An Instrument Pointing System (IPS) will be delivered later this year. This first unit has been given to NASA free-of-charge. A second unit will be bought in Europe by NASA.

---

* Member States of ESA are Belgium, Denmark, France, Germany, Ireland, Italy, The Netherlands, Spain, Sweden, Switzerland, and United Kingdom. Austria and Norway are Associate Member States and Canada has close associations with the Agency.

The total cost to European Member States of the Spacelab Development Programme amounts to 760 MAU, of which the major proportion (approximately 55%) was provided by Germany (Figure 4). Of the present ESA Member States, Sweden and Ireland did not contribute to the Spacelab Programme but Austria, an Associate Member State, did

THE ADVANCED SPACE TRANSPORTATION SYSTEM COMPRISES

- SPACE SHUTTLE
    - ORBITER
    - SOLID BOOSTERS
    - EXTERNAL TANK

- TDRSS SATELLITES

- UPPER STAGES
    - PAYLOAD ASSIST MODULES
    - INTERIM UPPER STAGE

- SPACELAB

- GROUND SEGMENT
    - TDRSS GROUND STATIONS
    - MOCR AND POCC
    - TRACKING STATIONS

Fig. 3. The STS Family

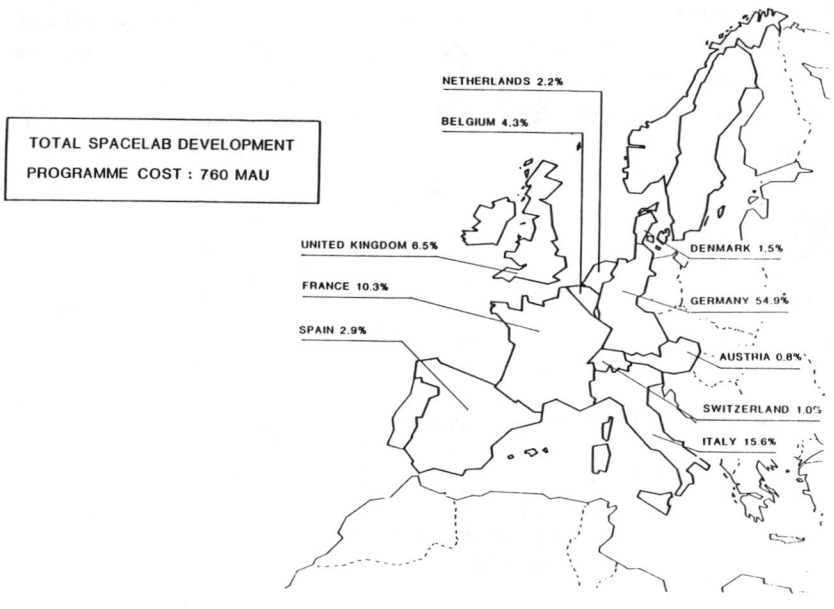

Fig. 4. Cost Breakdown

participate in it. The overall objective of the Spacelab Programme (Figure 5) was to provide a reusable orbital laboratory in which man-tended experiments could be operated much the same as on Earth.

The resulting Spacelab design (Figure 6) consists essentially of two basic parts. The pressurized module provides a "shirt-sleeve" environment for a crew of up to four scientists whereas pallet elements permit the direct exposure of instruments to space. Modularity ensures that the most suitable configuration can be chosen for a mission. This can be done by varying the size of the module and using up to five pallet elements. Three basic use modes are evident: module-only, pallets-only, and module and pallets. Variations on these basic modes can be imagined.

- LABORATORY AND OBSERVATIONAL PROVISIONS APPLICABLE TO ALL DISCIPLINES
    - MODULE
    - PALLET
    - SCIENTIFIC AIRLOCK
    - OPTICAL WINDOW

- NEAR-EARTH LABORATORY CONDITIONS
    - O_2/N_2 ATMOSPHERE AT GROUND PRESSURE
    - "SHIRT-SLEEVE" ENVIRONMENT
    - NORMAL RESOURCES, i.e. USABLE POWER, DATA HANDLING, THERMAL CONTROL, STRUCTURAL SUPPORT TO EXPERIMENTS
    - RACKS TO TAKE 19" TRAYS
    - STANDARD INTERFACES BETWEEN EXPERIMENTS AND SERVICES

- MANNED ATTENDANCE – SCIENTISTS AS PART OF CREW

- ACCOMMODATION OF LARGE AND HEAVY INSTRUMENTS AND FACILITIES

- POINTING CAPABILITY BETTER THAN ORBITER

- EASY AND LOW COST ACCESS TO SPACE

Fig. 5. Objectives of Spacelab Application

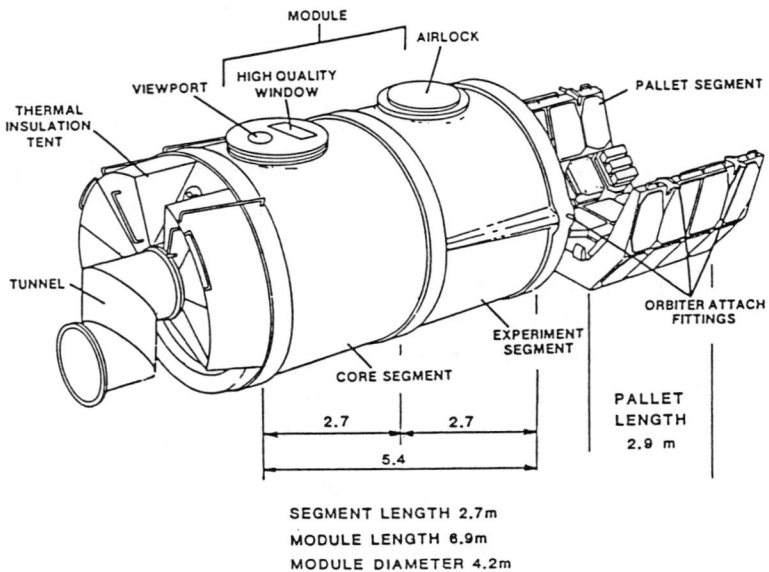

```
SEGMENT LENGTH 2.7m
MODULE LENGTH 6.9m
MODULE DIAMETER 4.2m
```

Fig. 6. Major Spacelab Elements

The Spacelab was produced in Europe with European funds by a European industrial team under the overall management of ESA. This team, drawn from all of ESA's Member States, was led by MBB-ERNO of Germany as main contractor. The development programme covered a period of eight years, with delivery of the Flight Unit in December 1981.

Spacelab is carried into orbit in the cargo bay of the Space Shuttle Orbiter, remains there during the flight and returns to Earth with it. The time spent in orbit is nominally seven days but may be anywhere up to ten or twelve days.

The first Spacelab flight, carrying a joint ESA-NASA payload took place over the period 28 November to 8 December 1983. The flight was extremely successful and demonstrated that Spacelab can be used in a variety of scientific and application disciplines. Some of the experimenters with equipment on-board Spacelab during that mission are at this Workshop and will talk about their results and experiences.

ESA has not only provided hardware for the STS family but plans to use the system in its future research activities (Figure 7). In fact, ESA-sponsored experiments or equipment will fly on all the Spacelab flights planned for next year, via. SL-3, SL-2, EOM-1, and the German D-1. Further in the future, ESA expects to partake in the International Microgravity Laboratory (IML) Programme promoted by NASA, providing facilities and experiments in the fields of Life Sciences and Materials Sciences. In addition, the Space Shuttle will be used for the launch and retrieval of EURECA (Europe's Free-Flying Retrivable Carrier) and will be used, when cost effective, for the launch of expendable satellites. Also, ESA has a core of three scientist-astronauts, all established experimental scientists who are available for the operation of experiments to be carried on flights that have ESA involvement.

The contribution of Europe to NASA's Space Station has not yet been decided. However, it seems logical that this should use the experience gained during the design, development, and manufacture of Spacelab. Hardware elements based on Spacelab and EUREKA are firm candidates for future cooperative ventures. Much has been learned from the Spacelab Programme, both from the point of view of technology and management. It is ESA's hope that its sound investment in Spacelab will pave the way for continuing ESA-NASA cooperation.

This Space Shuttle Experiment and Environment Workshop is an important event in a programme for the use of the Shuttle and Spacelab (and eventually the Space Station). Only if the environment of the Orbiter and Spacelab, either natural or induced, is well-known will the results of experimenters be meaningful. Further, the known environment must be communicated to future experimenters. ESA recognizes the importance of these factors and wishes to be a part of the overall Shuttle Environment Plan.

SPACE SHUTTLE ENVIRONMENT

o   SL-3
    -   REFLIGHT OF VERY WIDE FIELD CAMERA EXPERIMENT FROM SL-1

o   SL-2
    -   VERIFICATION OF IPS AND PALLETS-ONLY MODE
    -   ENGINEERING SUPPORT

o   EARTH OBSERVATIONS MISSION (EOM-1)
    -   REFLIGHT OF GRILLE SPECTROMETER AND METRIC CAMERA EXPERIMENTS FROM SL-1
    -   ESA MISSION SPECIALIST (CLAUDE NICOLLIER)

o   GERMAN D-1 MISSION
    -   FLIGHT OF SPACE SLED, BIORACK, IMPROVED FLUID PHYSICS MODULE
    -   ESA FLIGHT PAYLOAD SPECIALIST (WUBBO OCKELS), ESA BACK-UP PAYLOAD SPECIALIST (ULF MERBOLD)

o   ULYSSES (SOLAR-POLAR) WILL USE SHUTTLE-CENTAUR AS LAUNCH VEHICLE

o   IML
    -   ESA TO PARTICIPATE WITH MICROGRAVITY FACILITIES, e.g. BIORACK, ANTHRORACK, SLED, FLUID PHYSICS MODULE
    -   CAN PROVIDE PAYLOAD CREW MEMBERS

o   GERMAN D-2 (EUROPEAN MICROGRAVITY MISSION)
    -   ESA TO PROVIDE FACILITIES, e.g. SLED
    -   CAN PROVIDE PAYLOAD CREW MEMBERS

o   EURECA WILL USE ORBITER FOR LAUNCH AND RETRIEVAL (MULTIPLE MISSIONS PLANNED)

o   SPACE STATION - CONTRIBUTION NOT YET DECIDED BUT WILL USE STS FOR LAUNCH

Fig. 7.   Planned ESA Participation in Future STS Missions

SHUTTLE AND THE FUTURE

Isaac Gillam
Office of Space Flight
Headquarters, National Aeronautics and Space Administration

In order to talk about the Space Transportation System (STS) and the future, we have to look at where we are now, and then look at where it is we would like to go with the STS. I will start off by talking briefly about the goals and objectives of the program. These goals and objectives were first formulated about a year ago. They have been updated this year. We'll talk about goals and objectives and then we'll go into a comprehensive overview of the Shuttle program, customer services, Spacelab upper stages, expendable vehicles, and advanced programs.

The objectives of the Office of Space Flight are listed in Figure 1. The first objective of the program is for each mission to be conducted safely and successfully. Inquiries have been made on whether or not we are backing off on that objective to meet other goals such as schedules. The answer to this question is, no; it still is our number one priority. One of the new things that has been added to the list is the fourth bullet: To implement a management structure for the operational era. As you know, we are in transition from a program in which the Shuttle was the goal; this was the development of the Shuttle. We would like to move toward an era where the Shuttle is a tool. We are looking at various organizational structures, both within the Headquarters organization and within our field network to organize future streamlined operations.

I'd like to mention, as stated in the third bullet, that we are working hard on reducing the operational costs associated with using the system. The sixth bullet describes how an infrastructure will be provided for the Space Station. I'll talk more about that later in the presentation.

SPACE SHUTTLE ENVIRONMENT

In the Office of Space Flight (OSF) we talk about the Space Transportation System which uses both the Space Shuttle, the Spacelab, and the various upper stages to constitute a total transportation system. Briefly I'll talk to you about some of our new systems so they will be familiar to you.

- EACH MISSION—SAFE AND SUCCESSFUL
- MAINTAIN OPERATIONAL LAUNCH SCHEDULE
- REDUCE OPERATIONAL COSTS
- IMPLEMENT MANAGEMENT STRUCTURE FOR OPERATIONAL ERA
- COMPLETE STS DEVELOPMENT/EXPLOIT STS CAPABILITIES
- PROVIDE INFRASTRUCTURE FOR SPACE STATION
- MAINTAIN A HEALTHY INSTITUTIONAL BASE

Continued Space Leadership for the United States

Fig. 1. Objectives of the Office of Space Flight

The PAM-A is a Payload Assist Module. This is a payload assist module for ATLAS Centaur class payloads as is PAM-D for Delta class payloads. D-II is the growth version for payloads slightly heavier than Delta class payloads. We'll talk later about each one of these in more detail. We have the Inertial Upper Stage (IUS) which was developed by the Department of Defense (DoD) and the Centaur which is going to be used on the Galileo and International Solar Polar Mission (ISPM). I understand the latter is now called Ulysses. We have the Galileo mission in 1986 and another commercial stage called the Transfer Orbit Stage (TOS), which is between the class of the PAM-A and the IUS; and of course we have the Spacelab that you've heard about already today.

One hope of ours is to provide a transportation system that has transportation in depth. We're just getting started on this road; therefore, we don't have that capability right now. As seen in Figure 2, through the redundancy provided by the system we hope to incorporate more and more flexibility into the system. We plan to use a government team with a contractor and to take advantage of and exploit the capabilities provided by having "people in the loop."

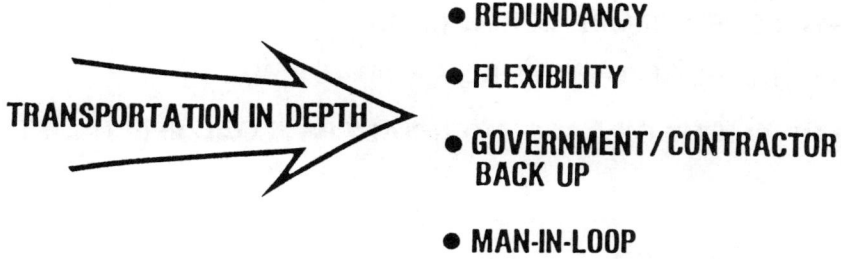

Fig. 2. Space Shuttle Program

Listed in Figure 3 are the "firsts" planned for 1984. The first flight of the Discovery has been delayed. The launch date will probably be the 27th, 28th, or 29th of August, depending upon a variety of factors. If we go on the 29th we will have about four or five slack days in the schedule. We are very sure that we can make the 29th, but we'd like to move that date up. Since it's a six-day mission, we would not like to launch on a day that would require landing on Labor Day, Saturday, or Sunday, which would complicate the process.

- 1st FLIGHT OF "DISCOVERY" ORBITER (41-D)
- 1st USE OF MANNED MANEUVERING UNIT (41-B)
- 1st LANDING AT KSC (41-B)
- 1st RMS GRAPPLE OF A FREE-FLYING SATELLITE (41-C)
- 1st SATELLITE REPAIR MISSION (41-C)
- HIGHEST ALTITUDE ORBIT - 250 NM (41-C)
- 1st AUTO-LAND LANDING (41-F)
- 1st SEVEN MEMBER CREW (51-B)
- 1st DEDICATED DOD MISSION (41-H)
- 1st FLIGHT OF A "COMMERCIAL" PAYLOAD SPECIALIST

Fig. 3. Special Events for 1984

We have used the manned maneuvering unit, have had the first landing at Kennedy Space Center (KSC), and have used the Remote Manipulator System (RMS) to grapple the Solar Maximum Mission (SMM) satellite. During the flight in which the highest altitude of 250 nautical miles was achieved, the first satellite repair mission was successfully accomplished. Automatic Computer Aided Landing was scheduled for 41-F. Since that mission has been combined into 41-G, the auto land will probably occur on a later mission and will probably not happen this year. The first seven-member crew will be on mission 51-B which is scheduled for December of this year and the first dedicated DoD mission is scheduled for the end of this year. The first flight of a commercial payload specialist is to occur on the next mission with Charlie Walker flying with the electrophoresis unit.

Our plan is as indicated in Figure 4. Four launches were planned in 1983 and eight in 1984. As you see from the buildup of orbiter deliveries, we expect the delivery of the Atlantis, which is the fourth Shuttle Orbiter in April of next year. It was originally expected in December of this year, but we decided to put in some of the modifications required before its delivery. At the same time, we were performing the modifications on Columbia to take out some of the Development Flight Instrumentation (DFI), e.g., some of the ejection seats, and some of the test equipment that was in Columbia. This removal is under way right now. We decided to use the time when we were looking at the problems associated with IUS and with PAM to get the orbiters in shape and ready so we would have four flight orbiters ready for 1985. We expect these problems to cascade into 1985, so we want to be sure that we meet the 1985 launch rate, and we also want to be sure that we are ready for the 1986 dual launch of Galileo and ISPM. The flight capability will be as indicated on Figure 4. The flight rate will probably be somewhat less because of less-than-expected demand. By 1988 we will be capable of flying 23 flights a year, and by 1989 we will be capable of flying 24, which is our goal.

18   SPACE SHUTTLE ENVIRONMENT

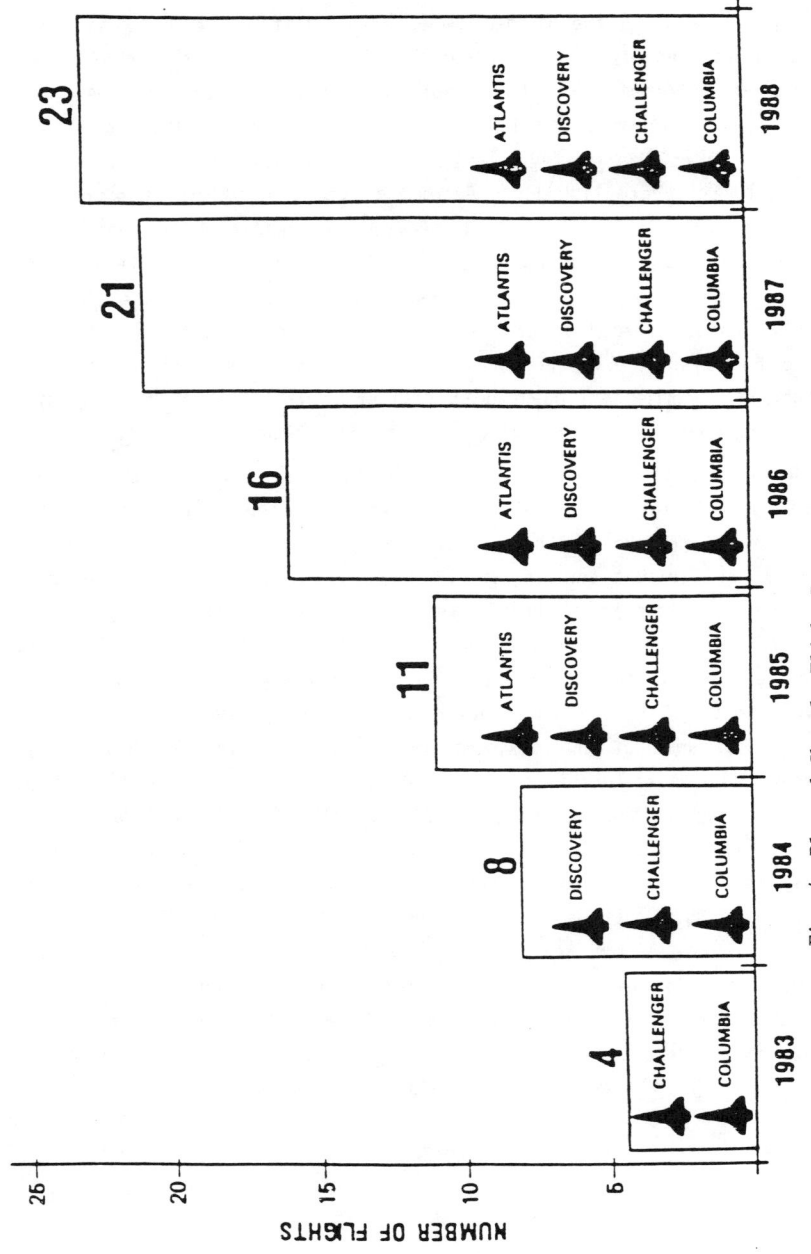

Fig. 4. Planned Shuttle Flight Rate

The basis of our planned capability growth that I just mentioned to you is indicated in Figure 5. We planned to reduce our turnaround capability to 35 days by the end of 1985. We have, in effect, already achieved a 35-day capability, if you take the minimum time required at each stage of buildup, but we haven't done that yet on a single mission. We are working to do that on a single mission with an ultimate goal of a 28-day turnaround on a single orbiter. We are working on increased flight operations efficiency, improved Space Shuttle Main Engine (SSME) performance, which will be dealt with in detail later. Ultimately, we'll have two launch pads at Kennedy and one at Vandenburg. The Vandenburg launch pad will be ready in October 1985, the second pad at Kennedy in 1986 in time for the Galileo/ISPM missions. We are adding to the program a third mobile launch platform which will also be available at Kennedy in 1986.

- **ADDITIONAL ORBITERS (FLEET OF 4)**
- **REDUCED TURNAROUND TIMES (30 DAYS)**
- **INCREASED FLIGHT OPERATIONS EFFICIENCY**
- **INCREASED PRODUCTION CAPACITY**
    - **EXTERNAL TANK**
    - **SOLID ROCKET BOOSTER REFURBISHMENT**
- **IMPROVED SSME PERFORMANCE**
- **TWO ADDITIONAL LAUNCH PADS (KSC AND VAFB)**
- **ADDITIONAL MOBILE LAUNCH PLATFORM**

Fig. 5. Bases of Planned Capability Growth

There is a planned profile for the performance capability of the vehicle. We currently plan to fly most of our missions at 104 percent of rated power. Using the 104 percent level, we found that we get more engine life with our current engine than if we fly it at a higher percentage of rated power. Although we are capable of flying at 109 percent, the engine is only certified for a certain number of hours at this level which is probably equivalent to about five flights. Galileo/ISPM will be flown at 109 percent. Because of additional payload performance requirements at Vandenburg, we have initiated a development program for the Filament Wound Case which lightens the cases of solid rocket boosters.

At the time Figure 6 was made, Columbia was undergoing modifications. Atlantis was planned for delivery in December 1984. We put it into its modification period. People have asked, "When you had a problem with Discovery, why didn't you change vehicles from Discovery to Challenger?" Well, Challenger is also undergoing modifications. By the time we go into fiscal year 1985, all the vehicles will be updated and capable of flying all the planned missions in the latest configuration. The first flight of the Discovery was scheduled for the time Figure 6 was made, in June 1984. It's now planned for takeoff at the end of August. This mission will be six days and will land at Edwards Air Force Base.

Looking at each one of the STS subsystems or systems I'd like to review where we stand with the orbiters, the engines, the Solid Rocket Boosters (SRB's), and the external tank programs. The highlight of the orbiter delivery program in terms of future programs is the scheduled delivery of Orbiter 104. We have, as indicated in the last bullet under the current status column of Figure 7, a set of structural spares in the works. We call them structural spares because, when we did the fiscal year 1984 budget, NASA went in for a fifth Shuttle Orbiter. A rather elegant compromise between yes and no was the idea of building a spare fuselage, spare wing, spare tail, spare whatever, without going for the assembly of these parts into a fifth orbiter. That keeps the production capability alive through the end of calendar year 1985, which gives us more time to look at the requirements and

OV-102	"COLUMBIA"	Undergoing Modifications
OV-099	"CHALLENGER"	Operational
OV-103	"DISCOVERY"	First Flight June 1984
OV-104	"ATLANTIS"	Planned Dec '84

Fig. 6. Approved Orbiter Fleet

CURRENT STATUS
- ELEVEN FLIGHTS CONDUCTED
- ORBITER PERFORMANCE HAS BEEN EXCELLENT
- TWO VEHICLES RE-FLOWN
- OV-103 UNDERGOING PREPARATIONS FOR MAIDEN FLIGHT
- OV-102 UNDERGOING MODIFICATIONS
- STRUCTURAL SPARES & LOGISTICAL SPARES PROGRAMS WELL UNDERWAY

FUTURE PLANS
- COMPLETE DELIVERY OF REMAINING PRODUCTION VEHICLE OV-104: DEC 1984
- UPGRADE OV-102 TO OPERATIONAL CONFIGURATION
- COMPLETE PROCUREMENT OF MAJOR STRUCTURAL SPARES
- POTENTIAL FIFTH ORBITER

Fig. 7. Current Status and Future Plans for Orbiters

needs for the production of a fifth orbiter. Therefore, the production capability will be maintained through the end of next year and we will have at least that amount of time to make a decision as to whether or not to complete the fifth orbiter. The potential fifth orbiter is listed down among our future options.

The SRB program is going along well, and its current status and future plans can be seen in Figure 8. One of the major items that we have to complete is the Filament Wound Case (FWC), which is the last bullet under the current status title. While continued development is still underway, its completion will give us a 4600-lb performance gain in terms of payload. That program is proceeding very well and on schedule. There have been little problems that crop up in the program but they are routine things that you would expect in such a development program. So far we have encountered no serious producibility problems. Many of our future plans are associated with the turnaround of each Shuttle for its next launch. We are tooling up and building up the capability to reprocess SRB's at Kennedy and to have a production rate and reprocessing rate which will give us a 24-flight-per-year capability by fiscal year 1987, which is two years in advance of the rest of the 24-flight-per-year capability.

CURRENT STATUS
- SUCCESSFUL COMPLETION OF FIRST 11 LAUNCHES
    - STS-4 REUSABLE HARDWARE LOSS
    - FIRST LIGHTWEIGHT SOLID ROCKET MOTOR USED ON STS-6
    - FIRST HIGH PERFORMANCE MOTOR ON STS-8
- HIGH PERFORMANCE MOTOR (HPM) FULLY QUALIFIED (3000 LB GAIN)
- RECOMPETITION FOR SRB FORWARD AND AFT SKIRT REFURBISHMENT/ ASSEMBLY/CHECKOUT IN PROCESS
- FILAMENT WOUND CASE (FWC) DEVELOPMENT UNDERWAY (4600 LB GAIN)

FUTURE PLANS
- PRODUCTION RATE INCREASE TO 24 FLIGHTS/YEAR BY FY 1987
- COMPLETE FWC STATIC FIRINGS/DEVELOPMENT PROGRAM
- FIRST FILAMENT WOUND CASE MOTOR FLIGHT IN OCTOBER 1985
- CONTINUE PRODUCIBILITY IMPROVEMENTS
- ACTIVATE NEW REFURBISHMENT SUBASSEMBLY FACILITY FOR BOOSTER ASSEMBLY AND CHECKOUT

Fig. 8. Current Status and Future Plans of SRB Program

The Main Engine Program is a program that has caused us a lot of difficulties. Those difficulties are associated with lifetime and utilization of the engine. As indicated in Figure 9, there is a phase II development program directed at increasing the pump life of our oxidizer and fuel pumps. This Phase II program is expected to be completed by 1986. Under the future plans column you see a phase III development program. It will be aimed at increasing the life of other selected components on the engine system so that by the 1987/1988 period, if it's required, we will be capable of flying the 109 percent engine as many as 20 flights and flying the 104 percent as many as 30 to 40 flights out of an engine.

CURRENT STATUS
- TWELVE FLIGHT ENGINES DELIVERED
  - FOUR RATED POWER LEVEL CONFIGURATION
  - EIGHT FULL POWER LEVEL FPL CONFIGURATION
- PREVIOUS YEARS PLANS FOR ADDITIONAL HARDWARE BEGINNING TO SHOW RESULTS
- HIGH PRESSURE PUMPS CONTINUE AS PROBLEM
  - HIGH MAINTENANCE
  - LIMITED NUMBER AVAILABLE
- PHASE II DEVELOPMENT DIRECTED AT PUMP LIFE

FUTURE PLANS
- COMPLETE PHASE II DEVELOPMENT OF HIGH PRESSURE PUMPS
- PHASE III ADVANCED DEVELOPMENT UNDERWAY
  - PROVIDE ENGINE WITH GREATER OPERATING MARGIN
  - INCORPORATES A PARALLEL COMPETITIVE ACTION
- STUDYING FEASIBILITY OF A NATIONAL ROCKET ENGINE SIMULATION FACILITY

Fig. 9. Current Status and Future Plans for Space Shuttle Main Engine

Another element of the Shuttle System is the External Tank program, which can be seen in Figure 10. We have undertaken a program to decrease the weight on the external tank. It was very successful, and we managed to take about 6000 lbs of weight out of that external tank. The lightweight external tank was flown on STS-7. Our primary activity now is a production capability buildup. It is similar to the one for the SRB's and will be available in advance of our requirement capability dated fiscal year 1989. With respect to the last bullet, we have had so much success on the program with our existing contractor, NASA has made a general policy decision that, after several years, we will recompete our contracts with a view toward cost reduction and other such factors. We have not yet decided whether or not we will include the external tank in such a recompetition program. As on most of our programs, we are moving toward incentive fee contracting because the system is developed now and it should be approaching a rather mature status in production.

PRESENT STATUS

- 16 FLIGHT TANKS DELIVERED (THRU 3/84)
- 11 TANKS SUCCESSFULLY FLOWN (THRU STS-11)
- 13 TANKS IN PRODUCTION FLOW
- INCENTIVE CONTRACT FOR 21 TANKS (THRU ET-60) SIGNED
- PRODUCTION BUILDUP TO 24 PER YEAR UNDERWAY

FUTURE PLANS

- CONTINUE PROGRAM COST REDUCTION/PRODUCT IMPROVEMENT ACTIVITY
- DELIVER FIRST TANK TO VAFB
- MOVE TOWARDS FIXED PRICE INCENTIVE FEE CONTRACTING
- DECIDE ON RECOMPETITION STRATEGY

Fig. 10. External Tank Program

Shown on Figure 11 is how production rates and capability rates all come together. This is our planned program, and we plan to reach the indicated capability in 1988. Since it doesn't look like the demand will require it, we believe that we have about a one-year cushion; however, the program elements are continuing on the basis of having the 24-flight-per-year capability by 1988. The potential 40-per-year capability will require significant investment in the neighborhood of 3/4 of a billion dollars in facilities and a fifth orbiter. It's just put up there for dream purposes, because there are not many of us who believe we will need to get beyond 24 in this decade. We may get to the requirement for 24 in this decade.

PRODUCTION RATES
(UNITS PER YEAR)

	1984	1986	1988	1990	POTENTIAL
• EXTERNAL TANK PRODUCTION	15	18	24	24	40
• SOLID ROCKET MOTOR PRODUCTION/ REFURBISHMENT	16	16	24	24	40
• SOLID ROCKET BOOSTER PRODUCTION/ REFURBISHMENT	16	18	24	24	40
• SHUTTLE MAIN ENGINE PRODUCTION/ REFURBISHMENT	4/3	4/4	4/4	4/4	6/8

PRODUCTION CAPACITY AND GROWTH AVAILABLE

Fig. 11. Production Capacity and Growth Available

SPACE SHUTTLE ENVIRONMENT

One of the areas that has created a great deal of concern both in the Congress and in the DoD is our logistics program which we have put in Figure 12. We are indicating that we are building up a significant logistics program. A significant amount of money is being invested in having an inventory of spares and other parts of the system available to support an assured launch capability, a launch-on-time capability, and whatever is necessary from a user's point of view. You can see our operational goal is 95 percent launch on time at 24 flights per year. This is the reason we are making such a massive investment in logistics.

SCOPE
- **4 ORBITER FLEET AND SUPPORTING FLIGHT HARDWARE**
- **SUPPORT KSC AND VAFB LAUNCH SITES**
- **SIGNIFICANT FUNDING ALLOCATED**
  FY 1984: $539 MILLION (1/6 Total Shuttle Budget)
  FY 1985: $649 MILLION (1/6 Total Shuttle Budget)
  FY 1983 - 1989: $3.2 BILLION
- **CURRENT MANPOWER**
  CIVIL SERVICE - 112
  CONTRACTOR - 1378

OPERATIONS POLICY
- **COORDINATED IN NASA/DOD INTEGRATED LOGISTICS PANEL**
- **MANAGEMENT TRANSITIONED TO KSC**

OPERATIONAL GOAL: 95% LAUNCH ON TIME AT 24 FLIGHTS/YEAR

Fig. 12. Logistics Program

Figure 13 is an indicator of our experience and where we are going on Shuttle flight turnaround. This is representative of the learning curve that we are anticipating, both in turnaround and in program costs. We are operating on about an 80 percent learning curve right now. The goal is to get about an 84 percent learning curve in terms of both performance and cost. We are well inside that band. The most notable exception was the first flight of Columbia with the Spacelab modifications. It took quite a bit more time in processing that vehicle for launch. We have had problems on the first launch of each vehicle. As you recall the first flight of Columbia was delayed a few days because of problems. The first flight of Challenger was delayed a few months with leaks in the engine system. Most recently we have had problems with Discovery. I might add that we haven't learned very much because there are no plans for dealing with such problems on Atlantis when it comes in.

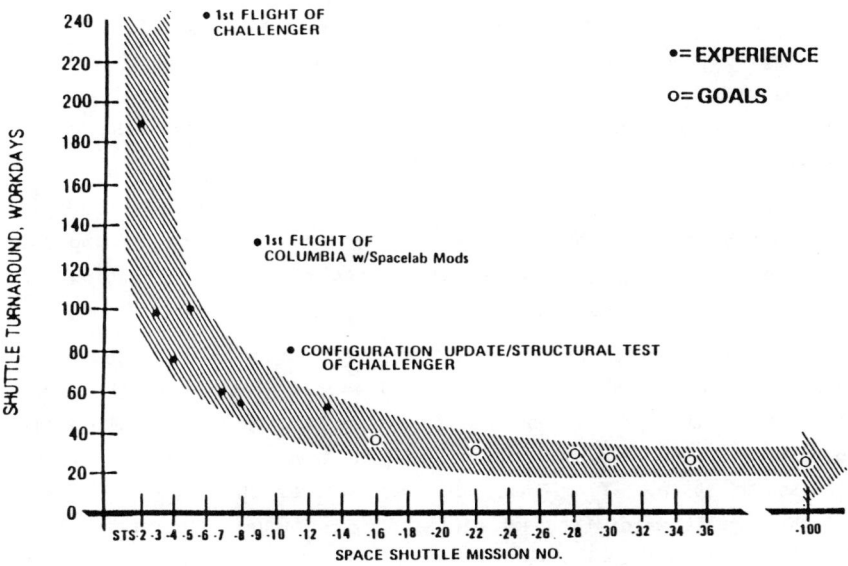

Fig. 13. KSC Shuttle Turnaround Experience and Goals

Figure 14 concerns organization and operations. That's an area to which we are devoting a great deal of attention. How we conduct the program is probably an area of importance to many of you. Mr. Beggs has appointed a working group, called the Shuttle Operations Strategic Planning Working Group, to work on the problem of organizing for the future. In organizing the Shuttle program, the agency is considering whether, in the long run, NASA should spin the Shuttle off to a private corporation or a government corporation, or retain it. They will be looking at these questions over the next three to four months. We have also established cost reduction goals for our Centers which we believe are reasonable. We expect the organizational elements to bring the program cost down over the next three or four years. We are looking at efforts to improve productivity, to achieve cost savings, and to increase the flexibility and efficiency in flight planning and operations. I'll go into some of the specifics when I get into the details of some other programs we have working. One secret is that if you have something which you want to fly, you should make it autonomous and avoid placing requirements on the orbiter system. We are working to provide more capability through the orbiter which will be independent of the basic orbiter systems to decrease the complexity of the integration process. We are also reviewing our contracting strategy.

We have a Division in the Office of Space Flight that we call Customer Services. The activities of this Division are growing and I'd like to review briefly the objectives of this organization and what they plan to accomplish as seen in Figure 15.

A goal was established in the President's National Space Policy announced on the 4th of July 1972. The goal stated that the first priority of the Space Transportation System Program is to have a fully operational, cost-effective Space Shuttle that will provide routine access to space. From that goal was derived the Office of Space Flight's goals which I showed earlier. This figure also represents a cascading of that goal into our customer-related activities. We want to provide our customers with a competitive launch service that is

- **DEVELOP MANAGEMENT STRUCTURE FOR OPERATIONAL ERA**
  - MAINTAIN COORDINATION/INTERFACE WITH SPACE STATION
- **IMPLEMENT AND MEASURE PERFORMANCE AGAINST COST GOALS**
- **CONTINUE CONTRACT CONSOLIDATION EFFORTS**
  - BOC AND SPC IMPLEMENTED
  - JSC OPERATIONS CONSOLIDATION UNDERWAY
- **EMPHASIZE EFFORTS TO IMPROVE PRODUCTIVITY AND ACHIEVE COST SAVINGS**
- **INCREASE FLEXIBILITY/EFFICIENCY OF FLIGHT PLANNING/OPERATIONS**
- **REVIEW CONTRACTING STRATEGY ACROSS PROGRAM**
  - RECOMPETE WHERE APPROPRIATE
  - MOVE TO INCENTIVE CONTRACTS

Fig. 14. General Management Strategy

GOAL
- FULLY OPERATIONAL, COST-EFFECTIVE SPACE SHUTTLE THAT WILL PROVIDE ROUTINE ACCESS TO SPACE

PLANS
- PROVIDE OUR CUSTOMERS WITH A COMPETITIVE LAUNCH SERVICE THAT IS BOTH RELIABLE AND FLEXIBLE
- PROMOTE THE USE OF SPACE
  - TO CIVIL GOVERNMENT, DOD, COMMERCIAL AND FOREIGN
  - USE A PRIVATE CONTRACTOR THAT IS APPROPRIATELY INCENTIVIZED
- PROVIDE SUPPORTING SERVICES
  - TO REDUCE RISKS; FINANCIAL, TECHNICAL AND INSTITUTIONAL
  - OTHERS TO SATISFY CUSTOMER NEEDS
- PROVIDE A CUSTOMER FRIENDLY ENVIRONMENT
  - REDUCE CUSTOMER COSTS THROUGH STREAMLINED OPERATIONS
  - INCREASE CUSTOMER SATISFACTION WITH NASA FOR REPEAT BUSINESS
- STIMULATE ADDITIONAL DEMAND BY PROVIDING NEW SERVICES
  - ON-ORBIT SERVICING AND REPAIR
  - OTHERS TO UTILIZE THE PRESENCE OF MAN
  - ASSIST IN THE DEVELOPMENT OF NEW RESEARCH, TECHNOLOGY DEVELOPMENT AND COMMERCIAL ACTIVITIES IN SPACE

Fig. 15. Customer Services

both reliable and flexible. We want to promote the use of space and are very active in the commercialization initiatives that are ongoing, mostly associated with materials processing. We want to provide some supporting services that meet our customers' needs in a customer-friendly environment. I'll give you an example of what I mean by a customer-friendly environment in just a few minutes. We want to stimulate additional demand by providing new services, and I will give you some examples of those kinds of new services very shortly.

Figure 16 shows the payload trends. This contains a comparison of payloads by category, or by class of payloads, which were scheduled in 1982 and 1983. The number of payloads in practically every category except one has gone up; 1983 was a good year. Remember, these are payloads, not spacecraft. They may include Get Away (GAS) canisters, Shuttle Student Involvement Program (SSIP), etc. Don't be deceived by including those classes and categories of payloads.

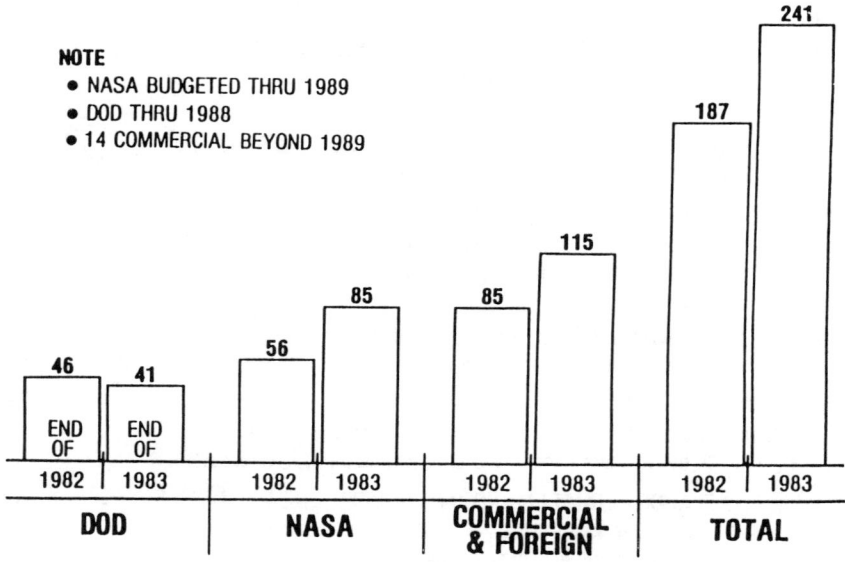

Fig. 16. Payload Trends

Figure 17 shows what we have in the way of planned flights, although these data are somewhat out of date. This chart shows a breakout of planned flights between NASA, the reimbursables or the commercial and foreign flights, and the DoD. We've included here the Vandenburg capability. When we talk 24 flights per year, we're talking a maximum of 20 out of Kennedy and an assumed four out of Vandenburg. The figure shows that it doesn't always work out that way.

	83	84	85	86	87	88	89
NASA							
KSC	1	3	5	6	6	6	6
VAFB	0	0	0	0	2	2	0
REIMBURSABLES							
KSC	3	2	4	5	7	7	6
VAFB	0	0	0	0	0	0	1
DOD							
KSC	0	0	2	1	5	6	6**
VAFB	0	0	0	2	0	2	3
SUBTOTAL	4	7	11	14	20	23	22
RESERVE *							
KSC	0	0	0	1	2	1	2
VAFB	0	0	0	0	0	0	0
TOTAL	4	5	11	15	22	24	24

* UNCOMMITTED INCLUDING RESERVE FLIGHTS
** ASSESSED FROM 10 (DOD INPUT) TO 6

Fig. 17. Planned KSC and VAFB Shuttle Flights

Our pricing policy can be seen in Figure 18, and this price is strictly for commercial and foreign customers. The Phase I pricing period as indicated was through fiscal year 1985. It was based on a 12-year average cost per flight as perceived. However, there were a few little anomalies that have resulted from the use of a 12-year average. At that time we planned 540 flights, we are now down to about 235. That was fairly optimistic planning. The full price for a full Shuttle on that basis was $38.5 million in 1982 dollars. The second phase of the pricing policy applies to flights between FY86 and FY89. The basis for this price is "out-of-pocket" costs. Actually, it's what most people would refer to as marginal costs, but its derivation yields a price slightly above the marginal cost. This price of the Shuttle is $71 million (in 1982 dollars).

• PHASE	I	II	III
• PERIOD	THRU FY 1985	FY 1986 THRU 1988	?
• POLICY	12 YEAR AVERAGE COST/FLIGHT	"OUT OF POCKET COSTS"	?
• FULL PRICE * (1982$)	$38,3M	$71M	?

PRICE FOR PHASE III IS CURRENTLY UNDER REVIEW

* ESCALATED TO PAYMENT DATE

Fig. 18. Shuttle Pricing Policy

Based on our learning curves and our projections, we would presume that the Phase III price, if we went to full-cost pricing in 82 dollars, would be about $88 million, if we go to full operating costs. There are people who believe that full-cost pricing means everything you can think of and therefore it should be higher. Then there are others who believe that full-cost pricing has to have some reason applied to it; so $88 million is a reasonable ballpark number.

Since you've heard a lot today already about Spacelab, I will just talk about a few "deltas" to the Spacelab program as we perceive it. Figure 19 is a variation on the figure which was presented earlier showing the various configurations of spacecraft. The thing that's been added at the bottom is the small carrier called Materials and Science Lab Materials Processing in space. It is the smallest of the Spacelab carriers, and it is a carrier that was added by NASA.

Figure 20 shows the Spacelab mission model. The title, Discipline Laboratories, describes that we have learned on Spacelab-1 that your integration problem goes down substantially if you go single discipline for each flight, as opposed to integrating many disciplines into a single flight. We are trying to convince people that the discipline laboratory concept is simpler and that flying a single discipline on each flight with a combination of experiments reduces cost.

Briefly, I would like to talk about Figure 21. The Get Away Special (GAS) handles one category of payloads and the Spacelab handles another category of payloads. We believe there is a gap between these two programs. We hope to fill the gap with the Hitchhiker program. There are actually three Hitchhiker programs, but it is difficult to distinguish between two of them. Hitchhiker-G is a Goddard program and it mounts along the side of the payload bay. Hitchhiker-M, for Marshall, is a Marshall Hitchhiker program which is an across-the-base structure. If you'll notice, the MPSS in the Spacelab area and the Hitchhiker in the Marshall area look a great deal alike. The Hitchhiker program is a concept, not of hardware, but a concept for consolidating the integration and paperwork in a manner that simplifies things greatly.

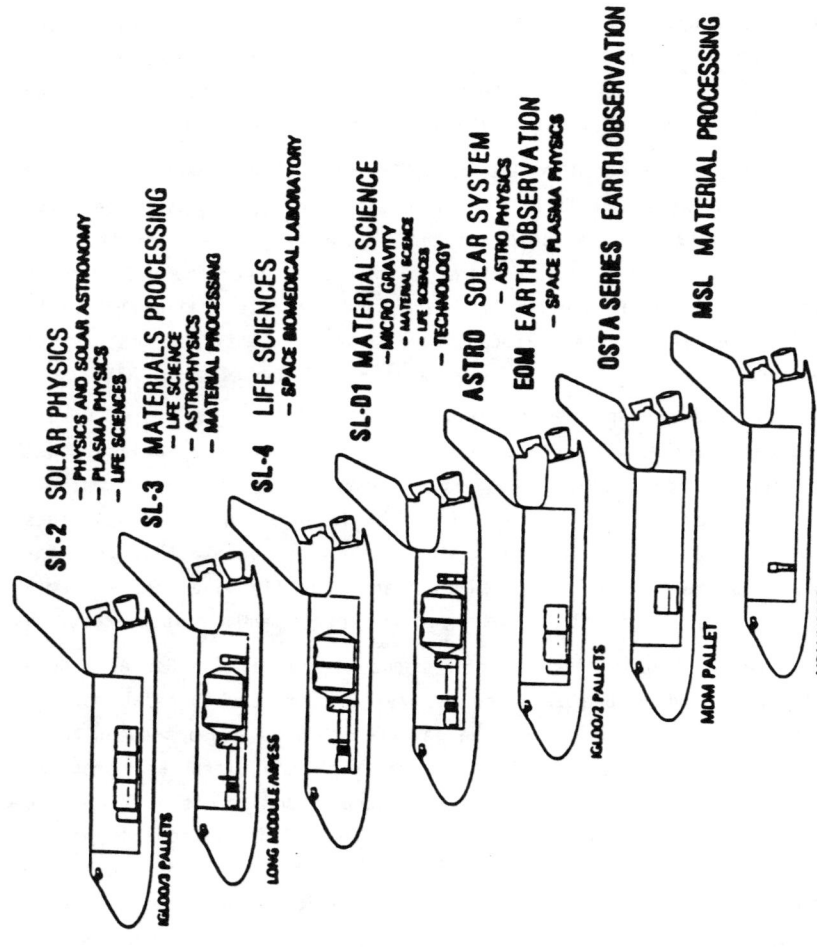

Fig. 19. Spacelab Major Missions

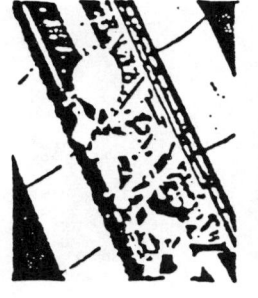

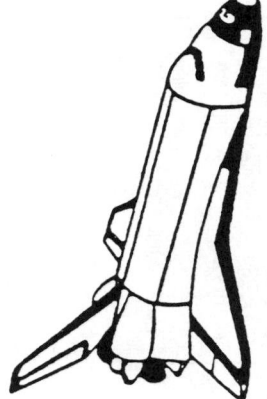

Fig. 20. Discipline Laboratories

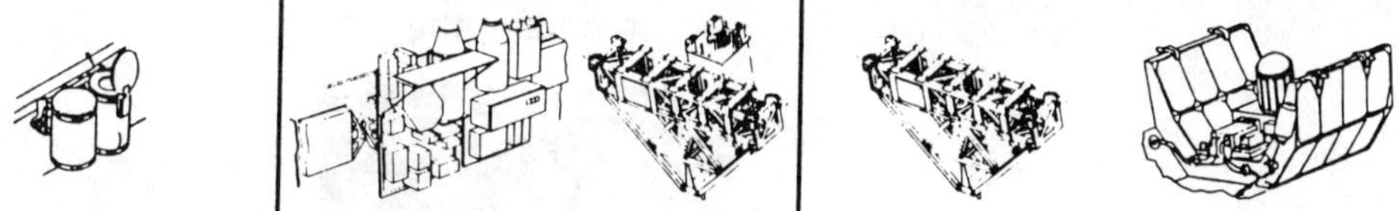

	GET AWAY SPECIAL (GAS)	HITCHHIKER GSFC	HITCHHIKER MSFC	SPACELAB MPESS	SPACELAB MDM PALLET
STRUCTURE	GAS CAN	PLATE 50x60	MPESS	MPESS	SPACELAB PALLET
SUBSYSTEMS	NONE, ON/OFF ONLY	FMDM	FMDM	FMDM	FMDM
CAPABILITY					
LOAD	200 LBS	750 LBS	3000 LBS	3000 LBS	5000 LBS
POWER	BATTERY	1500W	1500W	5000W	5000W
DATA	NONE	16KBS	16KBS	256KBS	256KBS
CONTROL	LIMITED	STD PANEL	STD PANEL	POCC	POCC
LENGTH IN BAY	SIDE MOUNTED	SIDE MOUNTED	40 INCHES	40 INCHES	113 INCHES
OPTIONAL SERVICES	VERY LIMITED	NO	NO	YES	YES

FMDM — FLEXIBLE MULTIPLEXER/DEMULTIPLEXER
MPESS — MISSION PECULIAR EXPT SUPPORT STRUCTURE
POCC — PAYLOAD OPS CONTROL CENTER

Fig. 21. STS Carriers

For example, on Hitchhiker we would like to have generic Program Interaction Plans (PIP's), wherein one only applies changes from the generic PIP which covers the things you're doing differently than were done on the general program plan. It cuts down on the paper work, safety requirements, etc.,--a lot of things that usually drive you, the users, crazy. The Hitchhiker concept is planned for your delivery of an experiment to Kennedy and six months later a delivery of your data from NASA. It's supposed to be a fast turnaround program. It's going to fly as a standby payload and, the way our manifest is going, it can expect ample flight opportunities. It will provide a great deal of flexibility and responsiveness to user requirements.

That's one of the new things which we are instituting. The third Hitchhiker program concerns NASA's plan to build only one Hitchhiker. There is a commercial version of the Hitchhiker being planned with slightly different capabilities. It would also have the increased flexibility of the NASA program, and NASA is trying to negotiate on a Joint Endeavor Agreement with the proposer. That would also be an across-the-base structure, but would be provided on a commercial basis as opposed to an in-house, NASA-provided service. The commercial Hitchhiker may or may not come to pass. We are involved in negotiations concerning that; we're not very far apart, but I don't know how successful these negotiations may be.

The Spacelab program will provide some benefits to the Space Station which can be seen in Figure 22. I won't dwell on these. For a while, NASA will continue operating expendable launch vehicles. They are operated by the Office of Space Flight, so we'll give you an overview of where that program is and where it's going.

BEFORE SPACE STATION FLIES, SPACELAB WILL HAVE:
- FLOWN 50 OR MORE MISSIONS
- COVERED ALL SCIENCE DISCIPLINES
- DEVELOPED PRACTICAL WAYS OF DOING BUSINESS
- PROVIDED OWN ESSENTIAL DATA BASE

SPACELAB LESSON LEARNED DISSEMINATION
- OPERATIONS AND UTILIZATION WORKING GROUPS
- INTERNATIONAL COOPERATION WORKING GROUP

USE OF SPACELAB
- TESTBED FOR PROTOTYPE SYSTEMS
- SIMULATION OF CRITICAL ACTIVITIES
- TRAINING

DEDICATED DISCIPLINE LABS
- DIRECT TRANSITION TO STATION MODE

POCC AUGMENTATION
- EVOLUTION TOWARDS STATION SUPPORT

Fig. 22. Spacelab Benefits to Space Station

NASA operates the vehicles which are shown in Figure 23. We have the Delta, which is the most active launch system. We have firm launch requirements through Delta 179, so we have about four more Deltas to launch before we complete the program. We will finish production of our Delta vehicles in the 4th quarter of this fiscal year. In other words, Delta will go out of production in October. The ATLAS Centaur production, however, will continue because the ATLAS Centaur has missions further downstream in FY87, both for the DoD and for the commercial user Intelsat. The ATLAS-F is basically used exclusively for the polar orbiting NOAA satellites. The ATLAS-H vehicles are being produced for the Air Force. The ATLAS-H will complete production in FY85. ATLAS-F is the refurbished version of the Atlases that were pulled out of the silos. There are about six or seven of those left for use. Finally we have the Scout. The DoD has requested a

DELTA
- FIRM LAUNCH REQUIREMENTS THROUGH MISSION NO. 179
- FINISH PRODUCTION AND DELIVERY OF 11 VEHICLES (THRU NO. 181) BY 4th QUARTER FY 1984

ATLAS—CENTAUR
- LOW PRODUCTION/LAUNCH RATE — ALL REIMBURSABLE (INTELSAT, NAVY)
- FUTURE LAUNCHES INCLUDE INTELSAT V AND FLTSATCOM THROUGH 1987 IN 4QCY 84), AND FLTSATCOM (IN 3Q85, 3Q86, 2Q87)
- NO FURTHER IMPROVEMENTS OR GROWTH ARE CONTEMPLATED AT THIS TIME

ATLAS—F
- CONTINUE CURRENT SUPPORT THROUGH CY 1988

ATLAS—H
- CONTINUE EXISTING PROCUREMENT OF VEHICLES FOR USAF

SCOUT
- OPERATIONAL — 15 PLANNED LAUNCHES THROUGH 1987 (14 FOR DOD) UNDER NASA MANAGEMENT

Fig. 23. Expendable Launch Vehicles Operated by NASA

two-year extension on our '87-end-date plan for the Scout. On this basis, it looks like Scout will continue in government use at least through '89, and in '89 we expect the DoD to write us another letter asking for an extension to '93. So Scout has been, and will be, around for quite a while.

Figure 24 is the flight schedule for the expendable systems. We still have the Automated Mesospheric Particle Tracer Experiment (AMPTE) scheduled for this week from Cape Kennedy on Delta. We have the NATO-IIID and the Galaxy-C to be launched. We have the two Geostationary Operational Environmental Satellies (GOES) scheduled for launch in 1986. The ATLAS Centaur program consists of the Intelsats and the DoD missions indicated as FLEETSATCOM (FSC). The ATLAS has NOAA missions scheduled through NOAA-I. We know the NOAA program is going to change, because they've had problems with both of the polar orbiting NOAA satellites.

	1983	1984	1985	1986	1987
SCOUT	HILAT	NAVY-21 AF-1 SAN MARCO	NAVY-22 NAVY-23 AF-2 AF-3 AF-4 AF-5	NAVY-24 NAVY-25	NAVY-26 NAVY-27 NAVY-28
DELTA	IRAS RCA-F GOES-F GALAXY-A ATT-A RCA-G GALAXY-B EXOSAT	LANDSAT-D^1 GALAXY-C AMPTE NATO-IIID		GOES-G GOES-F(179)	
ATLAS CENTAUR	INT V-F	INT V-G INT VA-A INT VA-B	INT VA-C FSC-F	FSC-G	FSC-H
ATLAS	NOAA-E	NOAA-F	NOAA-G	NOAA-H	NOAA-I
TOTAL	11	11	9	6	5

Fig. 24. Planned Launch Schedule for Expendable Launch Vehicles

The upper stages used with the Shuttle are indicated on the next chart, and I'll go through those individually. The kinds of upper stages are indicated along the bottom of Figure 25. The capabilities of those upper stages are compared to the ARIANE-1, -2, -3, and -4 vehicle configurations, and the band of performance capability of the NASA upper stages is indicated by the hash marks.

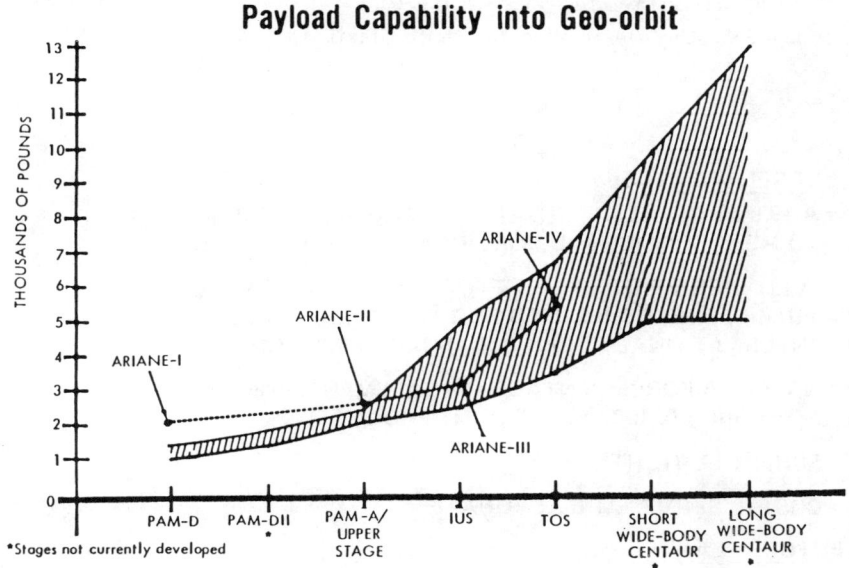

Fig. 25. Shuttle Upper Stages

The Centaur upper stage for Shuttle can be seen in Figure 26 and is proceeding well. There was a problem on the last ATLAS Centaur launch. The Centaur stage experienced a failure that caused the loss of the mission. An investigating team is looking at that failure. That failure was, as best as can be told right now, a mechanical failure in the tank system of the Centaur. We do not believe that there is a correlation between this failure and the Centaur-G or G Prime programs which are being built for the Galileo/ISPM, so we don't think, from what we know right now, that there would be a problem for these missions. The Centaur schedule is tight for making the Galileo/ISPM launches, but that's not a development or system schedule problem. That's a scheduling problem of getting all of the things walked through, and activities done at the Cape, without disrupting the rest of the STS schedule in time for those missions.

CURRENT STATUS
- A 19.5 FT STAGE DESIGNATED CENTAUR G BEING DEVELOPED TO MEET COMMON AIR FORCE/NASA REQUIREMENTS
- A LONGER VERSION (29.1 FT), DESIGNATED CENTAUR G PRIME BEING DEVELOPED FOR THE GALILEO AND INTERNATIONAL SOLAR POLAR (ISPM) MISSIONS
- NASA/AIR FORCE CENTAUR G AGREEMENT SIGNED NOVEMBER 9, 1982 FOR A JOINT PROGRAM
- SCHEDULE TIGHT
- COST RESERVES ARE ADEQUATE

FUTURE PLANS
- FIRST TWO FLIGHTS SCHEDULED FOR MAY 1986 (CENTAUR G PRIME)
- CENTAUR G VEHICLE TO BE OPERATIONAL BY MID-1987
- CENTAUR G VEHICLE PLANNED FOR VENUS RADAR MAPPER MISSION & TDRS-E IN 1988

Fig. 26. Centaur Upper Stage for Shuttle

The two G Prime missions are Galileo/ISPM. Centaur-G which is the second version of the Centaur will be ready by mid-1987. There is a question as to whether or not the Centaur-G vehicle will be used on the Venus Radar Mapper (VRM) or for the Tracking and Data Satellite System-E (TDRSS-E), and I'll just leave that because there are other systems available that can do those missions. Those competitive systems are available commercially, and we're looking at whether or not we should go competitive with procurement for those particular missions.

The upper stage Inertial Upper Stage (IUS) can be seen in Figure 27. This stage has had a flight problem, and it has delayed several missions. The most important mission, from your point of view, that has been delayed has been TDRSS. The investigating team has completed their failure analysis, and they have traced the failure back to a graphite seal. They have taken action to correct this problem, and we expect to have the IUS ready for flight later this year.

CURRENT STATUS
- DEVELOPMENT OF TITAN CONFIGURATION COMPLETED — FLOWN IN OCTOBER 1982
- DEVELOPMENT OF STS CONFIGURATION ALSO COMPLETED
  - SECOND STAGE FAILURE EXPERIENCED DURING LAUNCH OF TDRS ON STS—6 (4 APRIL 1983)
- FAILURE ANALYSIS COMPLETED — RECOVERY PLAN DEVELOPED AND BEING IMPLEMENTED

FUTURE PLANS
- SUCCESSFUL LAUNCH OF TDRS—B, —C, AND —D

CONCERNS
- STS MANIFEST IMPACT — TDRS SCHEDULE
- ADEQUACY OF "FIX"(TO BE DEMONSTRATED BY TEST FIRINGS

Fig. 27. Upper Stage Inertial Upper Stage (IUS)

The plans are to launch the TDRSS-B, -C, and -D on the IUS. The impact of having the IUS coming in all of a sudden, and having a high priority, will probably be an adverse one on the remainder of the missions that are scheduled there. At the time the figure was made, we had not completed the test firing program to determine the adequacy of the fix, but that has now been completed.

Payload Assist Modules (PAMs) are different because they are commercially developed upper stages. These particular upper stages, seen in Figure 28, were developed by McDonnell Douglas. There was NASA oversight of the developmental program.

CURRENT STATUS
- PAM—D COMMERCIAL DEVELOPMENT COMPLETED
- PAM—D ANOMALIES ON WESTAR AND PALAPA UNDER INVESTIGATION
- PAM—D OPERATIONS, INCLUDING SYSTEMS RELIABILITY, BEING ASSESSED
- PAM—DII COMMERCIAL DEVELOPMENT UNDERWAY

FUTURE PLANS
- CONTINUE SUCCESSFUL OPERATIONS OF PAM—D IN SUPPORT OF STS USERS
- PARTICIPATE IN MONITORING OF PAM—DII DEVELOPMENT AFTER OBTAINING NASA/MDAC AGREEMENT

Fig. 28. Payload Assist Modules (PAMS) Upper Stages

I must emphasize it was a commercially funded development. As you know, there were problems in the STS-10 mission where there was a failure of the nozzle in the PAM motors. McDonnell Douglas has gone through an extensive review and as of now they believe it was a materials problem associated with the nozzle that can be detected through additional nondestructive testing. Specifically, they believe they can CAT scan the nozzles and tell whether or not they can withstand the stresses that are induced on the nozzle during the first few seconds of firing. We plan to fly two PAMs on the upcoming mission. Those PAMs are Satellite Business Systems (SBS) and Telstar for AT&T. At last report, both of those missions are ready to go. McDonnell Douglas, Hughes, and the owners of the satellite are convinced that they have good motors and that they are ready for flight. We have continued successful operation of PAM with STS users and have participated in monitoring a world of the development of the PAM-D2. I might note that the problems with the PAM nozzle were with a carbon-carbon material; the PAM-D2 has a carbon phenolic material in the nozzle and we don't expect it to have the identical problems that were encountered with carbon-carbon.

McDonnell Douglas is planning a backup process of incorporating carbon phenolic into the PAM-D motor. That backup process may be available to users the first part of next fiscal year.

Another commercial development of a little bit larger stage is the Transfer Orbit Stage (TOS) (Figure 29). There is a group in Washington named Orbital Sciences Corporation that has raised venture capital funding to develop an upper stage. They call that upper stage the TOS.

It will be a pretty good size upper stage, much larger than the PAMs but much smaller than IUS and the Centaur. They plan to have an operational capability in December 1987. They have raised approximately $60 million to be able to proceed with that development.

CURRENT STATUS

- A NASA/OSC AGREEMENT FOR THE COMMERCIAL DEVELOPMENT OF TOS WAS SIGNED ON APRIL 18, 1983
- NEGOTIATIONS ON A SIMILAR AGREEMENT WITH BOEING ARE UNDERWAY
- ADDITIONAL PROPOSALS FOR COMMERCIAL DEVELOPMENT MAY BE FORTHCOMING
- COST AND PERFORMANCE TO BE COMPETITIVE WITH PROJECTED ARIANE-IV

FUTURE PLANS

- PROVIDE TECHNICAL MONITORING OF TOS DEVELOPMENT
- TOS OPERATIONAL CAPABILITY NO LATER THAN DECEMBER 1987

Fig. 29. Transfer Orbiter Stage (TOS)

We have received several commercial proposals (Figure 30). In May 1983, the President made the decision that after the phaseout of government operations of expendable launch vehicles, he would allow the private sector firms to operate these vehicles. We had received a proposal on Titan-34D; that proposal has subsequently been withdrawn. It was made in connection with the initial proposal by Space Transportation Corporation to purchase a fifth orbiter. They withdrew their proposal first to do the the Titan and second to do the fifth orbiter. We have since received a proposal from a group that was in Pittsburgh formerly known as the Cypress Corporation. It is now known as Astrotech International and we are discussing with them the possibilities of allowing them to finance and own the fifth orbiter as a commercial endeavor. Since these discussions are underway, I can't say much more about them. We have had a proposal from General Dynamics to operate the Atlas Centaur after 1987 when the government phases it out. We have negotiated an arrangement with a group in Washington called

Transpace Carriers, Incorporated, to turn over to them the Delta vehicle under certain terms and conditions on the first of October of this year. If those terms and conditions are not met, we probably will give them the option of taking over the program after 1986, after we've launched the last of the two geostationary weather satellites mentioned earlier. We've received numerous proposals and we have worked out numerous contracts with other companies on materials processing using the STS. Some of the companies are 3M, Microgravity Research Incorporated, John Deere, Dupont, etc. We've worked out what we call Joint Endeavor Agreements wherein they will do basic research using the Space Shuttle, the pallets, the Hitchhikers, the GAS canisters, and other capabilities that NASA will make available to them. In exchange NASA will provide them the facilities and the flight at no cost because they are expanding the basic research data base of the nation by providing the research results to NASA.

ELV'S

- TITAN 34D
- ATLAS CENTAUR
- DELTA
- NEW ELV'S (UNDER DEVELOPMENT)

STS

- MANUFACTURING/MATERIALS PROCESSING
- PLATFORMS FOR LEASE (INCLUDING FREE FLYER)
- TRANSFER ORBIT STAGE
- 5th ORBITER
- PAYLOAD PROCESSING
- OTHERS

Fig. 30. Commercial Proposals

There are platforms that are being conceptually designed for lease such as Leasecraft which is quite similar to the ESA EURECA. The Leasecraft will be a platform capable of an extended stay in orbit. They will lease space on the spacecraft, but we're not sure what the rates, terms, or conditions might be.

Earlier I mentioned the Transfer Orbit Stage, and I mentioned the fifth orbiter and our commercial proposal with respect to this orbiter. There is a new commercial payload processing facility in place just outside Kennedy. It came into being about May of this year, and it has a capability quite similar to that already existing on the Cape. This is designed to take care of the surge that is beyond the existing government capability, therefore relieving the government of the obligation of adding to its facilities.

Mike Sander mentioned earlier the Tethered Satellite System program (Figure 31). That program is temporarily within the Office of Space Flight. The newly-formed Space Station Office is looking at that program with a lot of "predatory-like" glances and we are expecting that the program will be transferred out of the Office of Space Flight into the Space Station Program within the next few weeks, but the figure indicates the status of the program. It's a very exciting small-scale program being conducted by the Office of Space Flight in cooperation with the Office of Space Science and Applications.

- DESIGN/DEVELOPMENT PHASE INITIATED JAN 1984
- INITIAL FLIGHT OF TSS ON SHUTTLE PLANNED FOR DEC 1987
- OBJECTIVES:
    - DEMONSTRATE USE OF TETHER FROM SHUTTLE
    - QUANTIFY INTERACTION BETWEEN TSS AND SPACE PLASMA IN PRESENCE OF ELECTRIC CURRENT CONDUCTED THROUGH THE TETHER
    - DEVELOP A REUSABLE FACILITY CAPABLE OF SUPPORTING A BROAD RANGE OF TETHERED SATELLITE APPLICATIONS

Fig. 31. Tethered Satellite System (TSS)

In summary, (Figure 32), we think we are moving right along with the STS toward a fully operational status. Progress comes slowly sometimes. Most of our plans and facilities are in place for the full operational capability. We are making progress in areas such as the Shuttle processing contract which is in place at Kennedy. We're moving toward having a similar contract in place at the Johnson Space Center. We're progressing in the areas of logistics and performance augmentation. We expect the Vandenburg launch site will be tested and proven as new elements of the STS next year. We're attempting to continue the cooperation necessary between NASA, the DoD, and industry for a total program success, and we are initiating programs to aggressively exploit the full capability of the STS and to provide an integrated link between the Space Shuttle and Station.

- **STS IS MOVING RAPIDLY TO OPERATIONAL STATUS**

- **MOST PLANS AND FACILITIES IN PLACE FOR FULL UTILIZATION**

- **BY 1986, SPC, LOGISTICS, PERFORMANCE AUGMENTATION AND VLS WILL BE TESTED AND PROVEN ELEMENTS OF STS**

- **CONTINUED TOTAL COOPERATION BY NASA, DOD AND INDUSTRY ESSENTIAL TO SUCCESS**

- **AGGRESSIVELY EXPLOIT FULL CAPABILITY OF STS**

- **PROVIDE INTEGRATED LINK BETWEEN STS AND SPACE STATION**

Fig. 32. Summary

In addition to these ongoing programmatic efforts, we have several advanced program study efforts underway which I'd like to go over with you just briefly to give you an example of the things that we're doing.

The objectives of our Advanced Program Office (Figure 33) are to support major changes in spacecraft design and use, by demonstrating capabilities that are available to them from the Space Shuttle; to proceed with the implementation of the appropriate infrastructure for accomplishing future space activities; and, finally, to define a long range future for major missions and goals. By long range I mean 1995 to 2005.

Figure 34 shows efforts to support major changes in spacecraft design and use. We are looking at these areas, some of which we've already demonstrated, and others in which the demonstrations are underway or activities are underway to support demonstrations. For example, we've already serviced the Solar Max Mission, but we have an orbital refueling demonstration coming up as an example of what we are doing in this area.

- **SUPPORT MAJOR CHANGES IN SPACECRAFT DESIGN AND USE**

- **PROCEED WITH INFRASTRUCTURE IMPLEMENTATION**

- **DEFINE LONG—RANGE FUTURE MAJOR MISSIONS/GOALS**

Fig. 33. Objectives of Advanced Program Office

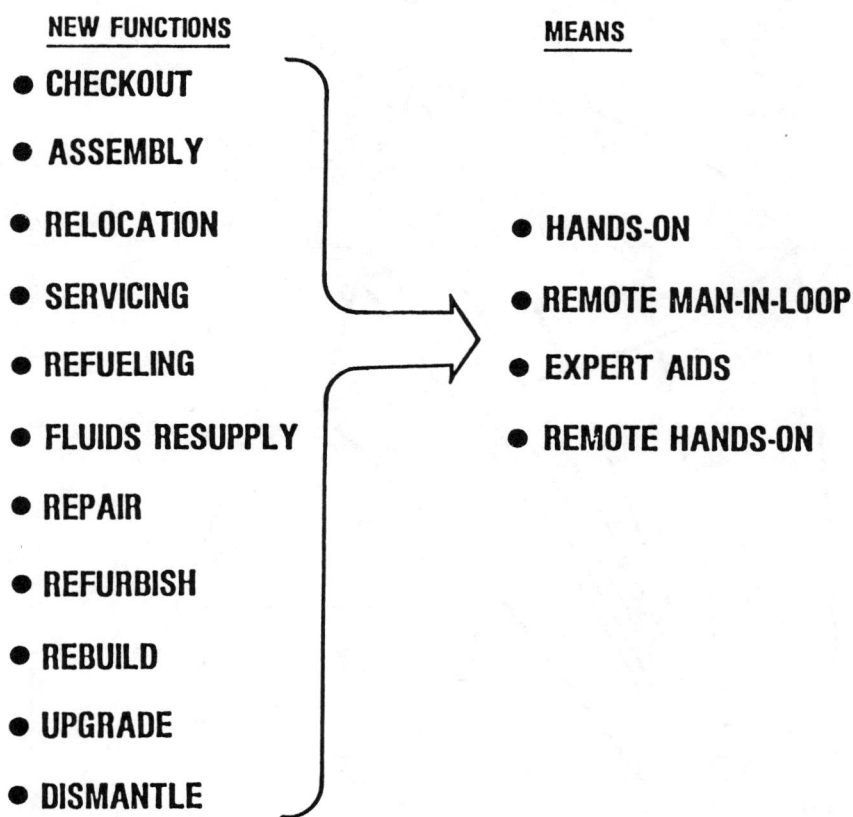

Fig. 34. Efforts to Support Major Changes in Spacecraft Design and Use

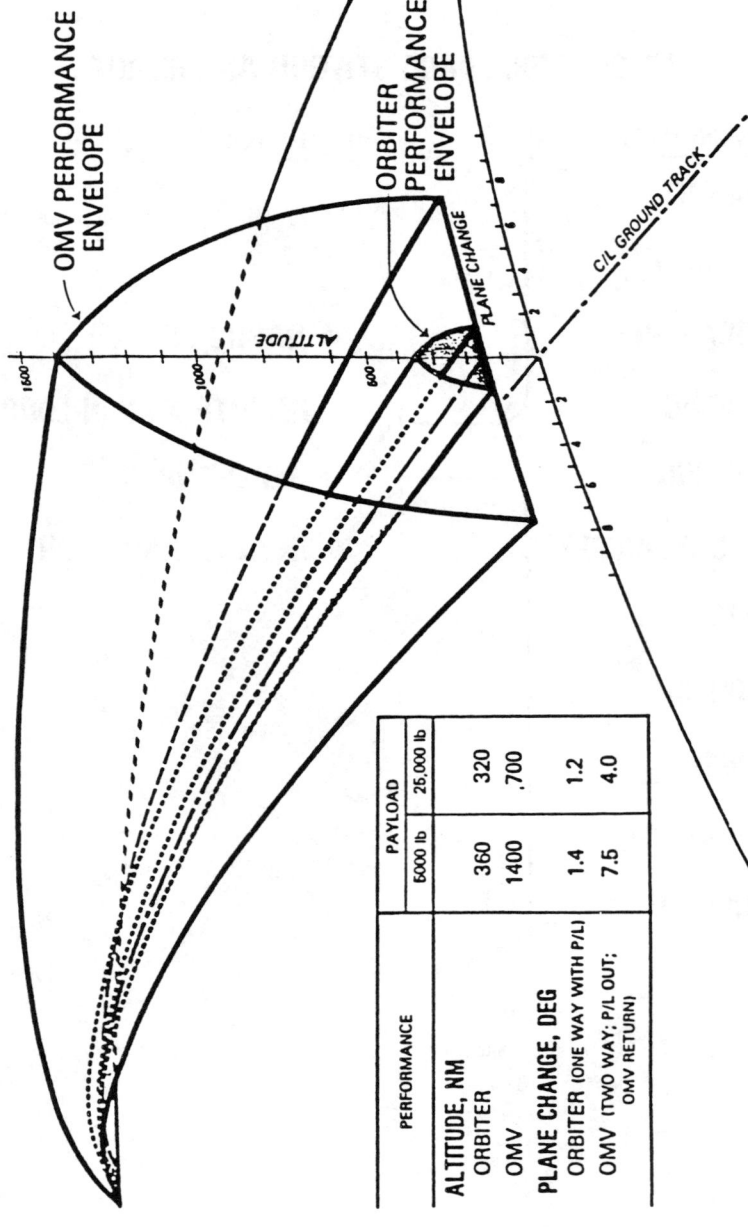

Fig. 35. Status of Phase B Activity of Orbital Maneuvering Vehicle

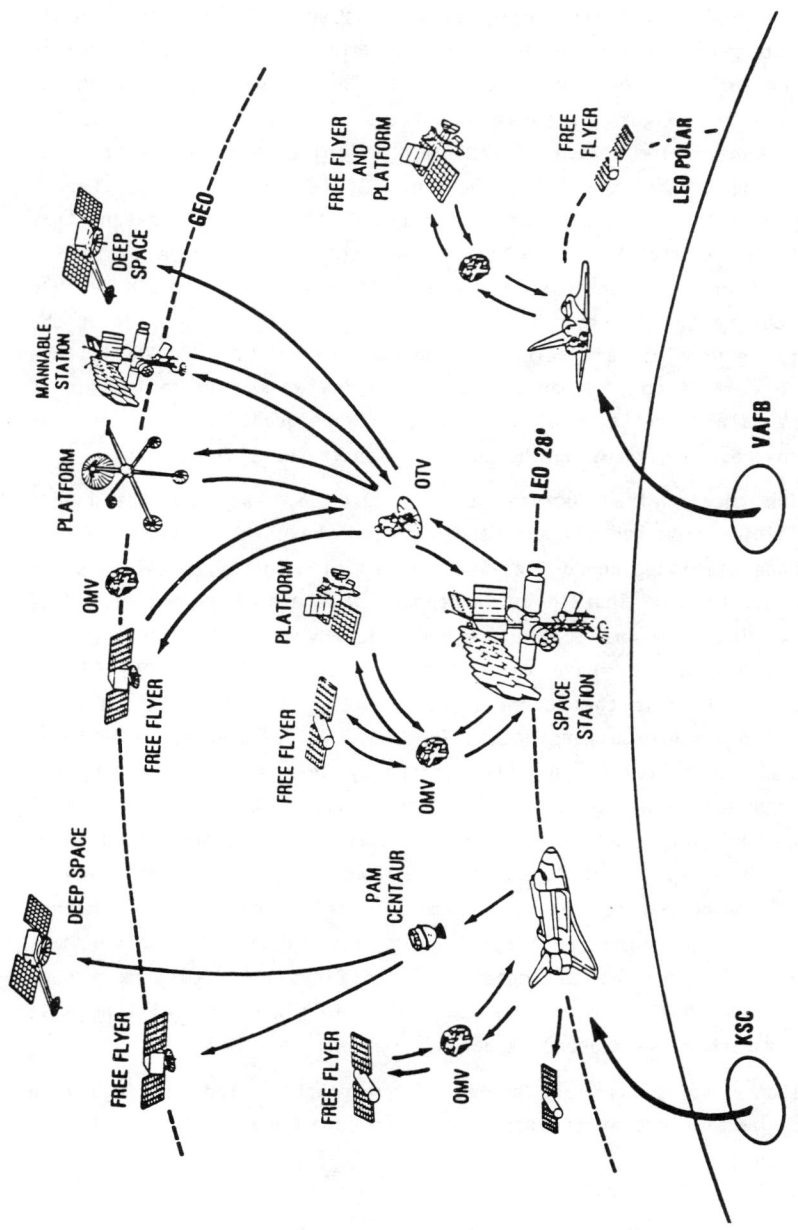

Fig. 36. Total Space Infrastructure

The orbiter refueling demonstration is scheduled for sometime in the next six or seven months. We are going to go out and actually transfer fuels in the payload bay of the Shuttle with an extravehicular activity. Figure 35 shows the status of Phase B activity for the Orbital Maneuvering Vehicle. The small triangle indicates the present maneuvering capability of the Shuttle. If we have an Orbital Maneuvering Vehicle in the payload bay of the Shuttle it would extend that capability to the larger volume indicated. For example, with the Orbiter Maneuvering Vehicle from the Shuttle orbit, you would be able to go out to the far left-hand side or the far right-hand side of the top of the envelope and return to the Shuttle orbit. The OMV is in a Phase B definition. We announced just a few weeks ago definition of the OMV as a potential FY86 new start. It has great possibilities, as you can see, for improving the capabilities in the system.

The total infrastructure is in Figure 36. We have the launch capabilities from Kennedy and Vandenburg, we have the Shuttle, we have the Space Station planned for 1992, and we plan to fly free-flyers and platforms from the Shuttle at Vandenburg in the initial phases. However, coming from the Station we have the OMV that will increase our capabilities, and we have the Orbiter Transfer Vehicle. The OTV is basically a vehicle that carries you from the low orbit Station altitude up to synchronous and back. That capability is anticipated to be a manned capability which will eventually lead to a capability for platforms and manned Stations at synchronous orbit. If you can have this by the year 2000, then with this infrastructure you can do anything in space you want to do for the next 30 years. By anything in space, I mean you can operate from this infrastructure to a manned lunar base, you can go to the asteroids, or you can go to a man/Mars mission. It provides the capability and flexibility to pick a goal for the year 2000. If this capability is in place, you will be able to accomplish those subsequent goals.

That completes a long series of viewgraphs. I'd like again to express my pleasure at the opportunity to come here and talk with you. Thank you very much.

SPACELAB SCIENCE IN THE NEXT DECADE

Robert D. Chapman*
Goddard Space Flight Center
National Aeronautics and Space Administration

Dr. Jeffrey D. Rosendhal, the Assistant Associate Administrator in the Office of Space Science and Applications, was going to come here and give an overview of the Agency's plans for using the Shuttle to do science during the next decade. We thought it would be a good way to set the stage for the activities during the rest of the meeting. It turns out that Dr. Rosendhal is unable to be here and he asked me to stand in for him. I am going to give you an overview of the Spacelab Payload Program as it is currently envisioned. I found it an interesting task preparing this talk because I had last been plugged into the activity about one year ago. The way things operate at NASA Headquarters, if you get plugged into an activity and understand it, then don't look at it for a year it's pretty amazing to see the changes after that length of time. That's the case with this program also. Some of the things that I want to tell you about have already been covered a little bit this morning, so we can go over that very quickly and get into the meat of the presentation.

Just to set the stage there are four missions that have already been completed: Spacelab-1, OSS-1, OSTA-1, and OSTA-2. That is the the message that's important for this meeting. We are beginning to develop an understanding of the Shuttle and how we can do science from it as a result particularly of these four missions.

The concept that I want to bring out is single discipline labs; this is the direction being taken by Agency planning, with the advice

---

* Now at NASA's Lyndon B. Johnson Space Flight Center.

of the scientific community. We have already heard this morning one of the user concerns that single discipline labs may not be the optimum way to do science. It might be better instead to have two or more disciplines on a single flight in order to take maximum advantage not only of the Shuttle pointing activities, but also of the astronauts and scientists that are on board to carry out the activities. Of course the bottom line is that everything that is happening on the Shuttle has a goal of learning to carry out science on Space Station. There is a planned evolution through a whole series of missions which will ultimately lead to the Space Station. I think that, as we talk among ourselves over the next few days, we have to keep in mind this evolution toward the Space Station. I have a feeling that, in the future, if this activity continues, we will become the Shuttle and Space Station Contamination Workshop.

Issac T. Gillam IV showed Figures 1 and 2 this morning, and he gave a very good overview of what they are all about, so I won't dwell on them. Figure 3 shows the status of some of the discipline laboratories that we are going to be talking about. You can see that they go all the way from the International Microgravity Lab (which has just completed a feasibility study) to the Materials Science Lab, the Space Science Lab, and the Astro I Mission which are in the implementation stage. The two missions that are farthest along, at least according to this chart, are the Shuttle Radar Lab and the Materials Science Laboratory.

The first Lab I will talk about is the International Microgravity Laboratory which in the not terribly distant past would have been called the Materials Processing Laboratory, with a first launch currently planned for 1987. It has an objective of doing low-gravity research in both the materials processing and the life science areas, and candidate facilities include both life sciences and materials processing. At this time the experiments have not been completely selected.

The Earth Observations Mission, which a year ago was called the Environmental Observations Mission, includes two types of observations. Solar irradiance observations look at the total energy output

R. D. Chapman 57

	1983	1984	1985	1986	1987	1988	1989	1990	1991
1. Space Life Sciences Lab			◁					◁	
2. Astro				◁	◁				◁
3. Solar Optical Telescope				◁				◁	◁
4. Shuttle Radar Laboratory		◁			◁	◁	◁		◁
5. International Microgravity Lab					◁		◁		
6. Payload of Opportunity Carrier		◁	◁◁	⋙	⋙	⋙	⋙	⋙	⋙
7. Space Plasma Lab				◁		◁	◁		◁
8. Shuttle High Energy Astrophysics Lab		◁	◁	◁	◁	◁	◁	◁	◁◁
9. Material Science Laboratory		◁		◁◁	◁◁	◁◁	◁◁	◁	◁
10. Shuttle Infrared Telescope					◁	◁	◁	◁	◁
11. Environmental Observation Mission				◁					◁
Multidiscipline Missions	SL-1 ◁	SL-3 ◁	SL-2 ◁						

Fig. 1. Discipline Laboratory Overview

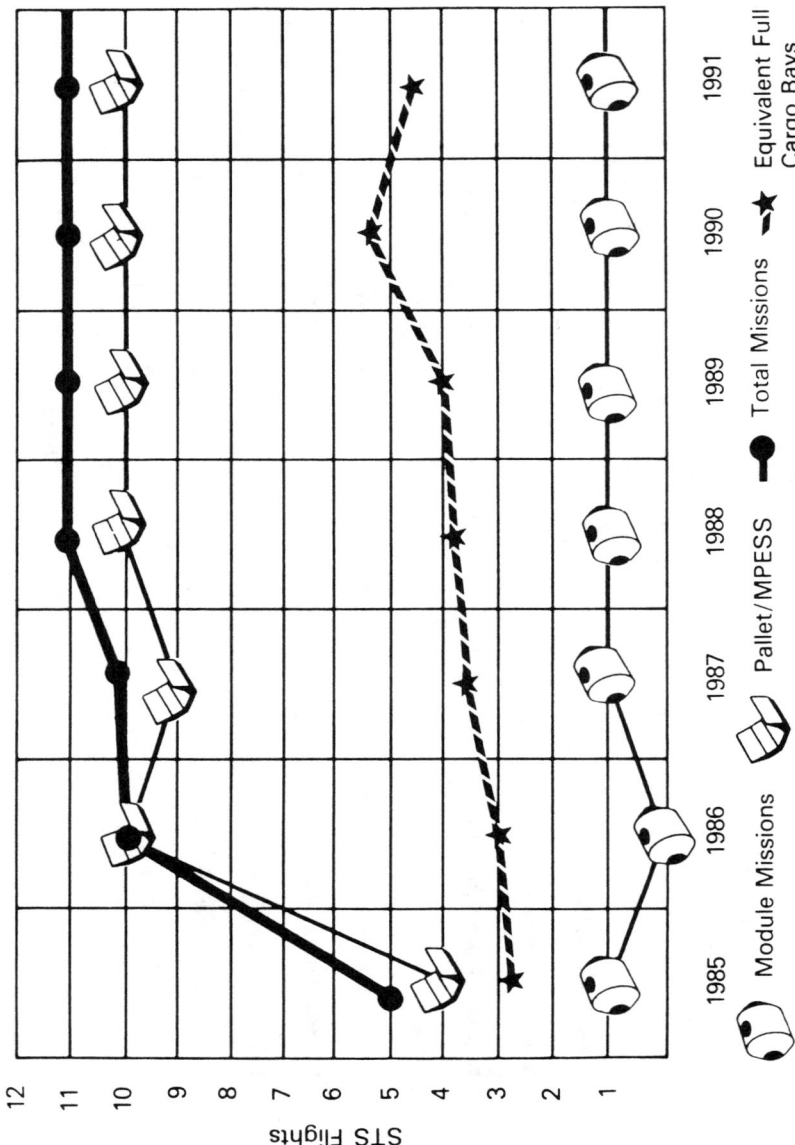

Fig. 2. Discipline Laboratories - Flight Model/ September 1983

Laboratory	Feasibility Study	Instruments Selected	Definition Study	In Implementation
1. Space Life Sciences Lab	▨	▨	▨	▨
2. Astro	▨	▨	▨	▨
3. Solar Optical Telescope	▨	▨	▨	
4. Shuttle Radar Lab	▨	▨	▨	▨
5. International Microgravity Lab	▨			
6. Payload of Opportunity Carrier	▨	▨	▨	
7. Space Plasma Lab	▨	▨		
8. Shuttle High Energy Astrophysics Lab	▨	▨		
9. Material Science Lab	▨	▨	▨	▨
10. Shuttle Infrared Telescope	▨	▨		
11. Environmental Observation Mission	▨	▨		

Fig. 3. Discipline Laboratory Status

from the Sun, both integrated over all wavelengths and as a function of wavelength. Instruments such as active cavity radiometers will be used in an attempt to get very high precision measurements of the solar output. It's interesting that back in the 20's and 30's, a lot of observations were made by Dr. Abbott of the Smithsonian Institution from the ground concerning the changes of the brightness of the Sun when sunspots rotated with the surface of the Sun. Few people had confidence in Abbott's results. Interestingly enough, the active cavity radiometer on the Solar Maximum Mission seems to be verifying Abbott's work. We want to continue modern observations and get higher precision and much more information about how the total radiation of the Sun changes with different solar activity conditions. In addition, the set of instruments on the mission will be looking at the Sun particularly during sunrises and sunsets in order to study the change in brightness of the Sun as it disappears behind the terrestrial atmosphere, which can be used to map the trace constituents in the Earth's atmosphere. The first launch is November '86, and will include five to ten experiments all concerning the Sun and its influence on the Earth's environment.

There's not too much that one can say about the Payload of Opportunity Carrier right now except that it was initially planned for a first launch in 1984. I think that, as late as we are in '84, this date may be a little optimistic. I don't know what the real launch plans are for this right now, but it does have the capability of putting small experiments in orbit quickly in order to respond to unusual opportunities. I expect it would include situations like astronomers wanting to respond to transients such as supernovae and similar things; it provides an opportunity to get instruments into orbit very quickly.

The Shuttle Radar Laboratory includes an instrument which is an outgrowth of SAR-A which was flown on the OSTA-1 mission and is going to consist of an imaging radar to carry out radar mapping along its ground track. This is a very slightly side-viewing radar which gives good images of geological features. The payload also includes a large format camera which produces simultaneous white light images and

Measurement of Air Pollution from Satellites (MAPS), which is a test of the feasibility of monitoring different kinds of air pollution from space. The first launch is scheduled August 1984 and it is to be flown on 12- to 18-month intervals.

The Space Life Sciences Laboratory was at one time called Spacelab-4. There are, in fact, a couple of principal investigators on this mission in the audience, and I'm sure they could tell you more about the mission than I can. The first launch according to the schedule is for January '86. Flight intervals are then 12 to 24 months. Twenty experiments will be on board. Another important user concern is to try to run 20 experiments with one-shift operation. I think operating such payloads is going to be quite an interesting challenge, and I'm sure it is something that will be coming up in the discussions in the future here. A major concern is effective operation of that many experiments with the number of payload people that are available.

The Materials Science Laboratory is planned to perform materials processing research in the low-g environment. This is actually an autonomous instrument package that is turned on and carries out a preplanned set of activities. The main package is the materials experiment assembly which has three different furnaces for doing different kinds of materials processing experiments. In addition, on each side are three get away special cans (GAS) with additional materials science.

The Shuttle High Energy Astrophysics Laboratory's first launch is planned for 1988. It's supposed to be launched once every 12 months initially. Three of them are the Large Area Modular Array of Reflections, the Diffuse X-Ray spectrometer and the Broad Band X-Ray telescope -- major facility instruments. In addition, there is something called the Cosmic Ray Nuclei Experiment. The Large Area Modular Array is an x-ray telescope for looking at very faint sources. It's about half an order of magnitude more sensitive than anything that's ever been flown. The Broad Band X-ray Telescope (BBXRT) will be doing spectrometry of point x-ray sources. The objectives are to survey the low-energy diffuse x-ray spectrum to observe cosmic x-ray sources.

The purpose of the Cosmic Ray Nuclei Experiment is to carry out a chemical analysis of heavy cosmic ray nuclei.

The Space Plasma Laboratory, which was at one time known as Spacelab-6, is a joint venture between Canada, Japan, the United States (NASA), and France. The basic idea of this package is to send a group of experiments into space which will actively experiment with the plasma environment of the Earth. There will be instruments such as radio frequency transmitters, sending long wavelength radio radiation into the environment, and particle beam accelerators, accelerating particles into the magnetosphere. Instruments on board will then look at the results of these unnatural perturbations of the natural environment. We have very little hands-on experience with plasmas on the scale that one sees in the Earth's environment. We are going to get more information, not only about the Earth's environment but also about plasmas on a scale that will be useful in magnetospheric physics and to the astrophysicists. I think a great deal of important plasma physics will come out of that, not only for the people that are interested in the Earth's environment but also for the rest of science as well.

The Astro Mission's first launch is planned for March 1986. I think we are marching along pretty well to achieve this date. There are four instruments on the package. The three major instruments are: the Ultraviolet Imaging Telescope which is being built by Goddard, an instrument being built at Johns Hopkins called the Hopkins Ultraviolet Telescope, and one that's being built at Wisconsin which is known by the acronym WUPPE (Wisconsin Ultraviolet Photometer Polarimeter Experiment). The fourth instrument which is a fairly recent addition to this payload is a small, simple Wide Field Camera. It turns out that the first launch of Astro will be when Halley's Comet is near to the earth and relatively bright. The project put on the wide angle cameras to let us get frequent wide angle observations of Halley's Comet.

The Solar Optical Telescope is slated for its first launch in 1990 with flight intervals of approximately 12 months. This major facility for solar physics is going to have very high resolution for looking at fine details on the Sun. The very best resolution that one can get from the ground is a quarter of an arcsec, while SOT will achieve 1/10 of an arcsec resolution. One may get quarter of an arcsec resolution observations on a few days of the year from the very best sites in the world. Typically one will achieve one arcsec or half an arcsec resolution on the standard, run-of-the-mill days. SOT will give somewhere between a order of magnitude and half an order of magnitude greater resolution on the Sun than has ever been achieved before. We know that there are things happening on the Sun on those scales and smaller. I think this mission is going to lead to a much more detailed understanding of the phenomena that take place on the quiet Sun, and hopefully someday on the active Sun on these scales. I say hopefully someday on the active Sun because the probability of catching a solar flare is relatively small from observations during a one-week mission. This is an instrument which is going to have to evolve into a Space Station experiment where it can stay in space for very long times in order to study solar flares.

Advanced Technology Missions have a first launch slated for June 1984, and a flight interval of 24 months. This mission will establish some of the space technology we are going to need for future spacecraft.

The Large Format Camera is a mission designed to look at the Earth with a camera, which is an outgrowth of the camera which was flown on Spacelab-1 from which we got extraordinarily interesting high resolution pictures of the ground. With a mission completely dedicated to a Large Format Camera, we'll be able to acquire a very large number of images to carry out synoptic high resolution studies of the Earth's surface.

Sunlab is an advanced instrument package for looking at a large number of solar phenomena. Slated for launch in July 1986, it will study magnetic fields on the Sun, velocity fields, etc.

In summary, the overall activity is leading toward a Space Station in order to get the long-term orbital studies which many of these missions require. In the process, it is providing opportunities for flights during this decade in order to maintain the science teams' interest during "free-flyer lull." This concludes a brief survey of some of the things the Spacelab Payloads Engineering Office is going to be putting forward during this next decade.

PEOPLE IN THE LOOP: SPACE OPERATIONS OF THE FUTURE

Byron Lichtenberg
Johnson Space Center
National Aeronautics and Space Administration

The launch of Spacelab 1 on Nov. 28, 1983 heralded a new era in the conduct of space science and operations. It marked the first voyage of a new space laboratory called Spacelab, designed, developed, built, and financed by the European Space Agency. It also marked the first flight of a new breed of space traveller called a Payload Specialist--a scientist selected to fly and conduct experiments in space by his/her science peers. In this case, the science peers, called Principal Investigators (P.I.'s) came from around the world and represented scientific experiments in five different science disciplines including Earth Observations, Space Plasma Physics, Astronomy and Solar Physics, Materials Sciences, and Life Sciences. The degree of crew involvement in operating these experiments varied from no involvement to complete responsibility for the set-up and conduct of the experiment, including real-time data verification from on board. We Payload Specialists were trained for five years in the laboratories of the P.I.'s to be able to make these types of data verification. The concept of operators on board that could communicate directly with their science peers on the ground was a new concept that had not been tried before.

Besides normal experiment operations, we also were trained in contingency operations so that we could diagnose and hopefully fix problems that might occur during flight. In fact, we had to perform a number of repairs and work-arounds to experiments in all the disciplines, but we were especially useful in the areas of life and materials sciences and Earth observations. These operations consisted of mechanical fixes to power supplies, unjamming stuck sample cartridges, patching software, substituting back-up equipment, and unjamming and rethreading a large format film magazine. Some of the fixes were purely engineering, but some were initiated by the Payload Specialist to improve data collection or experiment operation.

With Payload Specialists (laboratory qualified scientists) working as a part of the experimenter team, it is possible to enhance the design of science experiments. The PI can take advantage of our broad-based knowledge of the entire Shuttle system to facilitate cross-communication between various experimenters or Shuttle systems. One very specific example of this is the use of the video system to record materials science experiments during their operation. The crew was instrumental in making the capabilities of this system known to experimenters and helping to coordinate the use of this equipment with the various Shuttle project offices. This allowed real-time ground viewing of complex operations in the Fluid Physics Module which permitted interactive science advancement. The PI's were able to observe and comment during the operation and direct it almost as if they were there.

Although the Spacelab is fairly new (at this printing it has flown only once), and will undoubtedly change the nature of space experimentation, the great advance in space operations will come with the Space Station. The ability to communicate in real time with surrogates on board and also to have data available in both places will definitely change the manner of space operations. The Spacelab flights will not be as severely affected due to the short duration of the missions (ten days or so) which makes each minute very precious. Thus, onboard data analysis and interpretation is not cost effective. However, with the advent of powerful personal computers and advanced communications technology, the Spacelab will serve as a test bed for the Space Station and a new form of science called "telescience" coined by Prof. Peter Banks of Stanford University. This concept will use advances in technology to bring the ground-based scientist "closer" to his/her work by using teleoperators for remote controlled experimentation or by being in constant communication (voice, video and data) with remote operators (payload specialists) in space.

With the advent of this new technology, it is important that humans be utilized for the particular skills that they posess. It is clear that computers can do many of the routine process controls of an experiment like temperature regulation, data formatting, sequence control, etc., and that people are not well-suited for repetitive tasks or routine monitoring. The human should be used to provide higher level supervisory control, interactive decision making, and integration and synthesis of multiple inputs.

STS ENVIRONMENT MEASUREMENTS AND PRESENT PLANS

Robert L. Blount
Lyndon B. Johnson Space Center
National Aeronautics and Space Administration

The STS has participated in workshops of this type over the past few years in order to express the current status of STS environmental measurements and to be in a better position to respond to experimenter's environmental concerns. I intend to take advantage of the Henniker workshop to state very briefly what the STS program of environmental measurements has been, and to outline those areas where the STS has firm plans to make additional environmental measurements.

THERMAL ENVIRONMENT

Payload related thermal environmental measurements were made on STS-1 through STS-5. A total of 33 cargo bay measurements was made on each mission with a wide range of Orbiter thermal attitudes flown. Review of flight data indicated that the STS initial thermal models were conservative, but less than 40 degrees F conservative in all cases, and usually correlating within 10 degrees for the hot and cold cases. In the spring of 1984, the STS did release a new improved thermal model which more accurately predicted on-orbit conditions. Predictions using the new model were found to correlate within 5 to 10 degrees F for the vast majority of the cases evaluated. The STS has no further measurement plans in the thermal area and considers existing thermal models to be adequate for payload integration activity including future flights from the Western Test Range (Fig. 1).

CONTAMINATION

The STS contribution to the measurement of the contamination environment consisted of flying the Induced Environment Contamination Monitor (IECM) which was made up of twelve individual instruments on

SPACE SHUTTLE ENVIRONMENT

FLIGHT DATA

- 33 CARGO BAY MEASUREMENTS
- FLOWN STS-1 THROUGH STS-5
- WIDE RANGE OF ORBITER ATTITUDES

INITIAL MODEL COORELATION

- MODELS CONSERVATIVE (LESS THAN 40°F) FOR ALL CASES
- MODELS WITHIN 10° FOR HOT AND COLD CASES

MODEL UPGRADING - 1984 - 5 TO 10° CORRELATION

	390 NODE	SIMPLE
OPEN DOOR	ES-3-76-1 REV D	JSC 19540
CLOSED DOOR	ES-3-77-3 REV D	JSC 19692

WTR CONSIDERATION

- NO NEW UNIQUE THERMAL ENVIRONMENT

Fig. 1. LA/6 Mission Integration--Thermal

STS-2, -3, -4, and also on STS-9 as a part of the Spacelab mission. Column density and deposition rates were measured to be essentially as predicted, confirming that existing operational procedures are functional for contamination control. "Glow" was established as an environmental effect needing additional study. The STS did sponsor detailed science objectives on STS-5 and STS-8 to pursue the "glow" phenomenon. The STS has no additional glow measurement plans on the Orbiter, but will continue to work with the experimenters in order to complete the environmental definition and develop the needed operational constraints. The STS will continue a rigorous materials control program on the Orbiter and will attempt to prevent payload-to-payload contamination by careful attention to detail during the payload integration process. We do not consider that the Western Test Range will present a significant difference in the potential for the contamination environment when compared to KSC; by monitoring the launch facilities we can establish the degree of equivalence (Fig. 2).

FLIGHT DATA

o IECM (12 INDIVIDUAL INSTRUMENTS) ON STS-2, 3, 4, AND 9

MEASURED VS. PREDICTED ENVIRONMENT

o COLUMN DENSITY AND DEPOSITION RATES ESSENTIALLY AS PREDICTED
o CONFIRMED OPERATIONAL PROCEDURES ARE EFFECTIVE AT CONTROLLING CONTAMINATION
o VEHICLE "GLOW" AND ENVIRONMENTAL EFFECTS ON MATERIALS UNDER INVESTIGATION
  - STS SUPPORTED EXPERIMENTS ON ATOMIC OXYGEN EFFECTS
    AND SURFACE GLOW ON STS-5 AND STS-8

ADDITIONAL MEASUREMENTS

o STS HAS NO ADDITIONAL MEASUREMENT PLANS ON THE ORBITER, BUT:
  - WILL CONTINUE TO EXAMINE P/L DATA TO COMPLETE DEFINITION
    AND DEVELOP NEEDED OPERATIONAL TECHNIQUES
  - WILL CONTINUE ORBITER MATERIALS CONTROL PROGRAM TO LIMIT OUTGASSING
  - HAS INTEGRATED THE IOCM BY THE "FORM 100", "PIP", "ICD" PROCESS

WTR CONSIDERATION

o WTR NOT A PRIMARY DRIVER FOR CONTAMINTION DURING FLIGHT PHASES
o MONITORING LAUNCH FACILITIES CAN ESTABLISH COMPARABILITY TO KSC

Fig. 2. LA/6 Mission Integration--Contamination

ELECTROMAGNETIC ENVIRONMENT

The STS measurement program in this environmental area was primarily ground-based and made on Orbiter OV101. Those measurements were made prior to the OV101 drop test program and did result in revisions to the electromagnetic environment as reflected in the ICD2-19001. Additional electromagnetic measurements were made in the Shuttle Avionics Integration Laboratory (SAIL), primarily in the conducted emissions area. A limited number of measurements were made during the flight of OV102 using the PDP deployed on the RMS to gain spatial resolution. Measurements generally confirmed preflight predictions.

The STS does not expect the design electromagnetic environment to change as a result of flying the Space Shuttle out of the Western Test Range. We will update analytical models with additional data on a ground-based transmitter located at WTR.

There has been some concern in the user community directed toward the Orbiter Ku-band communication and radar system. We have developed a software mask which precludes direct beam exposure for payloads located within the payload bay. Payloads requiring complete protection may request Ku-band system "power down." There continues to be some user concern in this area, directed primarily at the reliability of the software mask in protecting payloads in the cargo bay.

The STS and users are also exposed to RF environments generated by ground-based transmitters which often exceed environments produced by the Orbiter or other payloads. Exposure to these ground-based threats are generally of short duration, and the exposure levels are not expected to be of a damaging nature for general avionic equipment; therefore, a design that can live with circuit upset from a short duration RF threat should prove to be acceptable (Fig. 3).

ENVIRONMENTAL MEASUREMENTS

- OV101 ELECTRIC AND MAGNETIC FIELDS (REV. TO ICD 2-19001)
- SAIL POWER BUS RIPPLE AND TRANSIENTS
- OV102 (PDP) GENERALLY CONFIRMED ICD2-19001

WTR CONSIDERATION

- ICD 2-19001 NOT EXPECTED TO CHANGE
- GROUND BASED XMTR DATA BASE UPDATE

KU-BAND SYSTEM

- CONTROL - MASK - PROCEDURES - TURNOFF
- SOME CONTINUE USER CONCERN

GROUND BASED XMTRS

- EXCEED STS PRODUCED ENVIRONMENT
- EXPOSURE DURATION USUALLY SHORT
- NOT CONSIDERED BY STS TO BE A DAMAGE CONCERN

Fig. 3. LA/6 Mission Integration--Electromagnetic

LOADS AND LOW FREQUENCY DYNAMICS

The STS flew nine accelerometers on the mid-fuselage measuring loads and low frequency dynamics on STS-1 through STS-5 (Figure 4). Additional accelerometer measurements were made by members of the payload community. In comparing flight data to predictive models, we find the response amplitude generally well below the design cases predicted by the model. There are some disparities in the frequency content of the flight data versus the existing models that we continue to live with at the present time. Users have suggested we needed additional data in order to refine our modeling techniques and to assure our methodology was as it should be. I am happy to report that the STS has added 10 accelerometers to OV099 since the last environmental workshop. These added instruments have flown one time, and will fly again in October of 1984. We also have 10 payload bay accelerometers authorized for installation on OV103 prior to its first flight out of WTR. This added instrumentation has always been anticipated as a requirement to evaluate possible differences in launch pad stiffness between the Eastern and Western Test Ranges. Engineering is underway to add accelerometer measurements to the payload bay area of OV102 and OV104, but again as in the recent past, those installations have been held up because of interference with the scheduled logistic flow plan for the two vehicles. The STS does intend to collect additional loads data to insure that our methodology is correct and to reduce the cost and time required for the loads verification cycle (Figure 5).

VIBRATION AND ACOUSTICS

Vibration and acoustic environmental data in the Orbiter payload bay has been measured on flights STS-1 through STS-5. Four microphones and 13 high frequency accelerometers were flown on each of the first five missions. The acoustic amplitudes were lower than predicted in the high frequency range, and random vibrations were slightly higher than predicted. In addition, higher level discrete tones were identified in the area of the payload bay vents. Changes in ICD1-19001 were made in order to reflect the above mentioned findings.

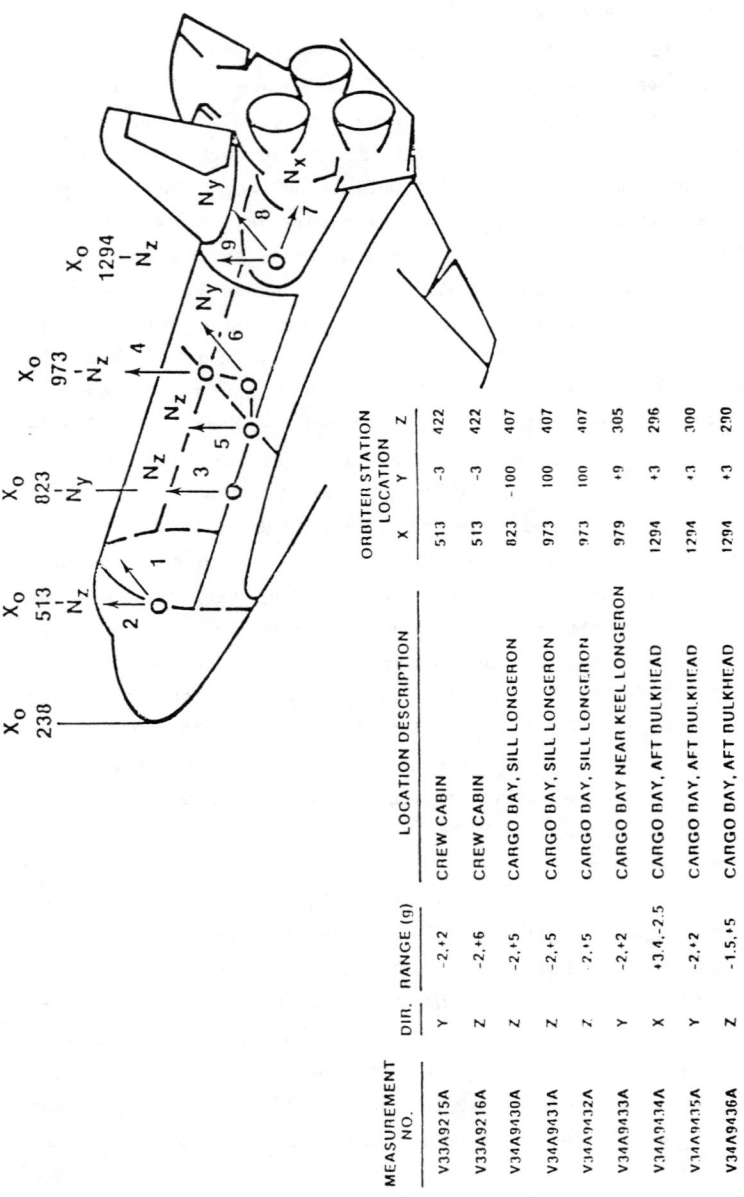

Fig. 4. Orbiter Low Frequency Accelerometers STS-1 through STS-5

FLIGHT DATA

- o Nine OV102 fuselage 0-20Hz accelerometers STS-1 through STS-5
- o Additional payload supplied accelerometers
  (OSS-1, SBS, GAS, IUS/TDRS)

DESIGN CASE LOADS VS FLIGHT LOADS

- o Response amplitudes generally well below design cases
- o Frequency comparison - excitation overpredicted in 30Hz range - underpredicted in 3Hz range for liftoff
- o Landing impact well below design case except for STS-3

ADDITIONAL MEASUREMENTS

- o 10 accelerometers now flying on OV099
- o Authorized for OV103 prior to first WTR flight
- o Additional measurements on OV102 and OV104 are TBD

PROGRAM GOALS

- o Verify basic loads methodlogy
- o Reduce cost/time of loads verification cycle

Fig. 5. LA/6 Mission Integration--Loads and Low Frequencies

Again, as in the case of loads and low frequency dynamics, the STS has responded to user requests for additional data, by installing additional measurements for vibration and acoustics in OV099, which has now flown one additional flight with many more flights anticipated. The STS intends to (1) characterize the payload bay vent noise, (2) establish large-volume payload effects if any, (3) better define acoustics in the forward third of the payload bay, and (4) evaluate possible differences between our various vehicle configurations. Also OV103 will be modified to include additional payload bay vibration and acoustic instrumentation prior to launch from the Western Test Range (Figure 6).

FLIGHT DATA

- o 4 Microphones, 13 H.F. accelerometers - STS-1 through STS-5
- o Additional accelerometers and microphones on payloads

MEASURED VS PREDICTED ENVIRONMENT

- o Acoustics lower than prediction in high frequency range
- o Random vibration slightly higher than predictions
- o Discrete higher level vent tones identified
- o ICD 2-19001 changed in accordance with above findings

ADDITIONAL MEASUREMENTS

- o 3 Microphones, 8 accelerometers now flying on OV099
- o 7 Microphones, 8 accelerometers authorized for OV103 prior to first WTR flight
- o Additional measurements on OV102 and OV104 are TBD

PROGRAM GOALS

- o Characterize vent noise and large volume P/L effect
- o Acoustics in FWD 1/3 of cargo bay
- o Differences, if any, between OV102, OV099, OV103 and OV104

Fig. 6. LA/6 Mission Integration--Vibrations and Acoustics

SPACELAB - NATURAL AND INDUCED ENVIRONMENTS APPLICABLE TO EXPERIMENTS

Alan Thirkettle
Manager, European Resident Team at Kennedy Space Center
European Space Agency

The Spacelab program has published a document entitled The Spacelab Payload Accomodation Handbook (SPAH). The contents are a compilation of the best information available to the program from both analytical and ground test data bases, but they do not include any iterations from flight test results. The first Spacelab flight - on STS 9 in November/December 1983 - was extensively instrumented, and a considerable body of environmental data was generated. This paper briefly summarizes the present contents of the handbook, describes the flight test data available from Spacelab 1 and recommends that, where appropriate, the flight test data be used to further refine and update the contents of the handbook.

The Spacelab program has issued an ESA/NASA jointly controlled document entitled "Spacelab Accommodation Handbook", (Ref. 1). This presentation describes briefly the contents of this document, the sources of the data contained therein, potential data derived since its publication, and recommendations for subsequent upgrading. Reference is made to Figures 1-18 attached.

Figure 1 notes the scope of the SPAH, the fact that the data are a mixture of analysis and ground test results, and the fact that the document is now published and distributed by the NASA/MSFC Spacelab Program Office. The subsequent presentation concentrates on two sections of the main volume, i.e., Payload Environment and Design Requirements for Experimenters.

Figure 2 shows the schematic development of the document as a function of time (and hence Spacelab development and maturity). The

SPACE SHUTTLE ENVIRONMENT

o <u>SPACELAB PAYLOAD ACCOMMODATION HANDBOOK</u>

- JOINT ESA/NASA PROGRAMME OFFICE DOCUMENT, EVOLVED THROUGHOUT D & D PHASE
- CONTENTS INCLUDE
  - o MAIN VOLUME - STS/SL OVERVIEW & SYSTEM DESCRIPTION
    - SL SUBSYSTEM DESCRIPTION
    ⟶ - PAYLOAD ENVIRONMENT
    - OPERATIONS
    ⟶ - DESIGN REQTS FOR EXPERIMENTS
    - SAFETY
  - o APPENDIX A, AVIONICS I/F (PHYSICAL, VOLTAGE, IMPEDANCE, SIGNAL CHAR.)
  - o APPENDIX B, MECHANICAL I/F (PHYSICAL, LOAD CAPABILITIES)
  - o APPENDIX C, ENVIRONMENTAL I/F (ATMOSPHERE, THERM. CONT, HEAT REJECTION)
- CONTENTS BASED ON ANALYSIS AND GROUND TEST
- AVAILABLE FROM

                J. THOMAS
                MANAGER SPACELAB PROGRAM
                NASA/MSFC
                HUNTSVILLE, ALA 35812

Fig. 1. Scope of the SPAH

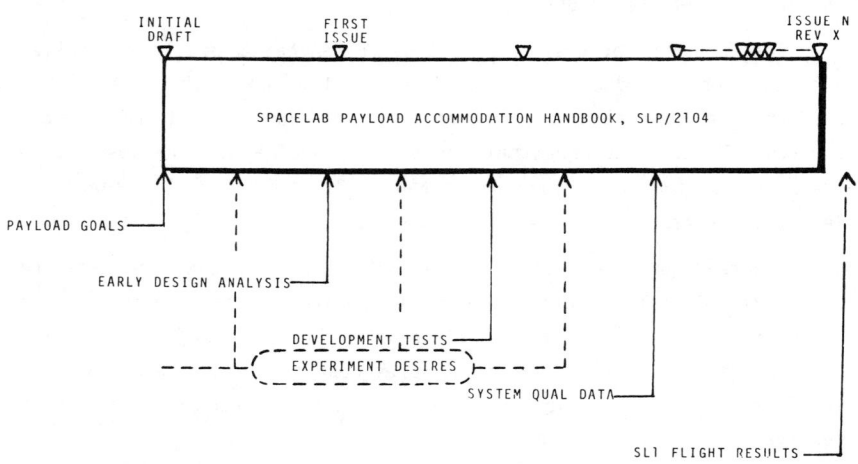

Fig. 2. Spacelab and Maturity

contents of the handbook were initially based on goals, refined by results of analysis, development, qualification and checkout testing. The currently published issue does not contain changes which might be appropriate as a result of the first Spacelab flight on board STS-9 in November-December 1983.

Figure 3 is the contents list for the section of the SPAH concerning Payload Environment (main volume, sect. 5). The data include information regarding mechanical, thermal, atmospheric, electromagnetic, cleanliness and contamination, on-orbit radiation and on-orbit meteoroid environments. The following six figures give example sheets taken from the handbook for a selection of the above environments.

Section 5 - PAYLOAD ENVIRONMENT

				Page
5.1	Mechanical Environment			5-1
	5.1.1	Flight Environment		5-1
		5.1.1.1	General Description of Mechanical Environment	5-1
			5.1.1.1.1 Static and Low Frequency Transient Accelerations	5-1
			5.1.1.1.2 High Frequency Excitation	5-1
		5.1.1.2	Sinusoidal Vibration	5-2
		5.1.1.3	Random Vibration	5-3
			5.1.1.3.1 Aft Flight Deck	5-3
			5.1.1.3.2 Module	5-4
			5.1.1.3.3 Pallet	5-5
		5.1.1.4	Acoustic Noise	5-6
			5.1.1.4.1 Module	5-8
			5.1.1.4.2 Aft Flight Deck	5-10
			5.1.1.4.3 Pallet	5-10
		5.1.1.5	Shock	5-10
			5.1.1.5.1 Module	5-10
			5.1.1.5.2 Pallet	5-11
		5.1.1.6	Acceleration	5-12
			5.1.1.6.1 Normal Mission/Emergancy Sequence	5-12
			5.1.1.6.2 On-Orbit Manuvers	5-15
			5.1.1.6.3 Orbit Atmosphere Accelerations	5-16
	5.1.2	Ground Environment		

Fig. 3. Table of Contents for SPAH Concerning Payload Environment

SPACE SHUTTLE ENVIRONMENT

		5 – 17
5.2	Thermal Environment	5 – 18
5.2.1	Flight Environment	5 – 18
5.2.1.1	Module	5 – 18
5.2.1.2	Airlock	5 – 19
5.2.1.3	Aft Flight Deck	5 – 20
5.2.1.4	Pallet	5 – 20
5.2.1.4.1	Launch/Landing Air Temperature	5 – 20
5.2.1.4.2	On-Orbit Conditions	5 – 22
5.2.2	Ground Environment	5 – 25
5.3	Atmospheric Environment	5 – 26
5.3.1	Flight Environment	5 – 26
5.3.1.1	Module	5 – 26
5.3.1.1.1	Pressure	5 – 26
5.3.1.1.2	Atmospheric Composition	5 – 27
5.3.1.1.3	Airlock Atmosphere	5 – 28
5.3.1.2	Pallet	5 – 28
5.3.1.2.1	Launch Sequence	5 – 28
5.3.1.2.2	On-Orbit	5 – 29
5.3.1.2.3	Re-Entry Sequence	5 – 30
5.3.2	Ground Environment	5 – 32
5.3.2.1	Integration Operations	5 – 32
5.3.2.2	Orbiter Cargo Bay	5 – 33
5.3.2.3	During Transportation	5 – 34
5.3.2.4	Terrestrial Environment	5 – 34
5.4	Electro-Magnetic Environment	5 – 35
5.4.1	Electrical Environment	5 – 35
5.4.1.1	Radiated Emissions	5 – 35
5.4.1.1.1	Orbiter Emissions	5 – 35
5.4.1.1.2	Spacelab Emissions	5 – 39
5.4.1.1.3	Combined Orbiter/Spacelab Electro-magnetic Environment	5 – 40
5.4.1.1.4	Launch Site Environment	5 – 40
5.4.1.2	Conducted Emissions	5 – 41
5.4.1.2.1	Narrowband Noise – deleted –	5 – 40
5.4.1.2.2	Broadband Noise and Spikes – deleted –	5 – 40
5.4.1.2.3	Transients – deleted –	5 – 41
5.4.1.3	Bonding and Lightning Protection	5 – 42
5.4.1.4	Electrical Surface Properties	5 – 42
5.4.2	Magnetic Environment	5 – 43
5.5	Cleanliness and Contamination	5 – 44
5.5.1	Flight Environment	5 – 44
5.5.1.1	Module	5 – 44
5.5.1.2	Pallet	5 – 44
5.5.1.2.1	Molecular Number Column Density (NCD) Predictions	5 – 44
5.5.1.2.2	Molecular Return Flux Predictions	5 – 44.2
5.5.2	Ground Environment	5 – 44.2
5.6	Radiation Environment	5 – 45
5.6.1	External Environment	5 – 45
5.6.2	Internal Environment	5 – 47
5.7	Meteoroid Environment	5 – 49

DCN 131

Fig. 3 (cont'd.)

Figure 4 presents a typical table giving random vibration environments for pallet mounted equipment as a function of payload mass. The data are based on the results of a full scale acoustic noise test performed on a flight standard full scale pallet at Orbiter payload bay launch levels. Mass/stiffness dummies representing payloads, subsystems and utilities were installed. Similar data are available for payloads mounted in Module racks, Module floor, Scientific Airlock, etc.

Table 5-4: Random Vibration Environment for Pallet Mounted Equipment

LOCATION	FREQUENCY	LEVEL
In plane motion Independent of mass loading	20 Hz 20 Hz - 150 Hz 150 Hz - 550 Hz 550 Hz - 2000 Hz 2000 Hz Composite:	$0.00003\ g^2/Hz$ +12 db/oct $0.1\ g^2/Hz$ -12 db/oct $0.0006\ g^2/Hz$ 7.82 g RMS
Input to Experiments mounted on Pallet Cold Plates at Cold plate interface		
Out of plane motion Cold plate loaded by ≥ 10 kg but < 20 kg	20 Hz 20 Hz - 135 Hz 135 Hz - 450 Hz 450 Hz - 2000 Hz 2000 Hz Composite:	$0.00054\ g^2/Hz$ +12 db/oct $1.0\ g^2/Hz$ -9 db/oct $0.011\ g^2/Hz$ 23.7 g RMS
Out of plane motion Cold plate loaded by ≥ 20 kg but < 40 kg	20 Hz 20 Hz - 120 Hz 120 Hz - 450 Hz 450 Hz - 2000 Hz 2000 Hz Composite:	$0.00054\ g^2/Hz$ +12 db/oct $0.7\ g^2/Hz$ -9 db/oct $0.008\ g^2/Hz$ 19.9 g RMS
Out of plane motion Cold plate loaded by ≥ 40 kg	20 Hz 20 Hz - 110 Hz 110 Hz - 450 Hz 450 Hz - 2000 Hz 2000 Hz Composite:	$0.00054\ g^2/Hz$ +12 db/oct $0.5\ g^2/Hz$ -9 db/oct $0.0057\ g^2/Hz$ 17 g RMS
In plane motion Cold plate loaded by ≥ 10 kg but < 20 kg	20 Hz 20 Hz - 150 Hz 150 Hz - 400 Hz 400 Hz - 2000 Hz 2000 Hz Composite:	$0.0001\ g^2/Hz$ +12 db/oct $0.35\ g^2/Hz$ -12 db/oct $0.0006\ g^2/Hz$ 12.0 g RMS
In plane motion Cold plate loaded by ≥ 20 kg but < 40 kg	20 Hz 20 Hz - 150 Hz 150 Hz - 400 Hz 400 Hz - 2000 Hz 2000 Hz Composite:	$0.0008\ g^2/Hz$ +12 db/oct $0.24\ g^2/Hz$ -12 db/oct $0.0004\ g^2/Hz$ 9.96 g RMS
In plane motion Cold plate loaded by ≥ 40 kg	20 Hz 20 Hz - 150 Hz 150 Hz - 400 Hz 400 Hz - 2000 Hz 2000 Hz Composite:	$0.0005\ g^2/Hz$ +12 db/oct $0.15\ g^2/Hz$ -12 db/oct $0.0002\ g^2/Hz$ 7.82 g RMS

Fig. 4. Random Vibration Environment for Pallet Mounted Equipment

Figure 5 shows the predicted temperature profile of the Orbiter payload bay during re-entry of a Spacelab mission. The data are based on a (NASA) coupled thermal analysis and are design requirement data for Spacelab. They are also the best available data known to the Spacelab program up to but excluding the Spacelab 1 flight.

Figure 6 shows extremes of pallet surface temperatures for various operational phases corresponding to a worst cold case and a worst hot case. The data are based on (ESA) detailed thermal analyses using Orbiter-provided external inputs. If a payload can tolerate these extremes it has no problem. If the extremes are beyond the capability of the payload, various options - such as mission-peculiar thermal analysis to narrow the extreme range, or on-orbit constraints - may legitimately be considered.

Figure 7 shows the change of cargo bay pressure, as a function of time, during re-entry. It is based on both a nominal descent profile of the Orbiter and pre-flight predictions.

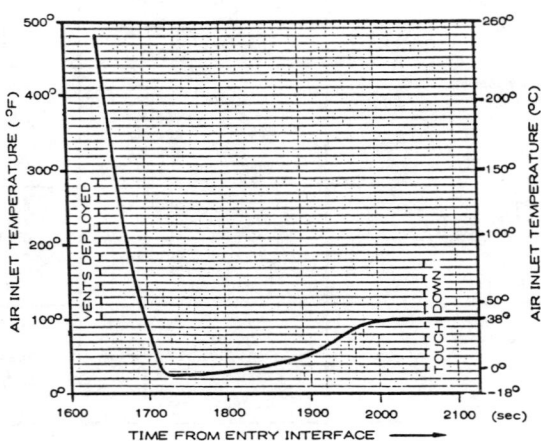

Fig. 5. Estimated Temperature Profile of the Cargo Bay Air During Descent

OPERATIONAL PHASE	T min. (°C)	T max. (°C)
Launch / Ascent	+ 4.5	+ 65.5
Orbit doors closed	− 50	+ 70
Orbit doors open	− 150	+ 120
Re-Entry	− 140	+ 110
Post Landing	− 110	+ 120

Fig. 6. Pallet Surface Temperatures

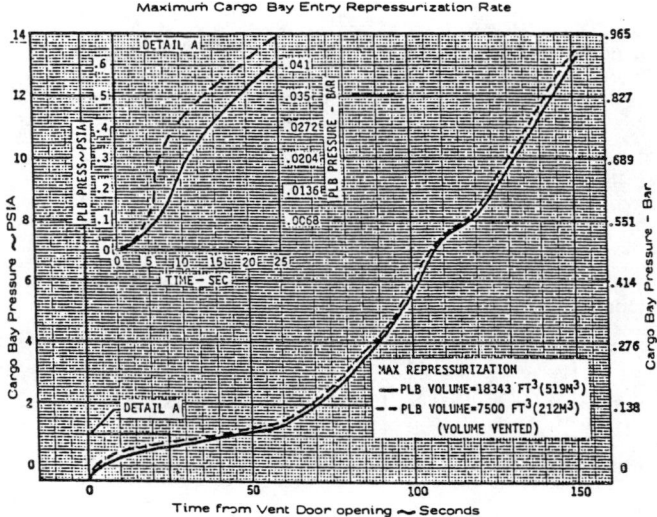

Fig. 7. Orbiter Cargo Bay Internal Pressure History During Entry for all Flight Modes

Figure 8 is an extract of handbook data on the electromagnetic environment emitted by Spacelab when in the Orbiter. It is based on predicted Orbiter levels as well as measured Spacelab levels. A full scale long module - one pallet, fully equipped with active subsystems was instrumented inside a Faraday cage. The predicted Orbiter EMC environment was simulated and measurements were taken at various locations within the Spacelab. Thus, the published data are a mixture of prediction and measurement.

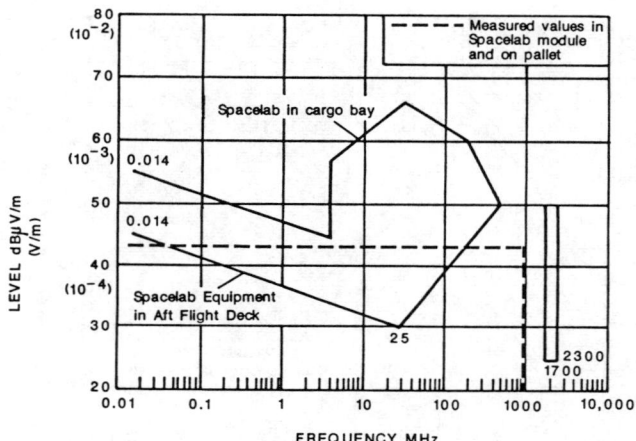

Fig. 8. Maximum Radiated Narrowband Emissions by Spacelab

Figure 9 defines the average total environment of micrometeoroid populations at representative Orbiter/Spacelab orbit altitudes. It is based on historical NASA data and resulting math models.

Particle Density   : 0.5 g/cm^3
Particle Velocity   :   20 km/sec
Flux Mass Models :

(1) For $10^{-6} \leq m \leq 10^0$ $\log Nt = -14.37 - 1.213 \log m$

(2) For $10^{-12} \leq m \leq 10^{-6}$ $\log Nt = -14.339 - 1.584 \log m - 0.063 (\log m)^2$

Nt = no. particles/m^2/sec of mass m
m = mass in grams

Defocussing factor for earth, and if applicable, shielding factor are to be applied.

Fig. 9. Meteoroid Environment

Figure 10 is the contents list for the section of the SPAH concerning Design Requirements for Experiments (main volume, section 7). The material does not cover environments applicable to experiments, but rather payload limitations necessary to maintain a suitable situation for Spacelab and/or the Orbiter. The limitations arise as a result of the environments which the payloads cause, i.e., cleanliness, EMC. This section also includes guidelines for using the environments of Section 5 (Figure 3) to verify payload flight worthiness, i.e. definiton of required safety factors.

SPACE SHUTTLE ENVIRONMENT

Section 7 - DESIGN REQUIREMENTS FOR EXPERIMENTS

			Page
7.1	Purpose of Design Requirements		7-1
7.2	Mechanical Design Requirements		7-1
	7.2.1	Experiment Mass and Volume	7-1
	7.2.2	Experiment Mounting and Installation	7-1
	7.2.3	Experiment Integrity	7-1
	7.2.4	Extension, Ejection, Deployment and Capture	7-2.1
		7.2.4.1 Emergency Retraction and Ejection	7-2.1
		7.2.4.2 Routine Ejection, Deployment and Capture	7-2.1
	7.2.5	Crew Interface	7-3
		7.2.5.1 General	7-3
		7.2.5.2 Loose Equipment Restraint	7-3
		7.2.5.3 Handholds and Handrails	7-3
		7.2.5.4 Equipment Transfer On Orbit	7-3
		7.2.5.5 Corners, Edges and Protrusions	7-4
		7.2.5.6 Area Closures	7-6
		7.2.5.7 Crew Applied Loads	7-6
		7.2.5.8 Controls and Displays	7-6
		7.2.5.9 Electrical Safety	7-6
		7.2.5.10 Labels for Caution and Warning and Emergency Use Items	7-7
		7.2.5.11 Crew/Equipment Interface for EVA	7-7
7.3	Thermal Requirements		7-8
7.4	Electrical Power Interface Requirements		7-8
7.5	Command and Data Handling Requirements		7-9
7.6	GSE Requirements		7-9
	7.6.1	Spacelab / Orbiter / Integration Center / Launch Site GSE	7-9
	7.6.2	Experiment Provided GSE	7-9
7.7	Environmental Requirements		7-10
	7.7.1	Natural and Induced Environment	7-10
	7.7.2	Experiment Induced Environment	7-10
		7.7.2.1 Accoustic Environment	7-10
		7.7.2.2 Electromagnetic Environment	7-10
		7.7.2.2.1 Bonding and Shielding Requirements	7-10
		7.7.2.2.2 Isolation and Grounding Requirements	7-10.1
		7.7.2.2.3 Conducted Noise Emission on Power and Signal Lines	7-12
		7.7.2.2.4 Static ("DC") Magnetic Field Emissions	7-20
		7.7.2.2.5 AC Magnetic Field Emissions	7-20
		7.7.2.2.6 AC Electric Field Emissions	7-21
		7.7.2.3 Ionizing Radiation	7-21
7.8	Test and Integration		7-22
	7.8.1	Test Requirements	7-22
	7.8.2	Integration and Checkout Requirements	7-22
7.9	Operational Requirements		7-22
7.10	Material Control Requirements		7-24
	7.10.1	Purpose of Material Control for Experiments	7-24
	7.10.2	Experiment Location and Associated Requirements on Specific Material Characteristics	7-24
		7.10.2.1 Orbiter Flight Deck	7-24
		7.10.2.2 Spacelab Habitable Area	7-24
		7.10.2.3 Spacelab Pallets	7-25
		7.10.2.4 Airlocks	7-25
		7.10.2.5 Sealed Containers	7-25
		7.10.2.6 Requirements Independent of Experiment Location	7-25

Fig. 10. Table of Contents for the Section of the SPAH Concerning Design Requirements for Experiments

			Page
	7.10.3	Material Control Program	7 - 25
	7.10.4	Test and Acceptance Criteria for Off-Gassing from Materials	7 - 25
	7.10.5	Test and Acceptance Criteria for Flammability Characteristics	7 - 26
	7.10.6	Test and Acceptability Criteria for outgassing under Vacuum	7 - 26
	7.10.7	Corrosion and Material Compatibility	7 - 26
	7.10.8	Outline for an Off-Gassing Test on "Black Box" Level	7 - 26
	7.10.9	Forbidden and Restricted Materials	7 - 26
	7.10.10	Stress Corrosion Cracking Materials	7 - 27
	7.10.11	Waivers and Deviations	7 - 27
7.11		Contamination and Cleanliness Requirements	7 - 29
	7.11.1	General Contamination Control Requirements	7 - 29
	7.11.2	Surface Cleanliness	7 - 29
	7.11.3	Contamination Inside the Module and/or Pallet	7 - 29
		7.11.3.1 Gaseous Contamination	7 - 29
		7.11.3.2 Particular Contamination	7 - 29
		7.11.3.3 Microbiological Contamination	7 - 29
	7.11.4	Contamination External to the Module and/or Orbiter	7 - 29
7.12		Caution and Warning Design Safety Requirements	7 - 30
7.13		Configuration Control of Experiments, Mission Peculiar Equipment and Privately Owned Spacelab Hardware	7 - 31
	7.13.1	Configuration Control of Experiments	7 - 31
	7.13.2	Configuration Control of Mission Peculiar Equipment (MPE)	7 - 31
	7.13.3	Configuration Control of Privately Owned Spacelab Hardware	7 - 31

LIST OF TABLES

Table		Page
7 - 1a	Minimum Design Safety Factors	7 - 2
7 - 1	Minimum Corner and Edge Radii for Loose Equipment	7 - 6
7 - 2	Recommended Continuous Current Versus AWG Wire Size	7 - 9

LIST OF FIGURES

Figure		
7 - 1a	Loose Items Penetration Hazard	7 - 2.2
7 - 1	Exposed Edges Design Criteria	7 - 4
7 - 2	Exposed Corner Design Criteria	7 - 5
7 - 3	28 V DC Bus Impedance Characteristics	7 - 12
7 - 4	Conducted Noise Limits for DC Power Lines	7 - 13
7 - 5	Conducted Inrush Transient Wave Form	7 - 13
7 - 6	400 Hz Bus Impedance Characteristics	7 - 14
7 - 7	Conducted Noise Limits for AC Power Lines	7 - 15
7 - 8	Conducted Noise Limits for RAU Flexible Inputs Used as Analog Inputs	7 - 16
7 - 9	Conducted Noise Limits for RAU Discrete and Digital Input	7 - 17
7 - 10	Common Mode Conducted Noise Limits for DC Power Lines	7 - 18
7 - 11	Common Mode Conducted Noise Limits for AC Power Lines	7 - 19
7 - 12	AC Magnetic Field Emission Limits	7 - 20
7 - 13	AC Electric Field Emission Limits	7 - 21

Fig. 10. (cont'd.)

Having briefly covered the scope and contents of sections of the SPAH, Figure 11 notes that Spacelab 1 - the results of which are <u>not</u> reflected in the handbook - was in fact a verification flight test and, as such, was extensively instrumented to measure, among other things, the actual environment. The flight of Spacelab 2 - scheduled at the time of the workshop to be April 1985 but now slipped to June 1985 - will be similarly instrumented. The next six figures indicate the scope of the Spacelab flight data.

FLIGHT TEST DATA

- SL1 WAS A VERIFICATION FLIGHT TEST. AS SUCH IT WAS HEAVILY INSTRUMENTED TO MEASURE ITS PERFORMANCE, ONE ASPECT OF WHICH WAS ITS ENVIRONMENT.

- VFI MEASUREMENTS TAKEN AND COMPARED TO PRE-FLIGHT PREDICTION, FOR EXAMPLE TEMPERATURES, ACCELERATIONS, NOISE.

- PLB "ATMOSPHERE" MEASURED BY IECM EXPERIMENT.

- IN ADDITION, RADIATION MEASUREMENTS WERE TAKEN.

- TOTAL DATA AVAILABLE AT MSFC.

- MORE TO COME ON SL2 (APRIL 1985).

- NO EMI DEDICATED EVALUATIONS, ALTHOUGH NO PROBLEMS HAVE EVER BEEN EXPERIENCED ON GROUND AT ERNO, KSC, NOR DURING SL1, EXCEPT WHEN EXPERIMENT VIOLATED SPAH BADLY.

Fig. 11. Summary of Results of Spacelab 1

Figure 12 indicates the sensors monitored on board the first Spacelab flight. Over 240, out of 363 sensors, measured directly the various environmental parameters listed.

In addition to the sensors of Figure 13, which were installed specifically for Spacelab 1, there are 579 operational sensors on board, which are permanently installed. Of these, 67 are devoted to environmental parameters.

Figure 14 shows the measured flight results of two temperature sensors which were mounted on the Spacelab 1 pallet, and a comparison with predicted values.

Figure 15 shows the measured flight results of two accelerometers which were mounted on the Module of Spacelab 1.

Figure 16 shows the results of an acoustic sensor mounted on the sill of the pallet of Spacelab 1, compared with the design requirement spectrum and the average results of Orbiter flights STS 2-5. It can be seen that the measured Spacelab results at this location are considerably lower than predicted.

Figure 17 is a summary of the measured results of Spacelab 1 in terms of Contamination and Radiation. The IECM (Internal/External Contamination Monitor) was flown aboard the pallet of Spacelab 1 and was operative during several phases of the mission. Various radiation detectors, active and passive, were flown in order to study the radiation levels of the South Atlantic anomaly and the horns of the radiation belts, to evaluate encountered isotopes, to derive the linear energy transfer spectrum and to study the effects on photographic film.

Figures 12-17 are all taken from the NASA Space Shuttle STS-9 Final Flight Evaluation Report, Volume II (Ref. 2). This document is not a complete assessment of the results and implications of the sensor measurements, but asserts only that the sensors and their recording system worked properly. Thus, at the time of this workshop, there is no published document which addresses the results of Spacelab 1 in terms of measured environments compared to analysis/ground test.

Fig. 12. VFI Flight Measurements

LOCATION	CODE	A ACCELERATION	B PHASE, ELECTRICAL	C CURRENT	D VIBRATION	E ELECTRICAL POWER	F FREQUENCY	G FORCE/STRESS/STRAIN	H POSITION/ATTITUDE	J LOGIC STATUS	K STIMULUS	L VELOCITY	M MULTIDATA	N RESISTANCE/CAMERA RADIATION	P PRESSURE	Q QUANTITY/HUMIDITY	R RATE	S SWITCH SCAN	T TEMPERATURE	U DEFLECTION	V VOLTAGE	W TIME	X DISCRETE EVENT	Y ACOUSTIC	Z COMPUTER STIMULUS CMD	SUBTOTAL
TUNNEL	140	3			3			5							1				20					1		33
MODULE FWD END CONE	160				3												2		8							13
CORE SEGMENT	200	18		7	12			10		8	41		3				2		27		10	4	10	2		154
VIEWPORT & WINDOW SEG.	260				3														1							4
EXPERIMENT SEG.	300	4						4											12							20
AIRLOCK EXP. SEG.	360				3			4																		7
MODULE AFT. END CONE	410																		6							6
PALLET UTILITY SUPT	460																		2							2
CONTROL PANEL	32V	13			5																		6	1		25
PALLET SEGMENT	600							7		2	19		1						47		8		1			85
PAYLOAD PLATFORM	900	3		1	3														7							14
NASA RI UTILITY KIT																										
SUBTOTAL		41		8	32			30		10	60		4		1		4		130		18	4	17	4		363

Fig. 12. VFI Flight Measurements

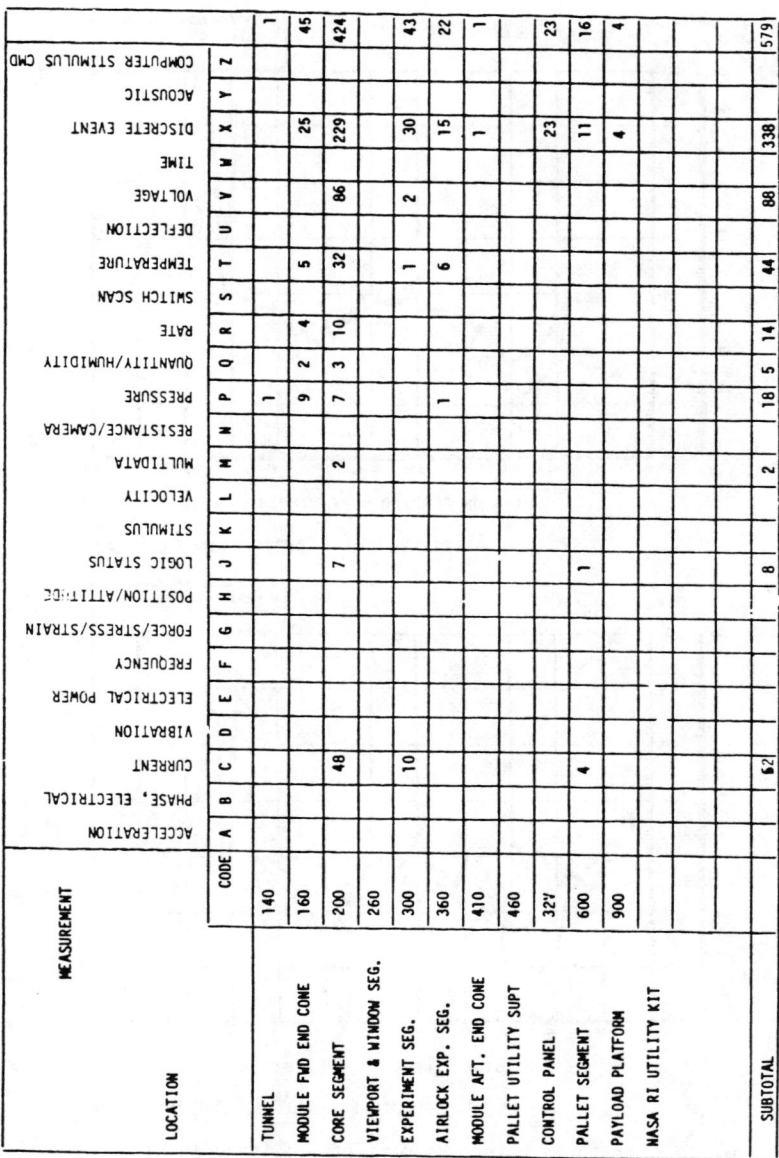

Fig. 13. Operational Flight Measurements

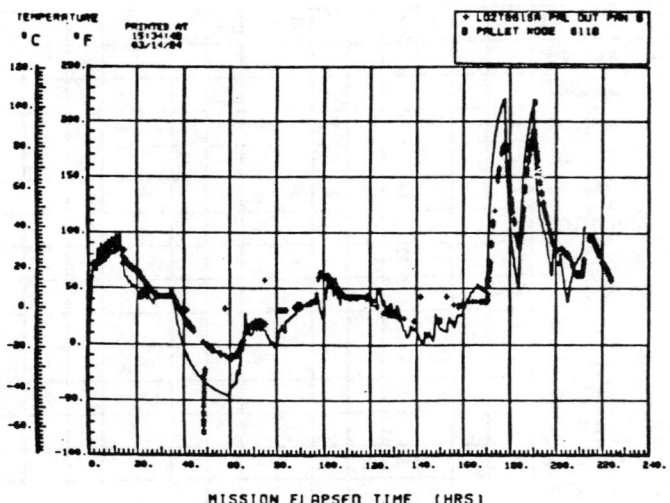

PALLET STRUCTURE, OUTER PANEL # 6 (PORT)

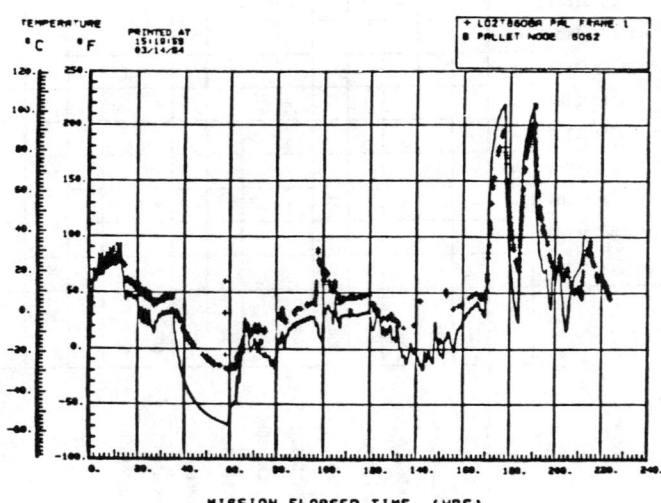

PALLET STRUCTURE, STBD FWD FACE MIDPT, FRAME 1

Fig. 14. Measured Flight Results of Two Temperature Sensors Mounted on Spacelab 1

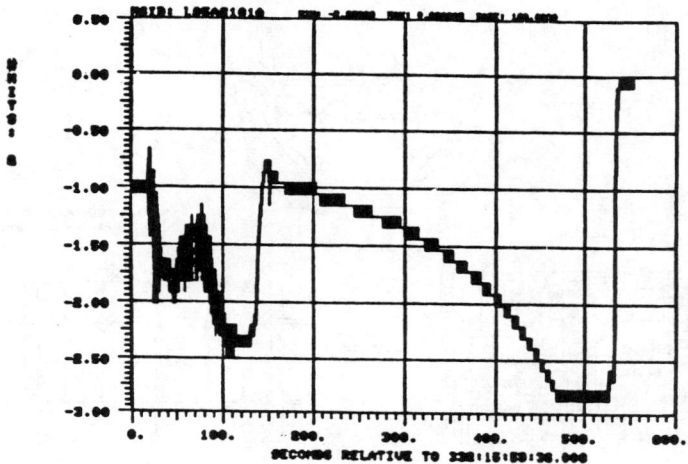

MODULE STABILIZING FITTING X-ACCELERATION
ASCENT TIME HISTORY

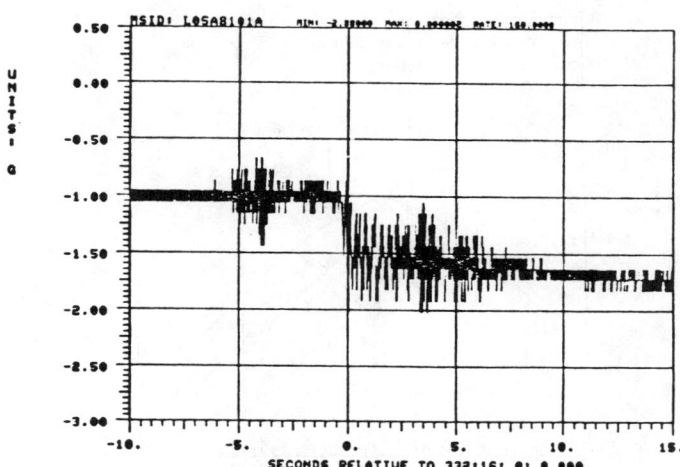

MODULE STABILIZING FITTING X-ACCELERATION
LIFTOFF TRANSIENT

Fig. 15. Measured Flight Results of Two Accelerometers Mounted on Spacelab 1

SPACE SHUTTLE ENVIRONMENT

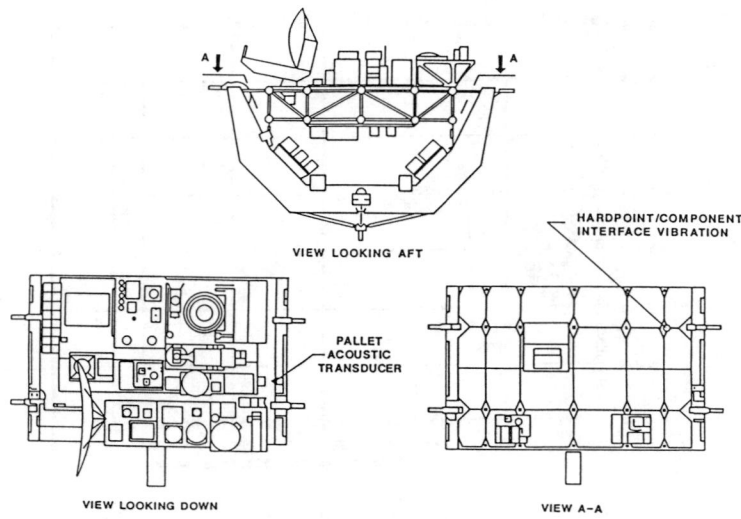

VIBROACOUSTIC MEASUREMENT LOCATIONS

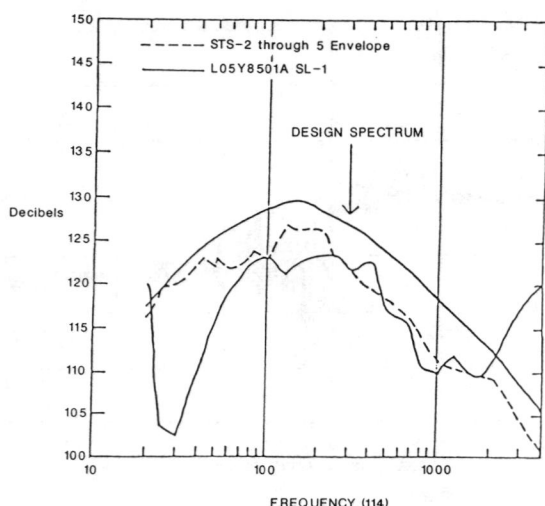

PALLET ACOUSTICS

Fig. 16. Pallet Acoustics

13.0 CONTAMINATION AND RADIATION

13.1 CONTAMINATION -The IECM was flown as part of the VFI system. The IECM engineering and scientific subsystems functioned normally during ascent and on-orbit until approximately 07/09:00:00 MET when an in-flight anomaly resulted in the loss of data on the IECM internal tape recorder for the duration of the on-orbit operations and the descent operations. The contamination-monitoring operations that were successful and are presently being analyzed include the following:
- A. Launch and ascent mode operations beginning at 00/00:00 MET
- B. On-orbit mode operations beginning at 00/00:15 MET and continuing until the anomaly discussed above occurred
- C. Mass spectrometer operations beginning at 00/05:00 MET
- D. Gas release and special mass spectrometer operations between 01/02:00 MET and 01/03:00 MET
- E. Passive sample exposure data for entire mission.

All the IECM flight engineering and science data were recorded on the IECM internal data recorder, and the anomaly was discovered only after the IECM was received at MSFC in late January. Based on partially complete flight data reduction, all instruments appeared to operate normally as scheduled throughout the above described operations until the flight anomaly occurred.

The S&E Failure Investigation Team concluded that the primary cause of the IECM anomaly was excessive temperature during a "hot soak" cycle which caused the IECM data acquisition and control system to malfunction. The thermal model has been modified to more accurately predict the environment during such hot cycles. Better predictions will prevent future operations in excessive thermal conditions. Key boxes within the IECM will be tested prior to SL-2 to ensure that there was no component damage as a result of the thermal stress on SL-1.

13.2 RADIATION -Analysis of the radiation detection samples from the SL-1 mission will be published under separate cover. Additional analysis results will be included in the VFT assessment. The following subsections present the status and preliminary results available at the time of publication of this report.

13.2.1 ACTIVE RADIATION DETECTORS -Quick-look data obtained during the flight and preliminary flight data show that both instruments operated properly during the flight except for approximately one day starting at 05/20:00 MET, when they were turned off to help reduce the temperature on the RAU 21 coldplate. Both South Atlantic anomaly passes and passes through the horns of the radiation belts were observed with good statistic.

13.2.2 PASSIVE RADIATION DETECTORS - Analysis by low-level gamma-ray spectroscopy using a high resolution Ge(li) detector has detected the isotope Co^{58} (E = 810 keV, $T_{1/2}$ = 77^d) from the nickel and cobalt samples. Averages fluxes of activating particles while in orbit can be derived as soon as counting is completed and the detector is calibrated. The neutron capture isotope, Ta^{182}, was detected in the tantalum sample, although the contribution of ground-level background neutrons must be accurately measured before the activating fluxes are determined. No activation products have been found in the vanadium or indium samples. This is consistent with the low level found in the other samples.

Fig. 17. Summary of Results of Spacelab 1 in Terms of Contamination and Radiation

13.2.2.1 THERMOLUMINESCENT DETECTORS - TOTAL MISSION DOSE - The thermoluminescent detectors (TLD's) in the three module passive radiation detectors (PRD's) have been preliminarily assessed. Preliminary net readings of about 0.1 rad in the module and about 0.2 rad in the pallet-mounted unit have been derived after subtracting the background unit readings. These results are lower than expected but within the predicted range. Postflight calibration tests and effect of thermal history will be included to derive accurate dose figures at each location of the detectors.

13.2.2.2 LEXAN AND CR39 HEAVY NUCLEI DETECTORS (LET STACKS) - These detectors have been partially processed (etched) at the University of San Francisco. The detectors appear to be in good condition. Analysis by microscope of the etch pits to derive the linear energy transfer (LET) spectrum has not yet begun.

13.2.2.3 NUCLEAR TRACK EMULSIONS - Nuclear track emulsions from the 3 module PRD's have been developed by the University of Washington, Seattle and returned to the University of San Francisco. The pallet-mounted PRD was returned to MSFC in mid-January and its emulsions have not been developed. A preliminary microscopic analysis indicates the emulsions are in excellent condition. The preliminary assessment is that the slow proton fluence in the module PRD's is somewhat lower than the preflight estimate of $10^5/cm^2$ from the South Atlantic anomaly. The background contains a large contribution of Compton electrons, probably from bremsstrahlung photons due to electrons in the north and south horns of the trapped radiation belts. Nevertheless, the total number of tracks and background is sufficiently low that cosmic ray experiments using emulsion may be flown on the shuttle with a similar altitude, inclination and flight time. Microscope measurements are continuing on the emulsions.

13.2.2.4 EFFECTS ON PHOTOGRAPHIC FILM - Photographic film samples packaged in the PRD consisted of 9 different film types. The types, all black and white, included three aerial films, two instrumentation films, one spectrum analysis film, one double negative motion picture film, one technical pan film, and one special film sensitive to ultraviolet wavelengths.

Each film container that flew within each of the PRD's, plus 2 ground truth containers that did not fly, contained 2 samples of each film type. One sample had a latent image implant consisting of a 20-step sensito/metric gray scale and a 3-bar 1951 A.F. resolution target. The second sample was unexposed before the Spacelab mission and received the same image implant upon recovery after the mission. After recovery, the 6 post-exposed samples of a particular film type were processed at the same time following a predetermined sequence and developing chemistry. Immediately upon completion of development of the post-exposed samples, the 6 pre-exposed samples were processed to the same sequence in the same chemicals.

At this time, density measurements of each step of the image of the 20-step sensito/metric gray scales have been made on test samples of eight of the film types. The remaining film type is the special film sensitive to ultraviolet wavelengths. The processed samples from this film type have severely obscured images of the step-wedge making density measurements unreliable if not impossible.

Fig. 17. (cont'd.)

The maximum effect that can be correlated with radiation dose so far is an increased optical density of about 0.2 in the most sensitive film carried in the pallet PRD.

The assessment of radiation effects on the film is preliminary at this time. A number of supplemental tests will need to be done before a firm assessment of radiation effects can be made. These tests include the following:

- A. Tests of the film types at a temperature-time profile matching the environment experienced during the mission so that the temperature effect can be assessed separately.

- B. Tests matching the time difference between pre-exposed and post-exposed latent image implant, to image development.

- C. Tests to determine density fluctuations within identical exposures on the same photographic emulsion and different positions along the film supply.

13.2.2.5 SUMMARY - All the radiation instruments performed as expected and the material and film samples were recovered in good condition. A number of postflight calibrations and measurements remain to be performed at this time and are in process.

Fig. 17. (cont'd.)

This concluding figure (Figure 18) summarizes the existence of the SPAH which contains a mixture of analysis, prediction and ground test data. It also indicates that there are more valid data available from the SL-1 flight (and yet more will come from SL-2). Obviously the user community should be aware of this availability. Feedback from the users concerning type, depth, format, etc. of data would be available to the Spacelab Program Office at MSFC.

The SPAH represents an honest attempt by the Spacelab program to document, as far as possible, real data rather than conservative design requirements. By comparison with other Orbiter carriers, e.g. MPESS, SPAS, and EURECA, it is of a much greater depth (due in part of course to the program maturity). However, via Spacelab 1 flight test data, a considerable amount of actual, measured, on-orbit data is available, some of which will relieve the impact of certain environments, some of shich may increase the impact, but in any case is of greater validity than ground data. Therefore, the flight test results should not only be published, but should be, where appropriate, the cause of an update of the SPAH to increase its validity.

o CONCLUSION

- ANALYTICAL AND TEST BASED DATA AVAILABLE IN SPAH
- FLIGHT TEST RESULTS FOR SL INDUCED ENVIRONMENT, AND LOCAL ORBITER ENVIRONMENT ARE AVAILABLE FROM SL1
- MORE OF THE SAME WILL COME FROM SL2

o QUESTIONS

- IS THE TYPE OF DATA THAT WHICH IS REQUIRED?
- IS IT IN SUFFICIENT DEPTH?
- IS IT PROPERLY AVAILABLE?
- IS IT PROPERLY FORMATTED?
- IF NOT, WHAT ELSE IS NEEDED AND HOW SHOULD IT BE OBTAINED?

Fig. 18. Summary of Existence of SPAH

References
1. Spacelab Payload Accommodation Handbook. Joint ESA/NASA Publication SLP/2104.
2. Space Shuttle STS-9 Final Flight Evaluation Report, Volume II. NASA Marshall Space Flight Center Report MSFC-RPT-1038, April 30, 1984.

THE IMPACT OF THE SHUTTLE ENVIRONMENT ON THE SCIENCE ON SPACELAB-1

Karl Knott
Spacelab Project Scientist
European Space Agency

The best test for the user-friendliness of the Shuttle environment so far has doubtless been the Spacelab-1 mission which was carried out in the period 28 November through 8 December 1983. On this mission, a total of 70 experiments from all known space disciplines has been flown. During the mission it was possible to determine the advantages and disadvantages of the Shuttle for the different disciplines and to obtain a first estimate on possible disturbances suffered by individual instruments.

In spite of the large number of experiments flown on this mission and the ensuing complexity of the mission architecture, this mission was quite successful. A number of new and relevant scientific results has been obtained. As reporting of science results is outside the scope of this paper, attention is drawn to the July 13, 1984 issue of "Science" magazine, which describes the results from individual investigations carried out on Spacelab-1. These are, of course, preliminary results, but they give a very good survey of what has been accomplished.

The objective of this report is to discuss briefly how the different disciplines represented on Spacelab-1 did cope with the environment of the Shuttle/Spacelab combination. To this end, the report briefly mentions what the different disciplines (Figure 1) wanted to accomplish and describes the degree to which the environment permitted them to do so.

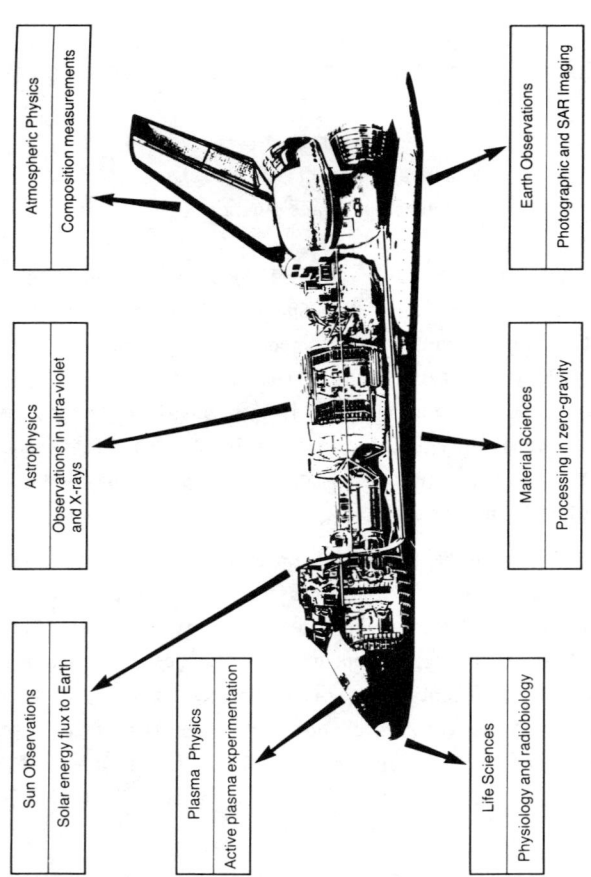

Fig. 1. Spacelab 1 - Scientific Disciplines and Main Objectives

Atmospheric Physics

The instruments of this discipline had a common objective, namely the study of the atmospheric composition on a global scale by remote sensing spectrometric techniques. The two largest instruments of the discipline were an Absorption Spectrometer and an Imaging Emission Spectrometer. Data from the former instrument are reported in a separate paper of the workshop so I would like to say a few words on the performance of the latter.

The Imaging Spectral Observatory (ISO) consists of five individual imaging spectrometers covering the range from 300 to nearly 13,000 Angstroms, e.g., from the UV to the near-infrared. The experiment was successful in determining the emission spectrum of the atmosphere, the so called "atmospheric glow". There were three environmental effects which influenced the performance of the instrument: (1) The optics of the UV spectrometer were coated with a thin layer of Osmium; the flux of atomic oxygen in the Spacelab-1 orbit (240 km altitude, 57 degree inclination) was sufficiently high that these coatings were eroded, and the sensitivity of the UV spectrometer was reduced accordingly. (2) Figure 2 shows a data sample from the ISO. Amongst many other lines those of carbon dioxide and carbon monoxide can be identified. At the altitudes where this data sample was taken, these constituents cannot be of natural origin and must be attributed to the Shuttle environment. The ISO investigator expects that natural and artificial constituents can be separated by a careful correlation of vehicle attitudes and vehicle-related events with background level and features within the spectrum. (3) Figure 3 shows another data sample from the ISO. It was obtained when the instrument was viewing into the velocity vector. The narrow field of view of the instrument ensures that protrusion of any orbiter or Spacelab parts into the field of view was excluded and the instrument is oriented towards dark space. The observation was made to detect possible emissions caused by atmospheric interactions with ram direction exposed surfaces. Figure 3 shows a significant enhancement of emissions at wavelengths above 6000 angstroms which must be attributed to the aforementioned surface interaction. The spectral characterization of this surface

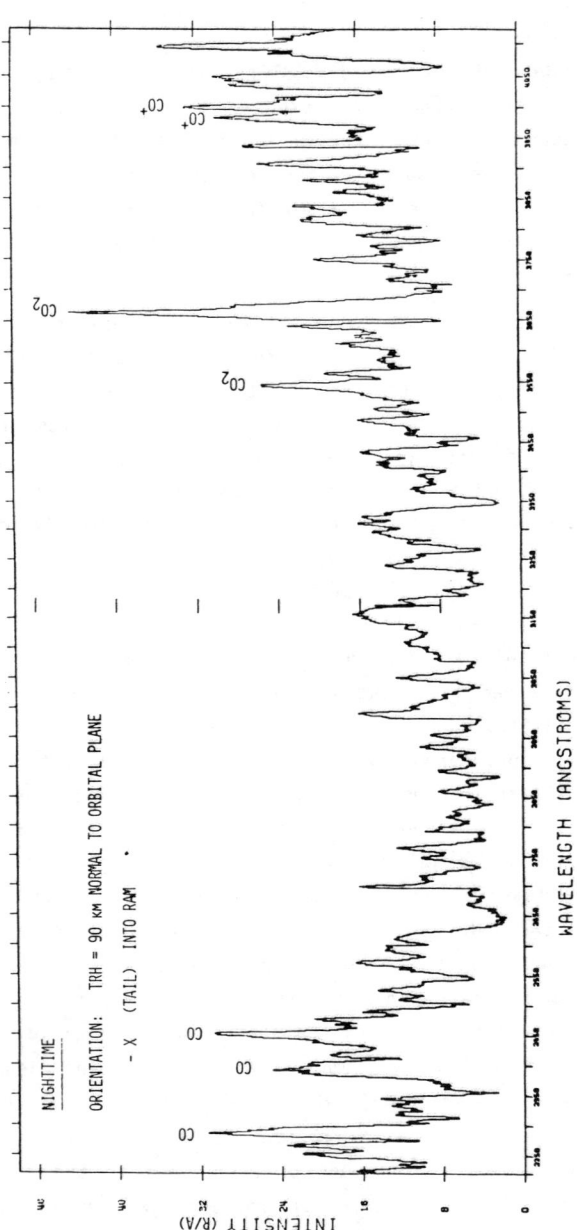

Fig. 2. Data Sample from the ISO

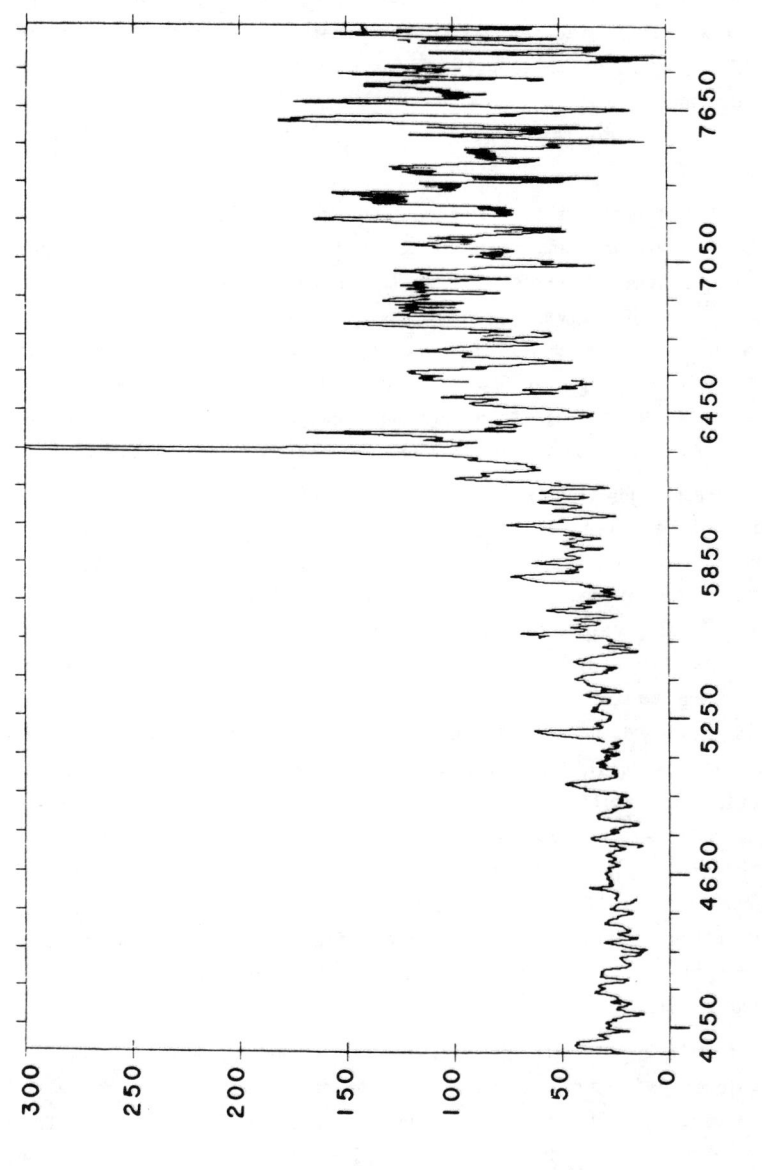

Fig. 3. Data Sample from the ISO

"glow" permits testing of existing models for the generation mechanism of this glow. It is interesting to note that the observations made on Spacelab-1 did not unambiguously prove models developed earlier for this phenomenon.

Plasma Physics

The aim of this discipline was to inject electron beams, ion beams, plasma blobs, and neutral gas clouds into the ionospheric plasma and to study the response of the natural environment to these stimuli. These objectives were largely met and several new and unexpected phenomena were observed. It was, for example, demonstrated that the whole Shuttle and Spacelab outer surfaces were covered by a bright glow when the electron gun was operated at high power, an impressive demonstration of the beam-plasma discharge phenomenon.

The plasma physics experiments encountered two effects which resulted from deficiencies in the environment. One is related to the discharge of the Shuttle/Spacelab combination when the electron beam emission was terminated. During the emission, the expected charging to potentials equivalent to the beam energy was observed. However, when the beam was switched off the charging did not, as one would expect, return to zero. Instead, an overshoot to a negative surface charge is observed before a return to zero is accomplished. This puzzling phenomenon (Figure 4) can only be explained by the fact that the Shuttle has insufficient conductive surface area. In fact, only the motor nozzles are conductive, which represents a comparatively small portion of the total surface area. In this situation it is to be expected that different surface areas charge and discharge at different rates. Thereby local electric fields are set up which are interpreted as negative surface charge by the diagnostic probes of the plasma experiment.

At certain points in the mission a systematic variation of the electron density in the immediate environment of the Shuttle has also been observed. The modulation occurs for electron densities in the range of $2-5 \times 10^5$ per cubic centimeter. The periodicity of the modulation is close to 10 seconds. This is clearly not a natural

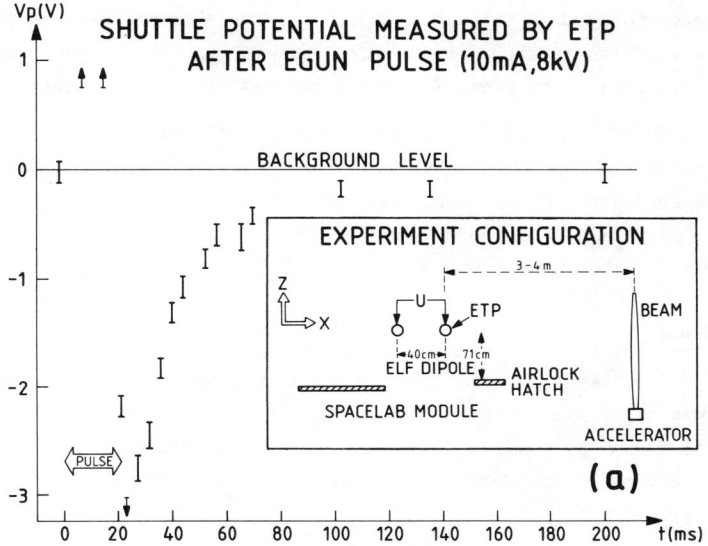
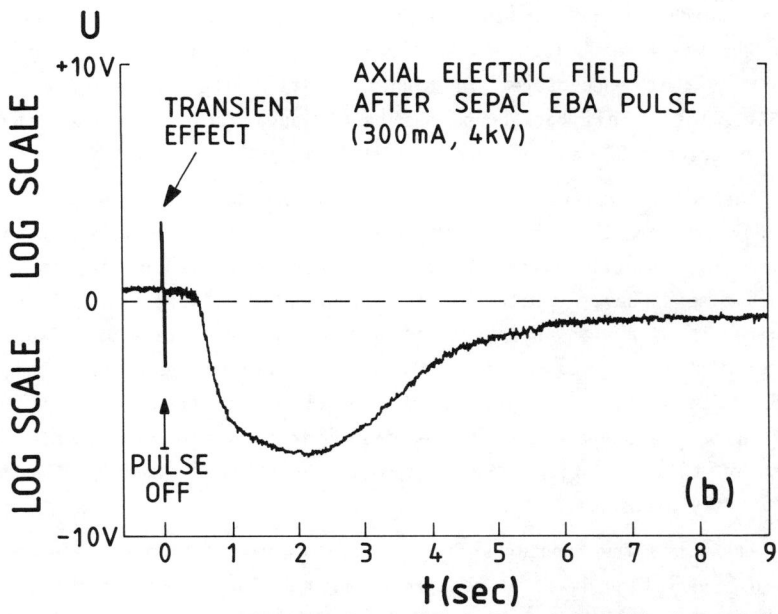

Fig. 4. Recording of Shuttle Potential Inferred from Electron Probe (top) and Potential Gradient near Surface (bottom).

phenomenon and is also not caused by the electron beam or any other plasma diagnostic emission. It is suspected, but not yet proven, that it has to do with thruster firings or flash evaporator operation.

In the area of DC magnetic and AC electromagetic cleanliness, the findings of the OSS-1 payload on STS-3 were confirmed on Spacelab-1. The geomagnetic DC-field was occasionally disturbed by parasitic fields. The AC background was sufficiently low to permit identification of natural wave phenomena.

Astronomy

Spacelab-1 carried three instruments for research in astronomy: A gas scintillation proportional counter for the observation of celestial X-ray sources, a UV telescope for individual star observations and a wide field UV camera for sky photographs aiming at the detection of extended but faint objects.

A result of the X-ray detector (Figure 5) was that the Spacelab-1 orbit, which--except for the South Atlantic Anomaly--does not enter into the van Allen belt, sees an X-ray background which is a factor of 5 better than encountered in higher orbits, like for example the EXOSAT orbit. This background has been mapped during the Spacelab-1 mission between 57 degrees geographical latitude.

The UV instruments suffered from the orbital conditions met in the November launch. The amount of orbital darkness occurring during a particular Shuttle flight depends, amongst other parameters, on the launch time of day. Orbiter safety considerations (more precisely the requirement of a transatlantic landing in daylight in case of a launch abort) dictated a launch time which led to marginal darkness conditions. In fact, the orbit was always close to the terminator and drifted on the seventh day of the mission into constant sunlight. In spite of this situation, the UV camera did obtain a number of good quality sky exposures.

All but a few exposures from the Far Ultraviolet Space Telescope showed a very high level of background light. The investigator has so far concentrated his effort on finding an explanation for this back-

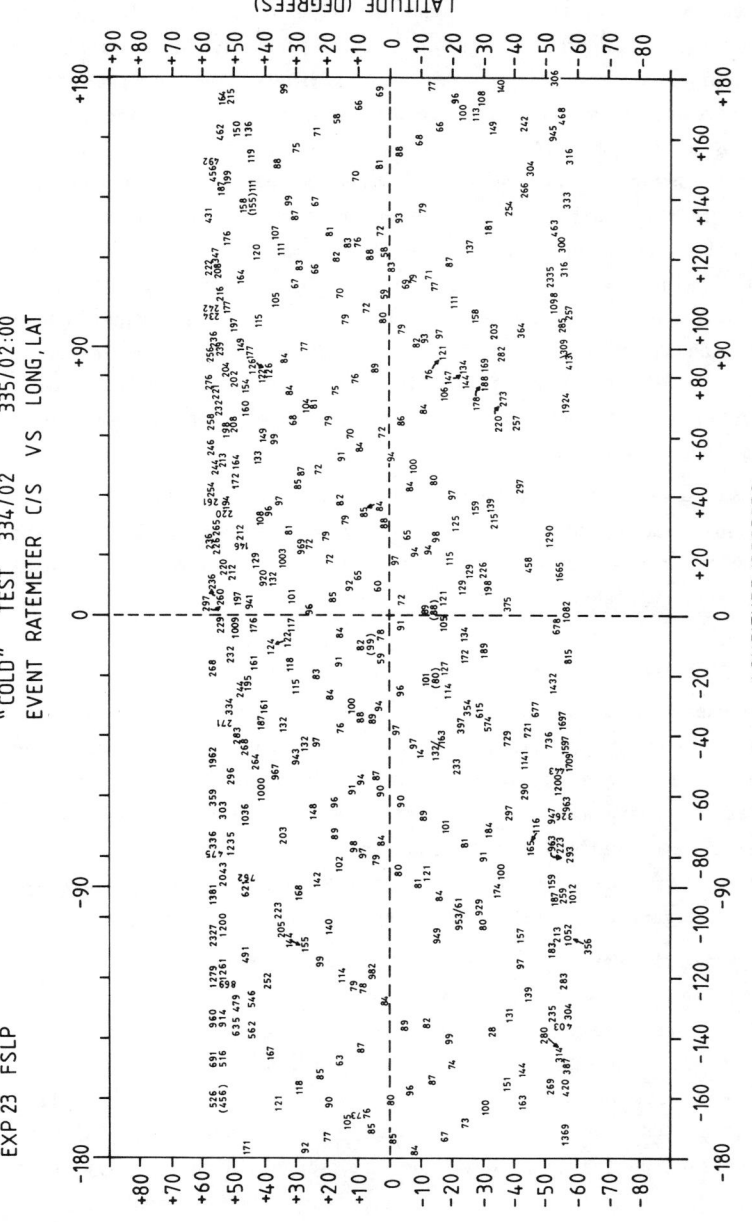

Fig. 5. Background Seen By X-Ray Gas Scintillator Spectrometer on Spacelab-1

ground and improving the instrument for the reflight. The primary source for the background light has been shown to be a particular airglow emission known as tropial arcs. Ninety-five percent of the interference is caused by this natural source of disturbance, e.g., not caused by the Shuttle environment. Tropical arcs appear close to the terminator and it is argued that this phenomenon was particularly active under the orbital conditions met in the November launch. There is indication for the existence of a second smaller optical contamination which might be caused by either Shuttle glow or light from the thruster firings. As a precaution for the reflight, the experimenter is replacing the film recording used on SL-1 by an opto-electronic system for the star imaging. This will allow data analysis on a frame-by-frame basis with the possibility to eliminate frames with sporadic high background light.

It was already realised before the launch of SL-1 that a November launch would not be ideal for the two UV experiments on board. The investigators concerned only agreed to a launch in November after NASA had granted a reflight.

Life and Material Sciences

The experiments of these disciplines were carried out inside the Spacelab module and required a maximum of crew interaction. The achievements in these disciplines must be attributed largely to the fact that on Spacelab-1 a scientific crew was both present to do experiments, and available to be experimented on. Already before the mission, it had been noted that Life and Material Science investigations are not very compatible from an operational point of view.

In many life science investigations the crew is exercising and moving inside the module quite intensively. At the same time, the material science investigators are anxious to protect the microgravity environment as much as possible. On Spacelab-1, this conflict of interest was resolved by scheduling a first series of critical life science experiments during the first days and only start material science processing on the third day. On the two following days, vigorous crew movement was prohibited in order to start crystal growth experiments under the best possible microgravity conditions. Once the crystals had grown to a reasonable size, life science experiments involving crew movement were resumed.

There was yet another conflict of interests between disciplines. The Life Scientists strongly opposed medication of the crew with anti-motion sickness drugs and had designed certain investigations which were motion sickness provocative to the crew. All other disciplines required the crew to be "top fit" to carry out their investigations. Again the time line was constructed in such a way that the conflicting discipline interests were respected to the largest possible degree. In any case, one lesson learned from the Spacelab-1 mission is that Material and Life Science investigations are difficult to interlace in a single mission.

In the Materials Science area, remarkable difficulties were encountered in controlling liquids in a microgravity environment. Almost all experiments in the Fluid Physics module suffered from extensive spreading of the test liquid over the contact surfaces. The anti-spread spray did not function as expected. In the mirror heating furnace, where a rod of silicon was remolten by the floating zone technique, more difficulties were encountered compared to previous ground processing. Finally there is a suspicion that some anomalies observed in the liquid cooling system of two furnaces might be attributable to the formation of air pockets and that the formation of these pockets were facilitated by the microgravity environment. This suspicion needs to be further investigated.

The microgravity environment met inside the Spacelab was in the expected range. Levels were of the order of 10^{-4} g on the average, with peak disturbances considerably higher than this average. Only a few cases were reported where such disturbances showed an influence on the material science products.

Earth Observations

Two instruments on Spacelab-1 were provided by this discipline: a Metric Photographic Camera and a microwave remote sensing experiment. The prime objective was Earth surface imaging by photographic and Synthetic Aperture Radar (SAR) techniques.

The metric camera was successful in proving the usefulness of Earth photography from space. It obtained overlapping pictures of 11 million square kilometers which are suitable for map production and map updates. The uniqueness of the metric camera flown on Spacelab-1 is that its picture is free of geometrical distortions and the distance between any two points can be determined with an accuracy of 20 m. Relative height can also be obtained by evaluating overlapping photographs with a stereo viewer. The maps which will be produced will have a scale of 1:100,000. The community claims that it would have taken years to collect the material obtained in a single Spacelab flight by conventional airplane photography.

The microwave instrument was designed to function in three different modes, the SAR-mode, the two-frequency modes in which the ocean wave spectrum can be obtained, and the passive radiometer mode. The instrument operated in X-band (9.65 GHz). Only a few results were obtained during the Spacelab-1 flight because of a failure in the high power amplifier (HPA). Post-flight failure analysis indicates that the failure was caused by a glow discharge between cathode feed and case of the TWT. In fact, the instrument was shown to work perfectly postflight in a normal atmosphere, but failed again in vacuum at a pressure of 10^{-5} Torr (in-flight conditions).

Thus, although only a few data in the radiometer mode were acquired during the mission, the flight was valuable from an instrument development point of view. The new possibility offered by Spacelab to retrieve the instrument and carry out post-flight failure analysis proved very valuable in the case of the microwave instrument. It is fair to say that this failure could have remained unexplained on a nonrecoverable free flyer.

Conclusion

The Shuttle/Spacelab combination represents a new space platform of considerable dimension. The local and artificially induced environment is expected to be different from the one met around conventional, low orbiting, free flying satellites. Useful measurements have already been carried out on earlier STS flights employing special equipment like the IECM. The Spacelab-1 flight offered a unique opportunity to test the functioning of instruments from all space disciplines in this new environment. In spite of the environmental signatures found in the data from a number of experiments, it appears not too diffficult to identify these signatures and separate them from the science data. The Spacelab-1 investigators welcome the plan to establish a data bank on the Shuttle environment and are prepared to provide their appropriate data to be included.

ONLINE SCIENTIFIC DATA BASES & DATA BANKS AVAILABLE FROM ESA-IRS

George Proca
Head, On-Line Services
European Space Agency-Information Retrieval Services

I. Mandate and Status of ESA-IRS

The mandate of the Information Retrieval Service of the European Space Agency (ESA-IRS) can be traced back to the time when the Space Documentation Service of the European Space Research Organization, forerunner of the ESA, entered into a generous Information Exchange Agreement with NASA.

The Exchange Agreement dates back to May 1964, following an exchange of correspondence between Dr. Hugh Dryden, NASA Deputy Administrator and Prof. Pierre Auger, Director General of ESRO. It has developed with the passage of time and is today one of the fundamental tenets of the European Space Agency's mandatory program on Documentation.

Similarly, the mandate of ESA-IRS has not varied. It was and still is a duty to provide the Agency, its Member States, and the Scientific and Technical Community at large with appropriate documentation primarily in support of aerospace programs.

On this basis, and often pioneering in the field of online retrieval of automated documentation, ESA-IRS has reached the status today where it is considered by customers as ranking first in Europe and second worldwide.

From an administrative point of view it may be useful to recall that ESA-IRS is one of the two activities currently present on the ESRIN site in Frascati, Italy, near Rome. The second activity on the site is the Earthnet Program Office. ESRIN, the smallest ESA Establishment, is organically reporting to the ESA Director General via the head of ESRIN, Mr. Francis Roscian.

The databank contains information regarding spacecraft components. As much information as possible is given for each component namely the name of all the projects in which it has been used, the procurement specifications, the quality approval, and many other facts. All the relevant Reports, Contractors Test Reports, and several others are also available therefore offering a detailed history of the past use and reliability of any spacecraft component.

Fig. 1.   SPACECOMPS-22 File Description

II.   Some ESA Data Banks

In the particular frame of this Space Shuttle Experiment and Environment Workshop, it is felt that an outline of a few factual data banks created by ESA would help in highlighting some of the characteristics of the ESA-IRS system.

1) For instance, the SPACECOMPS data banks (space qualified electronic components) provides the name of all the (European) projects in which it has been used, the procurement specifications, the quality approval, the contractor test reports, and several others offering a detailed history of the past use and reliability.

The subject coverage is shown in Figure 2 while the sources of the data bank in Figure 3. Not less than nineteen access codes are

Capacitors	Attenuators
Solar cells	Transformers
Inductors	Microwave devices
Connectors	Transducers
Piezoelectric devices	Quartz crystals
Relays	Diodes
Pyrotechnics	Transistors
Resistors	Thyristors
Isolators	Solders
Fuses	Instrumentation
Opto-Electronic devices	Integrated circuits
Batteries	Materials etc.
Filters	

Fig. 2.   SPACECOMPS-22 Subject Coverage

available for searching singly, or in combination (Figure 4). Outputs from searches may include tables (Figure 5) or graphic representation (Figure 6).

```
SOURCE A - Project parts list
SOURCE B - Qualified/Preferred parts list
SOURCE M - Manufacturers addresses
SOURCE N - CECC approved parts
SOURCE P - Construction analyses reports
SOURCE Q - Failure reports
SOURCE R - Miscellaneous reports
SOURCE S - Quality audits (DPA)
SOURCE T - Manufacturers test reports
SOURCE X - Radiation sensitivity reports
```

Fig. 3. SPACECOMPS-22 Sources

```
LI = LISTED IN                              LI = OTS-PPL
CO = COUNTRY                                CO = FRANCE
DA = QUALIFICATION                          DA = QUAL CERT,
     APPROVAL DATE                               72-06
LD = LOT DATE CODE                          LD = 7319
MN = MANUFACTURER NAME                      MN = SESCOSEM
PR = PROJECT                                PR = SPACELAB
RN = REPORT NUMBER                          RN = DPA 41I
     OR REPORT TITLE                        RN = CA 100
RO = REPORT DATE                            RO = 7612
     OR ORIGINATOR                          RO = TRW
SD = SCC CODE                               SD = 9100
     OR DESCRIPTION                         SD = I-C LINEAR
GC = GENERIC CODE                           GC = 5400
SO = SOURCE CODE                            SO = A
     (for list of codes see   ?FILE22)
PA = PARAMETER TESTED                       PA = VCE SAT
AS = FINAL ASSESSMENT                       AS = ACCEPTABLE
SS = SYSTEM OR SUB SYSTEM                   SS = AUTOPILOT
UR = USER                                   UR = AEG
PS = PROCUREMENT SPEC ORIGINATOR            PS = BADG
GS = GENERIC SPEC                           GS = MIL-C-20
DS = DETAIL SPEC                            DS = MIL-C-20/35

   or by direct select of the manufacturer type designation e.g.: SELECT SN5400J
```

Fig. 4. SPACECOMPS-22 File 22 Access Codes

85000816 SPACECOMPS
9101 IC-Linear, Operational Amplifier

Source	R (Miscellaneous Analysis);
	X (Radiation Test Reports)
Manufacturer Name	PMI-PRECISION MONOLITHICS
Country of Manufacturing	United States
Generic Code	OP08AJ
Mfr Type Designation	OP08AJ
Report Number	MIS194
Report Originator and Date	ESTEC; 8218
Specification Number	Commercial, full temp. range
Lot Date Code	7732
Tested Parameter	I bias; Vos; Ios; Open loop gain

REMARKS:
All 10 devices are marginal with respect to the spec. limits for Gain, Gainbandwidth product, and slew rate. These devices were not bought for Hi-Rel devices, but only full temperature range. This device is only irradiation hard up to 10 Krad. The sensitive parameters are Vos, Ios, Ibias and Gain. The four devices who had been chosen for irradiation test were within the specification up to 10 Krad. The OP08AJ cannot be considered 'rad-hard' for normal space application.

ANALYSIS REF: MIS 194				PARAMETER : Vos			
SPEC. LIMIT : 90 uV				UNIT : uV			
DEVICE S/N	# 1	# 6	# 9	# 10			
INIT. MEAS.	.6103	25.92	−9.28	−70.8			
Krad 1	.8544	30.55	−66.5	−.610			
Krad 10	1.830	64.20	−1.220	−62.3			
Krad 100	143.9	91.97	−104.5	−22.5			
Krad 200	507.7	567.2	−425.8	−375.2			

ANALYSIS REF: MIS 194				PARAMETER : Ios			
SPEC. LIMIT : 2 nA				UNIT : + pA			
DEVICE S/N	# 1	# 6	# 9	# 10			
INIT. MEAS.	148.0	770.7	21.18	11.90			
Krad 1	112.8	412.9	35.83	55.45			
Krad 10	139.7	312.9	72.35	62.25			
Krad 100	237.2	277.0	77.35	45.93			
Krad 200	203.4	140.1	60.69	6 069			

[Further tables are provided in the online reference]

Fig. 5. SPACECOMPS-22 Sample Record

87000151 SPACECOMPS
  9201 IC-Digital, Gate/Inverter
Source   X (Radiation Test Reports)
Manufacturer Name   RCA SOLID STATE
Manufacturer Type Designation   CD4001
Generic Code   CD4001
Report Number   GfW-TN53/08; HMI-B 248
Irradiation Source   Jet Propulsion Laboratories
Type of Irradiation   60-Co-gamma (radiation)
Report Originator   GfW; Hahn-Meitner Institut
Report Date   7744
Reference Document   RDC-Handbook Techn. Mem. 33-763
Tested Parameter   V threshold; tPLH; tPHL
Sample Size   10
Bias during Irradiation   VDD = 10V, All inputs = 10V
DATA SHEET VALUES: No data available
BIAS DURING MEASUREMENT: 1)ISS = 10uA, VDD = 10V 2)IDD = 10uA, VDD = 10V
         Sample size = 9 at 5x10E4 rad(Si)

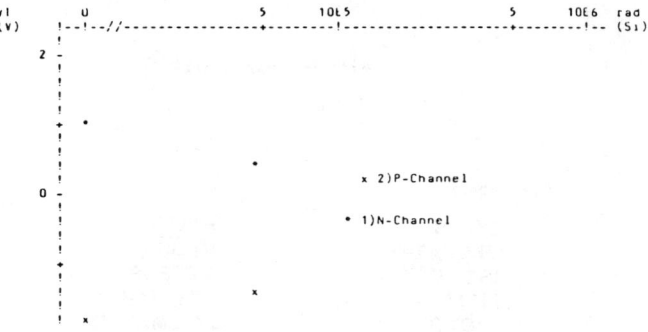

Fig. 6. SPACECOMPS-22 Graphic Representation

2) As regards SATELDATA, which is a data bank on satellite subsystems, mainly for mechanical, electrical, and electronic equipment, access is given to detailed data concerning equipment performance, characteristics, physical dimensions, etc.

The uses to which the data bank may be put are summarized in Figure 7. The subject coverage is given in Figure 8 while Figures 9 and 10 show samples of records. The important point to note here is the possibility offered by the ESA-QUEST retrieval software to specify a RANGE of numerical values for parameters of interest (Figure 11).

— Writing/evaluating equipment/subsystem specifications;

— Comparing design/performance parameters of existing, versus proposed equipments;

— Determining availability of similar equipment;

— Assessing technical input values for the execution of parametric cost models.

Fig. 7. SATELDATA-12 Scope

Currently loaded:

COS-B	HELIOS
EXOSAT	GEOS
ISEE-B	METEOSAT
ISPM	OTS/MAROTS
GIOTTO	ECS/MARECS

SPACE TELESCOPE (FOC)

To be loaded:

HIPPARCOS
OLYMPUS
.
.

Fig. 8. SATELDATA-12 Subject Coverage

```
MECHANICAL UNIT SAMPLE
```
81000013 SATELDATA
ANTENNAE Reflector Front Fed
Spotbeam SHF(Transmit)
Project ECS
Manufacturer Selenia
Mass 3.253 Kg
Dimension,length 620 mm (circular)
Dimension,width   620 mm (circular)
Dimension,breadth 625 mm
Dimension,diameter 620 mm
Bonding resistance ⟨1 ohm
Chemically clean Yes
Electrically bonded Yes
Feed single/multi Single
Frequency 10.95 to 11.7 GHz
Gain ⟩30.1 dB
Gain slope 0.002 dB/MHz
Interchangeable Yes
Life, design 7 yr
Life, storage 5 yr
Maintainable No
Material,Aluminium Yes (Honeycomb)
Polarisation,Circular No
Polarisation,Linear Yes(Dual-orthogonal)
Radiation compatible Yes
Radiation pattern Circular
Side lobe - 10 dB
Source document 1 SS-04-2-1,Iss4
Source document 2 EQ-04-2-3,Iss4
Surface finish, Corrosion protected Yes
Temperature,operating,max. 100 degC
Temperature,operating,min. -180 degC
Vibration frequency 5 to 2000 Hz
Vibration tested, random Yes
Vibration tested, sine Yes
VSWR transmit ⟨1.2
X-polar discrimination 33 dB
Year of technology 1978
Date of entry/update Aug82
Cross Ref 0663
File Ref AAG0663
Remarks: Three supplied

```
ELECTRICAL UNIT SAMPLE
```
85000071 SATELDATA
POWER Electronic Power Conditioner
High Voltage
Project ISPM
Manufacturer Thomson/CSF
Mass 2.2 Kg.
Dimension, length 300 mm
Dimension, width 115 mm
Dimension, breadth 83 mm
Analogue 90%
Bonding resistance ⟨7.5 mOhm
Chemically clean Yes
Digital 10%
Efficiency 75%
Electrically bonded Yes
EM compatible Yes
Hybrid Yes
Life, design 7 yr
Life, storage · 3 yr
Moments of Inertia(x,y,z)16.7Kg cm sq;96Kg cm sq;
                        107Kg cm sq
Overcurrent regulation Yes
Power out 53 W
Power regulation Yes
Pressurised Yes
Radiation compatible Yes
Redundancy,internal Yes
Regulated input Yes
Reliability 0.9
Source document 1 IS-EQ-TC-4470, IssC2
Source document 2 IS-IF-DS-0001, IssC1
Source document 3 IS-SR-DS-0001, IssC3
Source document 4 IS-EV-DS-0001, IssC3
Surface finish, Corrosion protected Yes
Surface finish, Painted Yes (Cuvertin)
Telecommand facility Yes
Telemetry monitored Yes
Temperature, non-op. max.   60 deg C
Temperature, non-op. min. -40 deg C
Temperature, operating, max.   45 deg C
Temperature, operating, min. - 5 deg C
Temperature, storage, max.   60 deg C
Temperature, storage, min. -45 deg C
Vibration frequency 5 to 2000 Hz
Vibration tested, random Yes
Vibration tested, sine Yes
Voltage in, range 28 V
Voltage in, tol 2%
Voltage out, range 5 V; 3200 V; 950 V; 60 V
Voltage out, tol 2%
Voltage, under-survival 10%
Date of entry/update Jun82
Cross Ref 0158
File Ref EAL0158
Remarks: Two supplied

Fig. 9. SATELDATA-12
Mechanical Unit Sample

Fig. 10. SATELDATA-12
Electrical Unit Sample

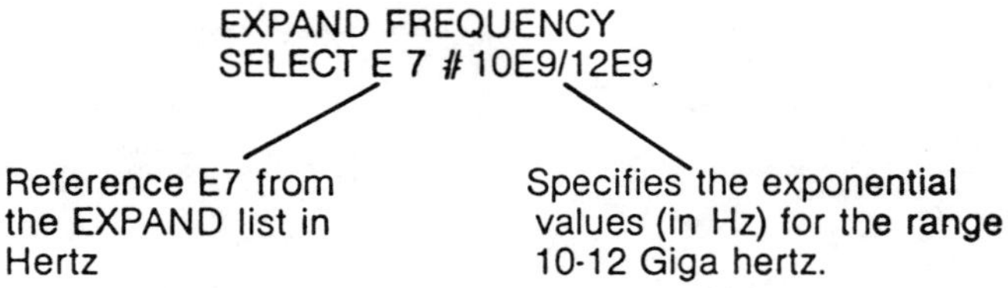

EXPAND FREQUENCY
SELECT E 7 # 10E9/12E9

Reference E7 from the EXPAND list in Hertz

Specifies the exponential values (in Hz) for the range 10-12 Giga hertz.

Fig. 11. SATELDATA-12 Range Searching

3) With LEDA 2, which is the catalog of imagery remotely sensed by the LANDSAT series of satellites and acquired by the Earthnet Fucino and Kiruna Stations, two new facets appear:

First, LEDA 2 is one example of a data bank which is physically residing on a computer different from the ESA-IRS mainframe. There are other such examples (Figure 12) of distributed data bases accessible by a variety of networks using different technologies.

Second, LEDA 2 is interrogated via a menu driven retrieval system which guides the customer in an orderly way towards an explicit and complete definition of his query (Figure 13).

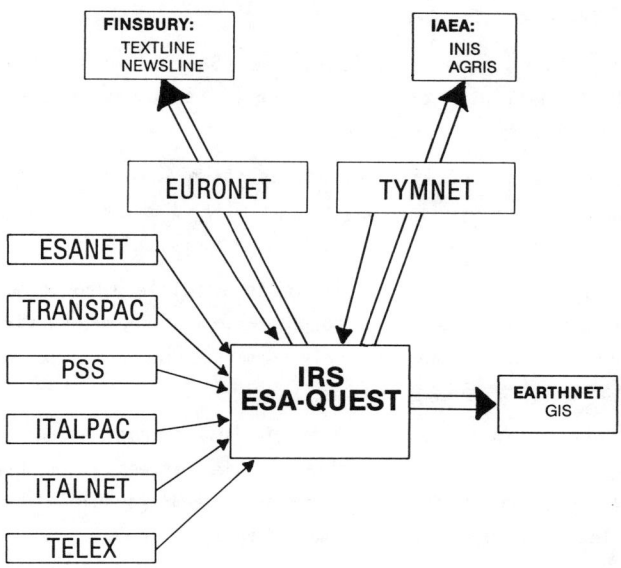

Fig. 12. Realtime Access To External Data Bases

```
ENTER GEOGRAPHIC SELECTION MODE. THE RESULTING GEOGRAPHIC AREA IS THE
SUM OF SUCCESSIVE SELECTIONS. YOU MAY LOOP ON THE FOLLOWING CHOICES :

AVAILABLE OPTIONS : RT TF PN RF PO SH LI DE SC CS AC SE

DEFAULT : SE >
PO

ENTER THE NEXT CLOCKWISE POINT OF POLYGON COORDINATES LAT,LON
(HUNDREDTHS OF DEGREE). AFTER THE LAST POINT ENTER <CR> :
4000,800+4100,900+4000,1000+

I NEED 10-20 SECONDS FOR GEOGRAPHIC COVERAGE COMPUTATION.

NUMBER OF LAST   SELECTED FRAMES :    4
NUMBER OF TOTAL SELECTED FRAMES :    4

ENTER GEOGRAPHIC SELECTION MODE. THE RESULTING GEOGRAPHIC AREA IS THE
SUM OF SUCCESSIVE SELECTIONS. YOU MAY LOOP ON THE FOLLOWING CHOICES :

AVAILABLE OPTIONS : RT TF PN RF PO SH LI DE SC CS AC SE

DEFAULT : SE >
```

Fig. 13. LEDA2-13

III. Other features of ESA-IRS

A very basic characteristic of the ESA-IRS retrieval system is the ability to deal with time series--or for that matter any series of data related to a uniformly growing parameter.

The statistical software package as shown in Figure 14 was originally conceived for use in conjunction with a commodities data base but may easily be adapted to suit other purposes.

It should also be noted that it is possible through a teletype compatible terminal to obtain a rough graphical representation of the time series obtained either directly from the data base or after data manipulation (Figure 15).

The data manipulation facility is--of course--enhanced by the possibility to create locally "private" time series. After downloading of a portion of a statistical database it is possible to replace locally selected values, e.g. for modeling purposes.

Clearly ESA-IRS also provides for more classical bibliographic searches often facilitated by the use of the QUESTINDEX function, see Figures 16 and 17, (a cross search facility amongst data bases) and the ZOOM function which extends to a variety of bibliographic fields (e.g. authors, corporated sources, titles, abstracts) and the well known .FREQUENCY command of the NASA-RECON retrieval language.

IV. Towards the future

From this brief survey it can be seen that the basic tools available from ESA-IRS, and currently used by more than 4000 customers, can be adapted and enhanced to allow the retrieval and manipulation of more specialized types of data such as those deriving from spacecraft or spacecraft experiments.

This is why there is close association with the studies currently undertaken with respect to the ESA project to set up a European Science Space Data Center. ESA-ESRIN is currently devoting considerable attention to the problems of data management and the additional services now required by the community of aerospace scientists soon to face the "data deluge" from existing and forthcoming space ventures.

COMMAND	ROUTINE NAME*	FUNCTION	INPUT PARAMETERS
V	(DEFAULT)	Calculates the statistics and tabulates the series belonging to the last set	
V	TABLES	Prints one or more series	
V	STATISTICS	Performs and prints elementary statistical functions (average, min. and max., variance and standard deviation)	
V	GRAPH	Represents a single series graphically by points	
V	PLOT	Represents up to 8 series graphically by indexing to a common base	
V	SEASONAL ADJUSTMENT	Calculates seasonal coefficients, the ciclo-trend component and residuals	- applied solely to the WKLY & Mthly series - RANGE must be min. of 4 years
V	AUTOCORRELATION	Performs the simple autocorrelation function and autocovariance up to n-lags	- LAG<No. of observations - 1<LAG<104 - by default: LAG = 1
V	MOVING AVERAGE	Interpolates through non-centred moving averages to n terms	- 1<No. of terms<No. of observations - by default: no. of terms = 2
V	GROWTH CURVES	Interpolates through linear, exponential and hyperbolic development curves	- by default: linear interpolation - by default: No. of forecasts = 0
V	POLYNOMIAL INTERPOLATION	Interpolates through linear, parabolical and cubic polynomials	- by default: Linear interpolation - by default: No. of forecasts = 0
V	MULTIPLE REGRESSION	Performs regression analysis	- the 1st set must be the dependent article
V	CORRELATION	Performs correlation analysis	
V	CROSS CORRELATION	Performs cross correlation analysis	- 1<LAG<104
V	EXPONENTIAL SMOOTHING	Performs the interpolation according to the exponential smoothing method	- No. of forecast>0 - 1<α<100
V	FORECAST	Performs forecasts from one to four weeks with, minimums and maximums	
V	PARITY DEVALUATION	Calculates the depreciation index of a currency in respect of another	
COMP	LOGARITHM	Performs logarithm operation	
COMP	EXPONENTIAL	Performs exponential operation	
COMP	SQUARE ROOT	Performs square root operation	
COMP	DIFFERENTIAL	Obtains the first differences in respect of any LAG	- 1<LAG<No. of observations - by default: LAG = 1
COMP	VARIATION	Obtains the % variations in respect of any delay (lag)	- 1<LAG<No. of observations - by default: LAG = 1
COMP	INDEX	Indexes according to any base	- 1<LAG<No. of observations - by default: LAG = 1
COMP	RIGHT SHIFT	Right Shift operations of n. periods	- 1<LAG<No. of observations - by default: LAG = 1
COMP	LEFT SHIFT	Left Shift operations of n. periods	- 1<LAG<No. of observations - by default: LAG = 1
COMP	ADDITION	Performs additions with constants	- by default the constant = 0
COMP	SUBTRACTION	Performs subtractions with constants	- by default the constant = 0
COMP	MULTIPLICATION	Performs multiplications with constants	- by default the constant = 1
COMP	DIVISION	Performs divisions with constants	- by default the constant = 1
COMP	(DEFAULT)	Performs the four operations, above, for a given time series	

* NOTE Any unambiguous abbreviation of the routine is valid e.g. TAB for Tables or STAT for Statistics

Fig. 14. Routines Provided by the Statistical Software Package

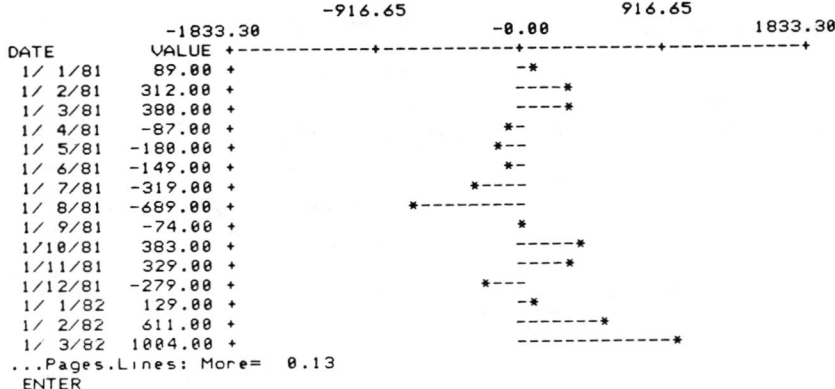

```
                      -916.65              916.65
            -1833.30          -0.00                1833.30
DATE      VALUE  +-----------+-----------+-----------+
 1/ 1/81    89.00 +                     -*
 1/ 2/81   312.00 +                     ----*
 1/ 3/81   380.00 +                     ----*
 1/ 4/81   -87.00 +                    *-
 1/ 5/81  -180.00 +                   *--
 1/ 6/81  -149.00 +                    *-
 1/ 7/81  -319.00 +                *----
 1/ 8/81  -689.00 +          *----------
 1/ 9/81   -74.00 +                    *
 1/10/81   383.00 +                     -----*
 1/11/81   329.00 +                     ----*
 1/12/81  -279.00 +                 *---
 1/ 1/82   129.00 +                     -*
 1/ 2/82   611.00 +                     -------*
 1/ 3/82  1004.00 +                     -------------*
...Pages.Lines: More=   0.13
ENTER
```

Fig. 15. Rough Graphical Representation of a Time Series

QT SPACE
TOPICS IDENTIFIED

QF SPACELAB

Base 01: NASA:1962-84,12
  2   2578 SPACELAB
Base 06: NTIS:1964-84,14
  2   727 SPACELAB
Base 08: INSPEC:1971-84,14
  2   366 SPACELAB
Base 14: PASCAL:1973-1984,05
  2   134 SPACELAB
Base 36: CONFERENCE PAPERS:72-84,05
  2   274 SPACELAB
Base 69: COSMIC 1984
  2   1 SPACELAB
Base 72: AEROSPACE DAILY 83.12-84.04
  2   106 SPACELAB
QUESTINDEX FUNCTION COMPLETED

Fig. 16. Search Using QUESTINDEX Function

```
QT·SPACE
TOPICS IDENTIFIED


QF SPACELAB ..., CONTAMINA?

Base 01: NASA:1962-84,12
   2     47 SPACELAB(F)CONTAMINA?
Base 06: NTIS:1964-84,14
   2     19 SPACELAB(F)CONTAMINA?
Base 08: INSPEC:1971-84,14
   2      5 SPACELAB(F)CONTAMINA?
Base 14: PASCAL:1973-1984,05
   2      0 SPACELAB(F)CONTAMINA?
Base 36: CONFERENCE PAPERS:72-84,05
   2      1 SPACELAB(F)CONTAMINA?
Base 69: COSMIC 1984
   2      0 SPACELAB(F)CONTAMINA?
Base 72: AEROSPACE DAILY 83.12-84.04
   2      3 SPACELAB(F)CONTAMINA?
QUESTINDEX FUNCTION COMPLETED
```

Fig. 17. Search Using QUESTINDEX Function

OVERVIEW OF ENVIRONMENT HANDBOOK

Lyle Bareiss
Martin Marietta, Denver, Colorado

It might be a little refreshing for you to know that I'm not going to present any more data. Instead, I have a program overview to present and I'd like to solicit support from the participants in three of the main panel sessions that we've covered so far in this workshop (i.e. particle/molecular contamination and surface interactions). The project we are working on supports Ed Miller, who is the monitor for this contract, and as you can see (Figure 1) the title is, "The Shuttle/Spacelab Contamination Environment and Effects Handbook." We call it the SCEE which is a little easier to say.

SHUTTLE/SPACELAB CONTAMINATION
ENVIRONMENT AND EFFECTS HANDBOOK

PROGRAM OVERVIEW

L.E. BAREISS
PROGRAM MANAGER
(303) 977-8713

ED MILLER, COR
NASA-MSFC

6-10 AUGUST 1984

Fig. 1. Title Page of SCEE

Throughout the discussion this week, we've achieved a good idea of the needs, desires and requirements for data bases. Figure 2 gives an idea of where I think we stand on a lot of the data base areas. That is, we have data strewn all over the place--on people's desks and in their memories--and right now, it's funneling into someone we call the end user. I think right now in the overall data base area, the end user is starting to be these panels, but we have not yet, in all areas, worked out the methodology to set up a central focal point where the end user actually goes to that focal point and draws out the useful information on his own.

We see the "Contamination Environment and Effects Handbook" as being that initial building block to input into a system such as the electronic data base that we've heard discussed here in the last couple of days. That allows everything to be funneled into one central focal point, and I think we all agree that data bases are well worthwhile.

At a top level, the objectives of the program that we are working are essentially threefold (see Table 1). The initial objective is to develop a comprehensive and accurate data base, primarily in the fields of molecular and particulate contamination and the surface interactions that relate to some of the new phenomena we've discussed tonight--surface glow, materials oxidation, erosion, and so forth. Our second main goal is to interpret the data, arrange them in a format that is easily usable and understandable to the common user.

Table 1. Objectives

o TO DEVELOP A SINGLE-SOURCE/COMPREHENSIVE DATA BASE OF SHUTTLE/SPACELAB CONTAMINATION ENVIRONMENTS

    - MAIN FOCUS: MOLECULAR/PARTICULATE/SURFACE INTERACTIONS

o TO INTERPRET/NORMALIZE AVAILABLE DATA INTO FORMAT WHICH CAN BE EASILY UNDERSTOOD BY USERS UNFAMILIAR WITH CONTAMINATION TECHNOLOGY

o TO INCORPORATE HANDBOOK INTO ELECTRONIC DATA RETRIEVAL SYSTEM AND DEVELOP USER FRIENDLY KEY WORD SEARCH SOFTWARE CAPABILITY

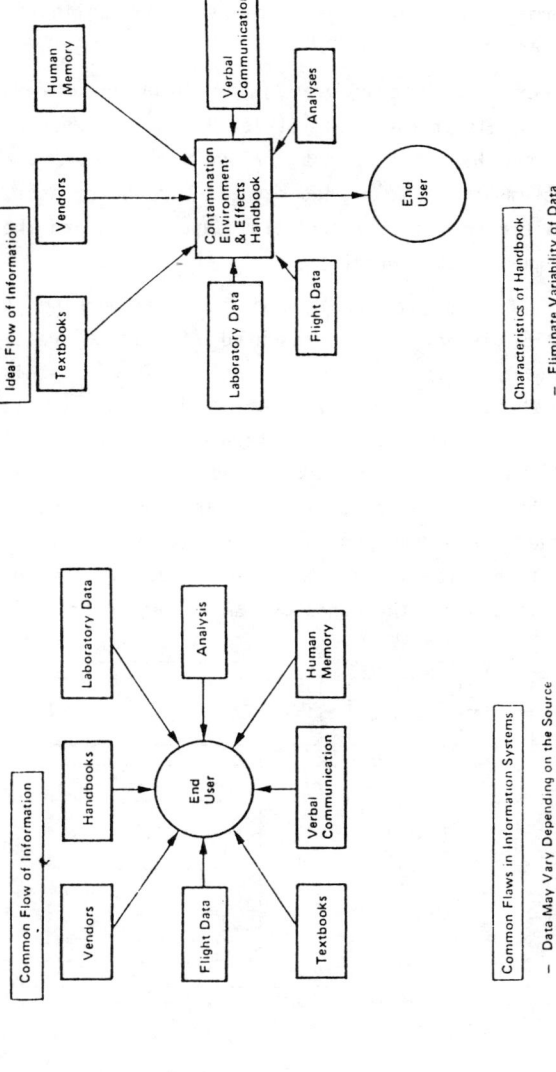

Fig. 2. Benefits of Contamination Environment Effects Handbook

I think we all understand that the technologies of "glow", "contamination", "loads", and "vibration" data are complex and variable. When presented to the designers and the payload users, the data need to be in a format which can actually be used and applied to specific projects and programs.

The third main objective of the program is to incorporate this particular Handbook into an electronic data retrieval system. Whether it will be identical to the Handbook or not, or whether it goes directly into the SCAN System is yet to be determined. In our project, we will at least have an electronic data retrieval capability, and the Handbook will be on a computer tape format.

Figure 3--an overall flow of the program that we are working is for the people with 20/15 vision. It's presented here at a pretty basic level. We are acquiring the data that will hopefully, at this forum and through personal solicitations, supplement the data base we have already accumulated. The data will be interpreted, and we will come out with a first hard copy handbook in approximately seven months. This handbook will be updated as new data are aquired and as comments are received back from the people who review it. In the meantime, we will not only be updating the hard copy, but also be developing a computer tape version of the Handbook and a data retrieval

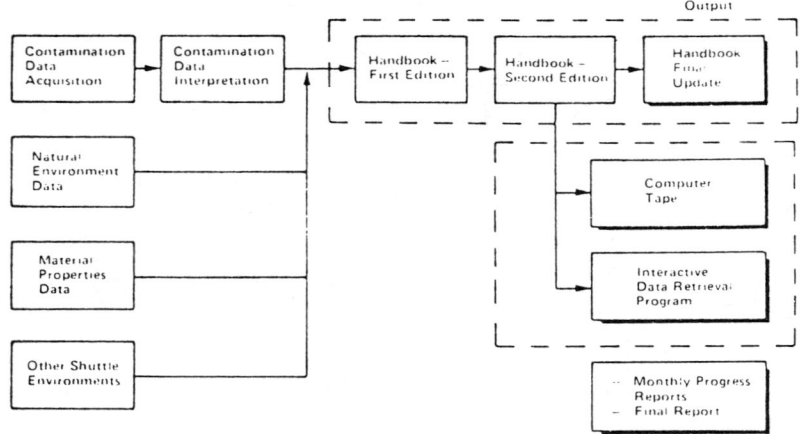

Fig. 3. Handbook Development Flow

system. Approximately 26 months after the go-ahead of contract, the final data acquisition system, final handbook, and the computer tape will be delivered. We will, therefore, have time to incorporate, hopefully, some of the new Spacelab II data we might be acquiring and also data that have yet to be published.

Our basic approach, (Table 2) as I indicated, will be to rely strongly on people such as you and the other people out in the industry and other agencies who have developed the data, have acquired and reduced them, understand them well, and have interpreted them properly. We are soliciting support from all of you folks, especially in the areas of contamination and surface effects. I want to guarantee you that any data we use in the Handbook will be properly documented and cited, so that due credit will be given to those who have provided the information and who have developed it. We are also conducting a formal literature survey, just to make sure that we do not have any holes in the data base system.

Table 2. Approach

o   RELY STRONGLY ON EXISTING DATA BASE DEVELOPED BY ASSORTED INVESTIGATORS THROUGHOUT INDUSTRY, ACADEMIA AND GOVERNMENT AGENCIES

- SOLICIT COOPERATION AND PROVIDE INSURANCE OF COMPLETE DISCLOSURE/CITATIONS OF DATA CONTRIBUTORS

- FORMAL LITERATURE SURVEY

o   DEVELOP CONTAMINATION DATA MATRIX COVERING ALL MISSION PHASES (PREINTEGRATION, INTEGRATION, ASCENT, ON-ORBIT, DESCENT, POST-LANDING, DEINTEGRATION)

- IDENTIFIES ALL AVAILABLE DATA FOR EACH PARAMETER OF INTEREST

- IDENTIFIES AREAS OF INSUFFICIENT OR INADEQUATE DATA

- IDENTIFIES AREAS WHERE FUTURE DATA AVAILABILITY IS ANTICIPATED

The next major step in working the program will be to develop a contamination data base matrix (Figure 4). That is essentially a management tool which allows us to make sure we don't have any holes left

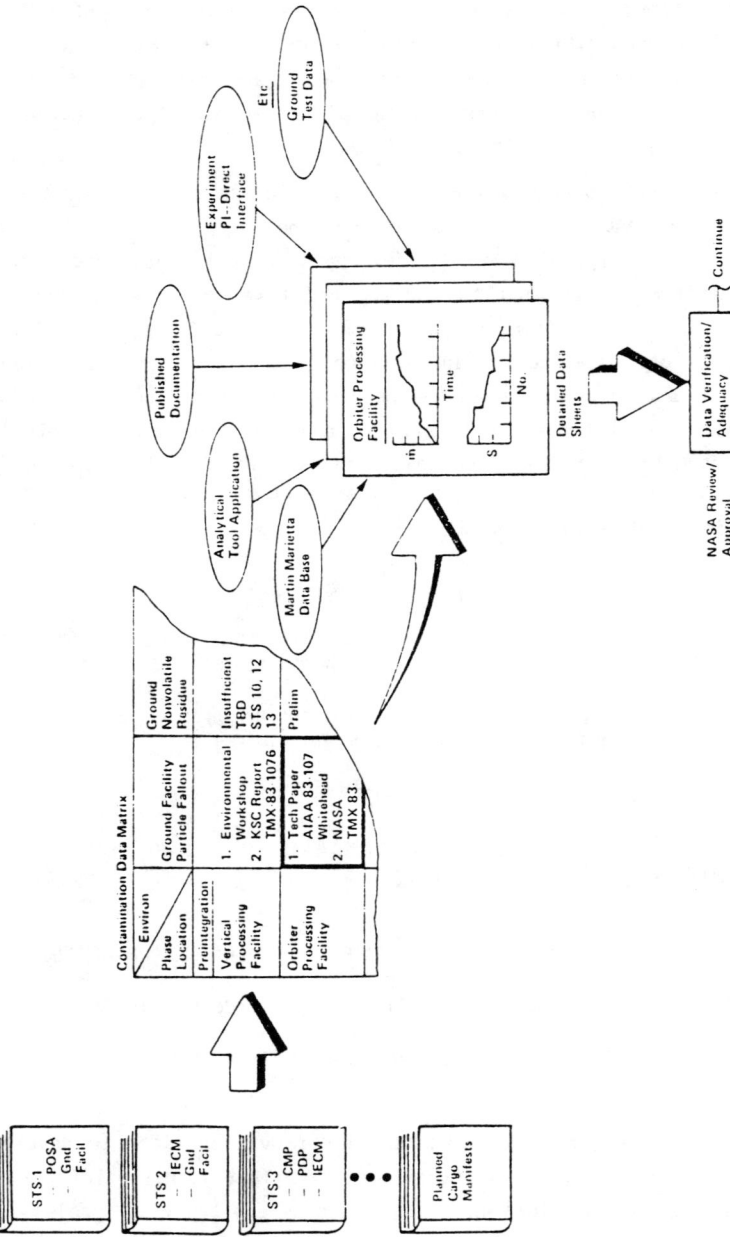

Fig. 4. Handbook Data Aquisition System

in the data base, to make sure we're covering all the major parameters of interest, and all of the major mission phases we are interested in. The matrix allows us to identify the parameters, it identifies the holes, and gives us an idea of where other work is being done and where future data may be obtained. Figure 4 gives you an idea of how that data matrix works. Of course, we do have the available data out in the real world. The data base matrix (which I envision to be quite an extensive document in itself) essentially includes the various phases of missions and locations where the data have been acquired, and the primary parameters, such as in this case, ground facility particle fallout, nonvolatile residue, and across the top - the other parameters we are monitoring. This will be backed up by the actual hard copy data publications in whatever format they come in from the various sources that we've identified, as well as others on this chart. After that, the data will be evaluated for accuracy (to the level of detail that we feel we should take it), and for its applicability to the Handbook (Table 3). We will address "conflict of

Table 3. Approach (cont.)

O   CONDUCT DATA INTERPRETATION/ANLAYSIS OF COMPLETE DATA BASE

- EVALUATE IN TERMS OF EFFECTS TO VARIOUS CLASSES OF SYSTEMS AND INSTRUMENTS

- EVALUATE IN TERMS OF PROTECTIVE MEASURES FOR EACH ADVERSE ENVIRONMENT

- DATA CONFLICT RESOLUTION

O   DEVELOP HANDBOOK IN USER FRIENDLY FORMAT

- USER'S GUIDE

- SAMPLE PROBLEM(S)

- EXTENSIVE BIBLIOGRAPHY

O   CONVERT HANDBOOK TO ELECTRONIC DATA BASE SYSTEM

- COMPUTER TAPE FORMAT

data" items using on a fairly straightforward ground rule. If we receive data from several sources on the same parameter, and they are all credible sources, all sets of data will be presented. We will not be the judges as to who is right, but as time goes on, we hope that these are resolved by additional information, analysis and testing. We therefore do not anticipate being the judge and jury of what data are right or who has the right approach, unless we have a consensus from the rest of the technical community.

The Handbook will be developed in as user-friendly a format as possibly can be done. We will have one section that is specifically a user's guide, which walks the user through the Handbook and allows an individual to pick it up and use it without being totally familiar with these technologies. We will include sample problems and an extensive bibliography after each section, which will provide the user with additional contacts if more information is required from what is presented in the Handbook.

As I indicated, we will convert to a computer tape format, and we will evaluate, in the selection process, what kind of data acquisition/retrieval system we should use. Figure 5 is just a little cartoon which indicates that, up front in the program, we will develop

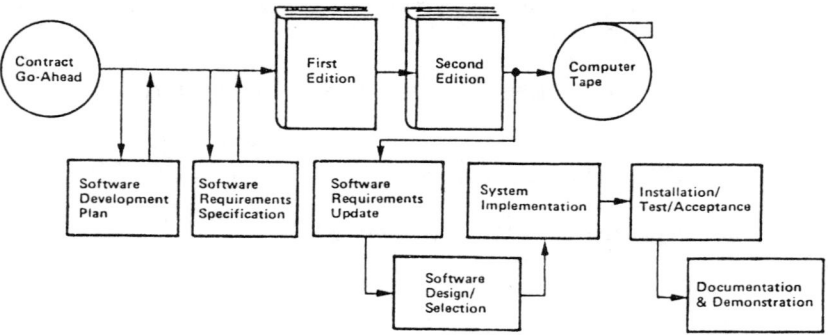

Fig. 5. Software Development Flow

the software requirements. This will help dictate the format of the hard copy handbook, and it will definitely dictate the format of the software that is developed later in the program. Once we complete the second edition, we will have a good idea of what the software will look like for the data retrieval system (Table 4).

Table 4. Approach (cont.)

0   DEVELOP INTERACTIVE DATA RETRIEVAL SYSTEM

   -   SYSTEM REQUIREMENTS SPECIFICATION

   -   TRADE STUDY TO OPTIMIZE UTILITY

   -   SOFTWARE DEVELOPMENT/IMPLEMENTATION

   -   DEMONSTRATION AND TRAINING

0   MAINTAIN AND UPDATE HANDBOOK

   -   FUTURE SHUTTLE/SPACELAB MISSION DATA

   -   DATA CURRENTLY NOT PUBLISHED OR AVAILABLE

We will use a standard software development process as shown in Figure 6. This will involve developing the software requirements, optimizing them for user utility followed by basic software develop-

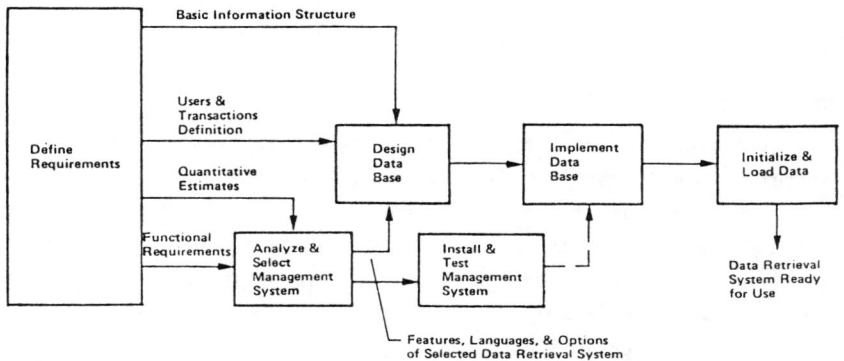

Fig. 6. System Requirements in Software Development

ment/implementation, and training and demonstration for the users at Marshall Space Flight Center and possibly at other locations.

After initial Handbook development, we will continue to maintain it. We will update it with new incoming data and data that have not yet been published or that individuals are not currently ready to release. This will be continued throughout the period of performance of the contract period.

Table 5 presents the program schedule. It is a 28-month activity, with the first draft released after about the seventh month. That will provide NASA with time to review it before it is officially released. As I understand, it will be released by MSFC out of the Government Printing Office at that point in time. The second edition will be released approximately 18 months after go-ahead, with the computer tape and software available in that same time period.

Table 5. Handbook Schedule Overview

o   CONTRACT DURATION - 28 MONTHS

o   CONTRACT START - JULY 1984

o   FIRST HANDBOOK EDITION - MONTH 7

o   SECOND HANDBOOK EDITION - MONTH 18

o   COMPUTER TAPE - MONTH 18

o   FINAL UPDATE - MONTH 26

I believe that as the week has progressed, I have been able to introduce almost everyone to what we are attempting to accomplish. Therefore, some of the items on Table 6 have already been addressed because I wanted to make sure that a few of them got discussed while we were here. Since I have now brought this up formally tonight and you are now aware of what's going on, I would be very interested in inputs or any suggestions on content, format, and arrangement of the Handbook. Inputs should be from a user's viewpoint as well as from a supplier's viewpoint. The available data that we are continuing to solicit are desperately needed. My last chart (Table 7) contains my name and address. Anyone who has additional information/data that they would like to see included in a Handbook such as this (with the proper citations) is encouraged to participate.

Table 6. Workshop Discussion Items

O   SUGGESTIONS ON HANDBOOK CONTENT, FORMAT AND METHODS OF DATA PRESENTATION

O   SHUTTLE/SPACELAB CONTAMINATION DATA AVAILABILITY

- PUBLISHED LITERATURE SOURCES/ANTICIPATED PUBLICATIONS

- PLANNED DATA ACQUISITION ON FUTURE MISSIONS

- ANALYTICAL TOOL DESCRIPTIONS

O   ELECTRONIC DATA RETRIEVAL SYSTEM SELECTION

- APPLICABILITY OF PLACING HANDBOOK ON SYSTEM SUCH AS "SCAN"

- IMPORTANT CAPABILITIES

Table 7. Solicitation for Support

O   DATA IN REDUCED AND/OR PUBLISHED FORM SHOULD BE SENT TO:

LYLE BAREISS, MS-M0487
MARTIN MARIETTA AEROSPACE
BOX 179
DENVER, COLORADO   80201

O   PARTICIPANTS/CONTRIBUTORS WILL BE CITED IN APPROPRIATE HANDBOOK SECTIONS <u>AND</u> WILL BE INCLUDED IN HANDBOOK DISTRIBUTION LIST

One final comment regarding the data retrieval system for which some decisions have yet to be made: We are now familiar with the SCAN system and its capabilities. Whether this Handbook would go directly onto a system such as that or whether it will be input in a different format is to be investigated later.

My final plea is, "Send us your data." I do guarantee that you will be properly cited for your participation and also included on the distribution list for the overall Handbook when it is completed. If there are any questions, I'd be happy to answer them now.

QUESTION: How does this Handbook relate to the Data Bases being discussed at this workshop?

To address this in more detail, an example may help. An earlier speaker tonight from the "loads" subpanel had some questions as to how all these data were going to get pulled together, and who's going to control it and maintain it. I think the Handbook is an example of how that can be done. It's this kind of activity that provides a focal point to bring the data together. I don't see the panel doing all that kind of activity in all areas. But with this kind of organization, I think we are going to be able to bring that data base together and, at least at a frozen point in time, have what's most current and most agreed upon by the community. All environmental data bases may not fall together perfectly, but I think, in these three or four major environmental areas which we're addressing in the Handbook, that this can't be anything but a benefit to enhancing the overall electronic data base system.

QUESTION: Is this a Handbook or a data base?

It potentially could be both. I see it initially as being a separate Handbook, with appropriate segments being included in the overall SCAN data base at a later date.

QUESTION: What about the data management system search capabilities?

That's part of that tiny flowchart (Figure 6) that I discussed earlier. The design specification really has to include those kinds of decisions.

QUESTION: Doesn't Martin Marietta already have a pretty complete data base developed?

To a certain extent that's right, although there is new data in these areas coming out all the time. This activity, then, is supplementing the already existing data base.

1985 AIAA MEETING: SHUTTLE ENVIRONMENT AND OPERATIONS II

Billy McCormac
Lockheed Palo Alto Research Laboratory

(Ed. note: With Dr. McCormac's permission, we are including here an updated version of the meeting announcement he made at the workshop.)

NASA really views AIAA meetings as a good way of getting the general aerospace audience educated into what's going on. The meeting on Shuttle Environment Operations, which is to be held in Houston on November 13-15, 1985, is to allow anyone who wishes to feel free to attend and try to present different levels of information on Shuttle environment and operations. This is how Shuttle affects the environment, integration, operations, etc. The organization is basically two half days of plenary sessions. Last time in Washington we had seven plenary session speeches which went quite well. That leaves time for a maximum of four concurrent sessions throughout the rest of the three days which will produce a total of sixteen sessions. My goal is to take any suggestions. Don't hesitate to write me or call me if you have some.

The call for papers will go out in the February issue of Aerospace America. After we receive those back, the Program Committee and the Session Chairman will organize them up to the limit of our capability. We picked Houston because more people agreed that is harder to find out the happenings at JSC than any other center; therefore, it seemed logical to go to JSC.

Hopefully this location will greatly increase interaction with the people at JSC. The call for papers is listed on the next page. Please note that a continuation of the Henniker Conference will be held in the evenings of November 13 and 14 at Houston.

AIAA Meeting

Shuttle Environment and Operations II

November 13-15, 1985 Houston, Texas
NASSAU BAY HILTON-CLEAR LAKE AREA

CALL-FOR-PAPERS

Abstract Deadline: April 18, 1985

The theme of this meeting is to discuss all aspects of the Shuttle environment and operations, and to present the latest results and experience. This is a full three-day meeting, with plenary sessions the first two mornings and up to four concurrent sessions the rest of the time. The plenary session speakers will be invited. The other sessions will be a mixture of invited and contributed papers.

The objectives of the meeting include discussion of:

* The environment in and around Shuttle that is created by Shuttle and which may affect the missions of Shuttle payloads.
* The environmental effects of Shuttle operations.
* The integration and deintegration procedures as they bear on contamination, operations and the success of payloads.
* The operations, data flow and real-time data processing and control.
* The effects of the environment on Shuttle launch and landing.

This is also a call for papers for presentations at a workshop that will be conducted in the evenings of November 13 and 14. The objectives will be to carry on the discussions that took place at the Henniker Conference, August 5-10, 1984. These sessions will provide a forum for evaluating information from recent Shuttle flights and/or reviewing the requirements and concerns of the scientific community on the Shuttle environment on payloads.

PRELIMINARY PROGRAM

November 13, 1985

    AM: Plenary   JSC
                         ESA
                         HQ NASA

    PM: Shuttle Environment - Gases/Glow: Chrm., T. D. Wilkerson
            Shuttle Environment - Vibration/Acoustics
            Shuttle Operations and Control
            Commerical Uses: Isacc Gillam

    Night Workshop: Chrm., M. Lauriente

November 14, 1985

    AM: Plenary   Mission Operations
                  Air Force Shuttle Operations: Brig. Gen. Cromer
                  Operations & Checkout: J. Ragusa
                  Shuttle Program Status

    PM: Shuttle Operations and Control
            Shuttle Environment - Particulates
            Shuttle Environment - Ionized and Nonionizing Radiation: Chrm., J. Reagan
            Spacelab and the Environment

    Night Workshop: Chrm., M. Lauriente

November 15, 1985

    AM: Shuttle Environment - Material Loss: Chrm., L. Leger
            Shuttle Environment - Loads and Dynamics
            Launch Site Processing & Integration: Chrm., J. Ragusa
            Vandenburg Facilities

    PM: Shuttle Environment - Electromagnetic Interference
            Shuttle Environment - Thermal and Humidity
            Shuttle Environment - Orbiter Motion
            Environment Effects of Shuttle Operations: Chrm., W. W. Vaughan

ABSTRACTS:

The Program Committee will review all abstracts. Submit one copy of a full one-page abstract with the "abstract submittal form" at the end of the AIAA bulletin. THIS FORM MUST CONTAIN THE FULL ADDRESS AND TELEPHONE NUMBER AND THE PREFERRED SESSION. DO NOT SUBMIT MORE THAN A ONE-PAGE ABSTRACT (only the first page will be reproduced for review by the Program Committee). The deadline for submission of abstracts to the General Chairman is a postmark of April 18, 1985. Early submission will be appreciated. Authors will be notified of the action taken on their proposed papers by July 3, 1985.

Publications: Authors of accepted papers (not Workshop presentations) will be expected to submit photo-ready manuscripts for publication to AIAA Program Coordinator by September 5, 1985. A proceedings of these manuscripts will be distributed to the participants at registration.

Speakers are also encouraged to submit papers on their subject to the appropriate AIAA journal.

Comments, suggestions, questions and ABSTRACTS should be directed to the General Chairman: Dr. Billy M. McCormac
Lockheed R & DD
D91-30/B202
3251 Hanover Street
Palo Alto, CA 94304
(415) 424-2816

Program Committee:

Lyle Bareiss Martin Marietta	George Paulikas Aerospace Corporation	A. T. Stair AFGL
Nobuki Kawashima ISAS, Tokyo	James Ragusa NASA KSC	William W. Vaughan NASA MSFC
Michael Lauriente NASA GSFC	Donald G. Rea JPL	John T. Viola AFSD
Lubert Leger NASA JSC	J. B. Reagan Lockheed R&DD	J. B. Weddell Rockwell
Glynn Lunney NASA JSC	Frank Redd Utah State University	Thomas D. Wilkerson University of Maryland
Andrew Masley TRW	John D. Rehnberg Perkin Elmer	M. Trella ESTEC
Jack Matsushita NRC Canada	Michael Sander NASA Hq.	M. Bignier ESA

THE NATURAL ENVIRONMENT

Paul Robinson
Jet Propulsion Laboratory
California Institute of Technology

The prospective panel of experts on the natural environment is made up of the following people:

E.S. Stassinopoulos (Chairman, Goddard Space Flight Center)	-Modeling/Application
A. Vampola (Aerospace Corporation)	-Trapped Particles
P. Robinson Jr. (Jet Propulsion Laboratory)	-Internal Charging SEU Events
A. Hedin (Goddard Space Flight Center)	-Atmospheric Environment
P. McNulty (Clarkson University)	-Radiation Effects, SEU
J. Lockwood (University of New Hampshire)	-Chemical Reactions
W. Vaughan (Marshall Space Flight Center)	-Atmospheric Effects
B. Cour-Palais (Johnson Space Flight Center)	-Micrometeorids

The prime objective of this panel is to define for you the upper atmospheric constituents, the micrometeor environment, and the radiation environment that you would typically see on a Shuttle flight. Since I alone am representing this committee in this conference, I'm going to concentrate on the area that I understand best. If you have questions on the other areas, ask them and I will find an answer for you.

You've heard a little about the upper atmosphere already from Dr. Garrett's panel that is dealing with surface interactions. You saw charts such as Figure 1, where you find the concentration of various ion species <u>versus</u> altitude as a function of solar cycle. I'm not really going to dwell too much on that; one of the important affects of that, of course, is erosion.

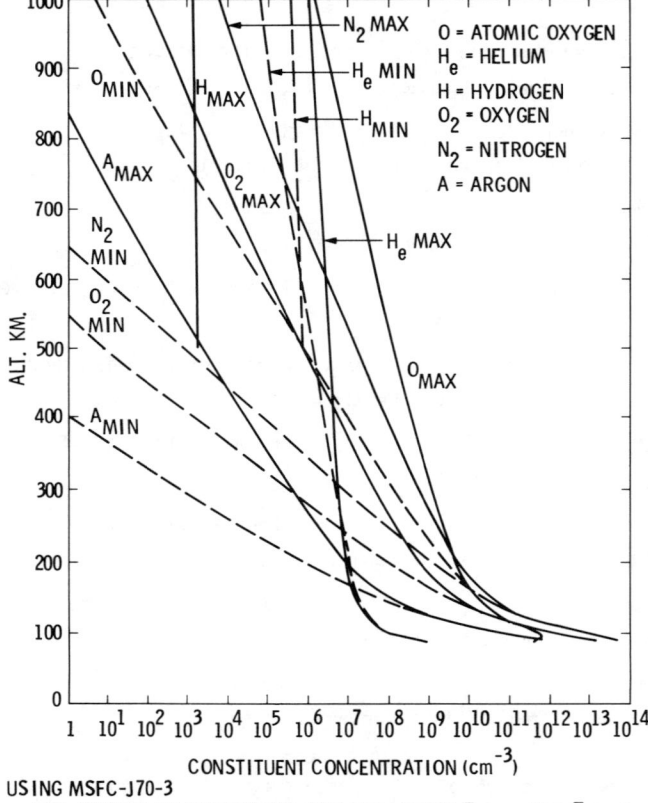

Fig. 1. Constituent Number Density vs. Altitude for Extremes of Solar Conditions

The part I understand best, and can speak to you a little more intelligently about, is radiation. Just to make myself clear, the kind of particles and things that we are talking about are as follows:

PHOTONS
    X-rays   (Synchrotron Radiation)
    Gamma Rays   (Cosmic Rays)

CHARGED PARTICLES
    Electrons   (Planetary Radiation Belts)
    Protons   (Planetary Radiation Belts)
    Alphas   (Packaging)
    Heavy Ions   (Cosmic Rays)

    NEUTRONS   (Nuclear Reactors, Weapons, RTG's)

We are talking about photons: x-rays or gamma rays. Some examples include: charged particles from either planetary radiation belts or from radioactive decay in electronic packaging, heavy ions from cosmic rays or solar flares, and neutrons from reactors or RTG's (Radioactive Thermal Generators). Stassinopoulos from Goddard can produce fairly thick volumes on exactly what the natural environment is on the average. You need to be aware that the natural radiation environment is highly variable. During a 7-day Shuttle mission, an average number isn't necessarily what you want or what you will see. Figure 2 shows log particle energy versus the log of the flux for a number of different kind of environments. Since I'm from JPL and we like to go to planets like Jupiter, there are some Jupiter data on there but there are also some Earth data as well. You'll notice that on these scales that you have events that are low and events that are high. We are trying to make a point. When you are concerned about a radiation environment, there could be a single event that happens during the time of your flight that just overwhelms any average you might have. You will also notice that we are interested in very extensive ranges of both energy and flux.

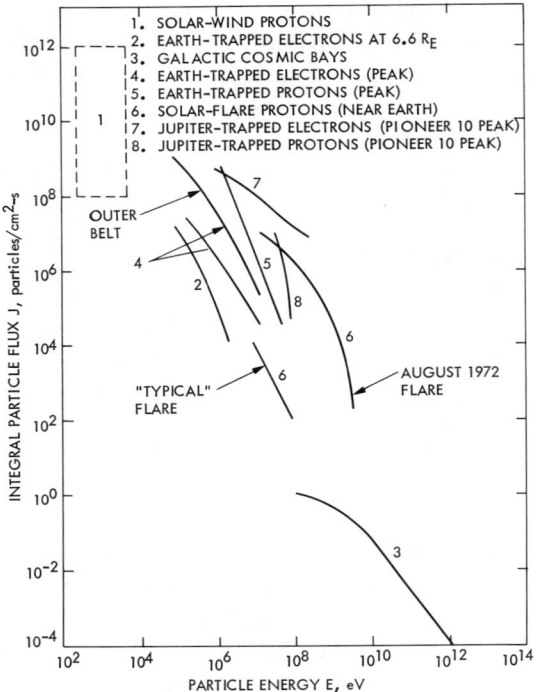

Fig. 2. Energy Spectra for Several Types of Radiation

I've got a number of different particles which I have described for you briefly, and they vary quite rapidly. So what; i.e. what do they do? There are several kinds of reactions that each of these various particles undergoes. Photons will undergo photoelectric processes, Compton scattering, and pair production. Charged particles will lose energy by ionization or by scattering, and can have a number of nuclear reactions. Neutrons also have nuclear reactions which include elastic and inelastic scattering and transmutation reactions. You need to consider the effect of these reactions on your experiment or your instrument.

Figure 3 shows some typical radiation effects. I am going to talk tonight about only two. One is logic upsets and the other is internal charge buildup on spacecraft electronics.

ENVIRONMENTAL FACTOR	EFFECT
1. COSMIC RAYS	- SOFT ERRORS - LOGIC UPSETS (DISALLOWED STATES), LATCHUP - SENSOR BACKGROUND*
2. SOLAR PARTICLE EVENTS	- SOFT ERRORS - LOGIC UPSETS - SENSOR BACKGROUND - DOSE TO COMPONENTS - DOSE TO ASTRONAUTS
3. TRAPPED RADIATION	- SOFT ERRORS - LOGIC UPSETS - SENSOR BACKGROUND - DOSE TO COMPONENTS - DOSE TO ASTRONAUTS - SOLAR CELL DAMAGE - INTERACTION WITH LARGE ION THRUSTERS+ - THICK DIELECTRIC CHARGING
4. SPACECRAFT CHARGING	- EMI, ESD - THERMAL CONTROL DEGRADATION - OPTICAL SENSOR DEGRADED - INCREASE OUTGASING - MATERIAL DEGRADATION
5. PLASMA INTERACTIONS	- INCREASE OUTGASING, - MATERIAL DEGRADATION - ENHANCED CONTAMINATION - HIGH VOLTAGE

6. SOLAR EM RADIATION (X-RAY, UV, VISIBLE, IR)	- SENSOR BACKGROUNDS
	- RESTRICTIONS ON EXTRAVEHICULAR ACTIVITY OF ASTRONAUTS (?)
	- SURFACE CHARGING-DISCHARGING
	- MATERIALS MODIFICATIONS
	  1) WINDOW DEGRADATION
	  2) S/C MATERIALS
	  3) DEPOSITED OUTGASSING PRODUCTS
	- ENHANCED OUTGASSING EFFECTS
	  1) INCREASED EMISSION
	  2) INCREASED RETURN VIA IONIZATION AND ELECTROSTATIC ATTRACTION
	- SENSOR BACKGROUNDS
	- EYE DAMAGE TO ASTRONAUTS
	- INDIRECT EFFECT VIA CHANGES IN UPPER ATMOSPHERE
	- HEATING
	  - GENERAL SPACECRAFT HEAT BALANCE
	  - SPECIAL TELESCOPE DESIGNS REQUIRED

*SENSOR BACKGROUND INCLUDE:
- SCINTILLATION AND CERENKOV RADIATION IN OPTICS
- CHARGE DEPOSITION IN CCD'S
- ENERGY DEPOSITION IN OPTICAL, X-RAY, PARTICLE COUNTERS, ETC.

+AFFECTS BEAM DISPERSION, RADIO SCINTILLATION, ETC.

Fig. 3. Radiation Effects

In electronics today you can cause a flip-flop in a computer memory to change state by the passage of a single particle. We have been calling that a single event upset (SEU). I find it particularly interesting and poetic that genes in people or animals respond to a single cosmic ray which causes mutations. Electronics doesn't want to be left behind biology. We've reduced the size and increased speed of electronics to the point where that can happen in electronics as well. We have indicated in Figure 4 that if a single heavily ionizing cosmic ray passes through a sensitive region of a chip of an IC, it can deposit enough charge in the off state of that flip-flop to cause the IC to flip. It doesn't damage the electronics at all. It still works just the way you thought it did, but now a one is a zero or vice versa. It can also happen with protons.

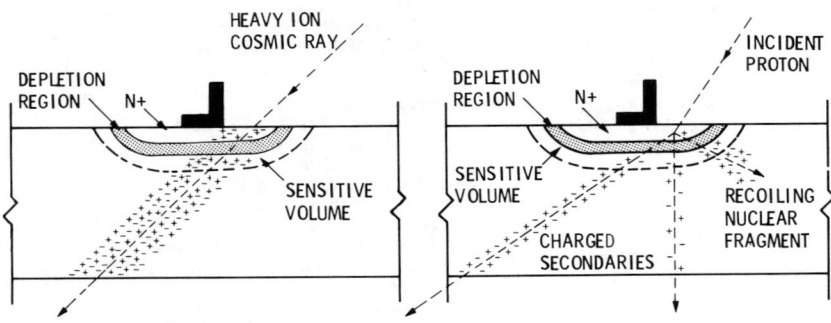

Fig. 4. Single Event Upset Mechanism

In an electronic part you get upsets for heavy ions if the ionization path through the material deposits a minimum charge and you don't get any single event upsets (SEU's) below this charge. This is expressed in Figure 5 as a cross section (i.e., a probability of that event occurring as we have shown here).

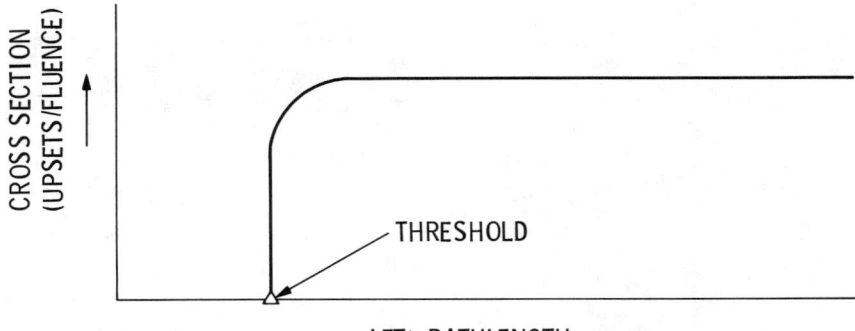

- MINIMUM AMOUNT OF CHARGE REQUIRED FOR CHANGE OF STATE DETERMINES THE THRESHOLD
- UPSET RATE DETERMINED BY GEOMETRIC AREA AND THICKNESS OF SENSITIVE REGION

Fig. 5. Simplified Single Event Upset

Various technologies that people are using are tending to become more and more sensitive to this particular effect. Figure 6 shows this trend in technology. A little later when we begin to talk about the data base, we will propose our intention to not only contribute to data on the environment, but to put in data on the kind of technology being used--which is also important. This way a system designer can come up with something he can work with. The point is, as technology improves you are going to be more susceptible to this kind of effect. From an engineering point of view one needs both an accurate description of the environment, and the sensitivity of the system to the environment. Once the engineer knows this, he then tries to design his system in a variety of ways to account for this particular problem. He can develop hardened parts, he can change the orbit that he wants to go on, etc.

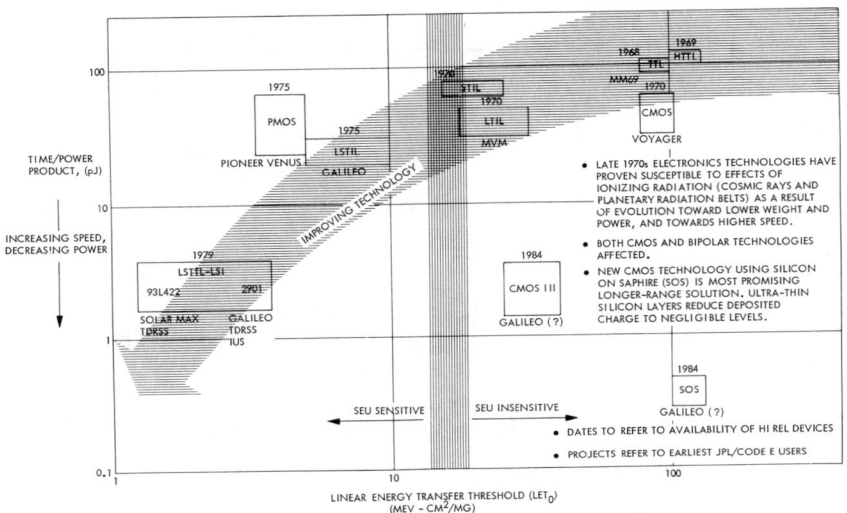

Fig. 6. Trends in Integrated Circuits Towards Higher Speed, Lower Power, and Increased Single Event Upset Sensitivity

At JPL we are acutely aware of this process. For Galileo we had a number of different environments to design to, all of which could cause single event upsets. That's the kind of process I think that we, as providers of this data base, have to have in mind so that the users can take our models of the environment, fold them together with useful data on sensitivities, and come up with a design that works for them.

Another physical phenomenon that's already implied in the environment estimates given, but you may not be aware of, is internal charging. On a new combined NASA-DoD mission, the CRRES satellite, we will fly an instrument called the Internal Discharge Monitor to measure these effects. These effects become important when an insulator isolates a volume such that the electric field across the insulator exceeds the breakdown voltage. This rule of thumb is shown in Figure 7. On CRRES we will investigate this phenomenon as a function of material and geometry. The geometries we are using are shown in Figure 8.

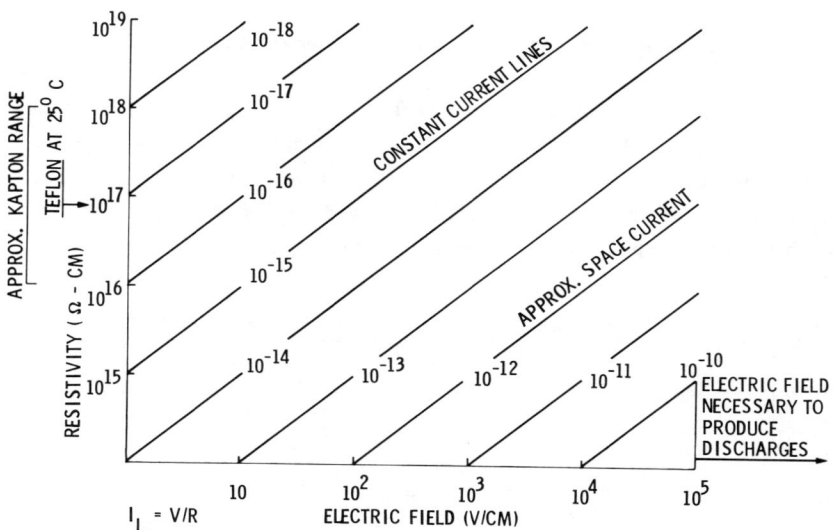

Fig. 7. Internal Charging is More Likely in High Resistivity, High Radiation Situations

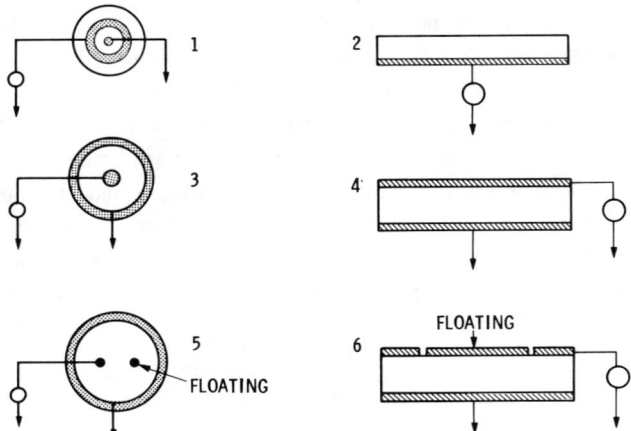

Fig. 8. Geometric Configuration for Internal Discharge Monitor

For Galileo, the most important engineering concern was volumes of isolated metal. This can occur quite naturally with ungrounded wires in cable bundles, or floating volumes of metal on circuit boards. An example of our design rules is shown in Figure 9. This kind of data is based on a series of experiments.

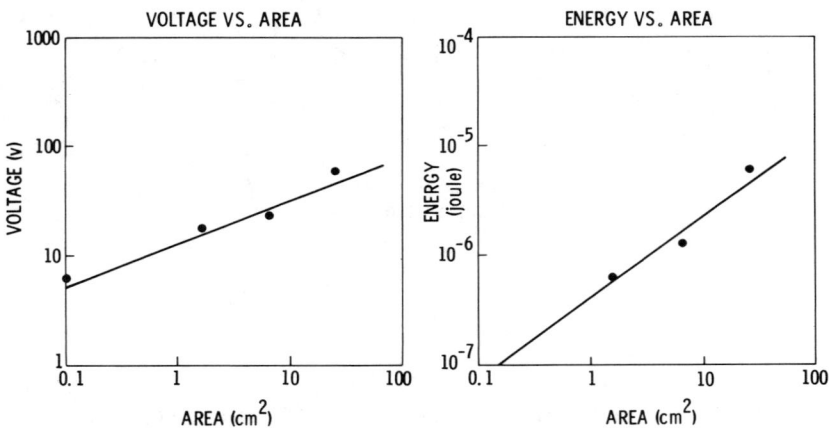

Fig. 9. Internal Charging Design Rules for Galileo

We have been investigating various kinds of dielectrics and floating conductors to quantify energies, voltage peaks, etc. that we expect to see for given configurations. This gives half the solution. The second half of the problem is the sensitivity of the electronics. For the Internal Discharge Monitor we investigated a number of typical circuits, and found maximum sensitivities as shown in Figure 10.

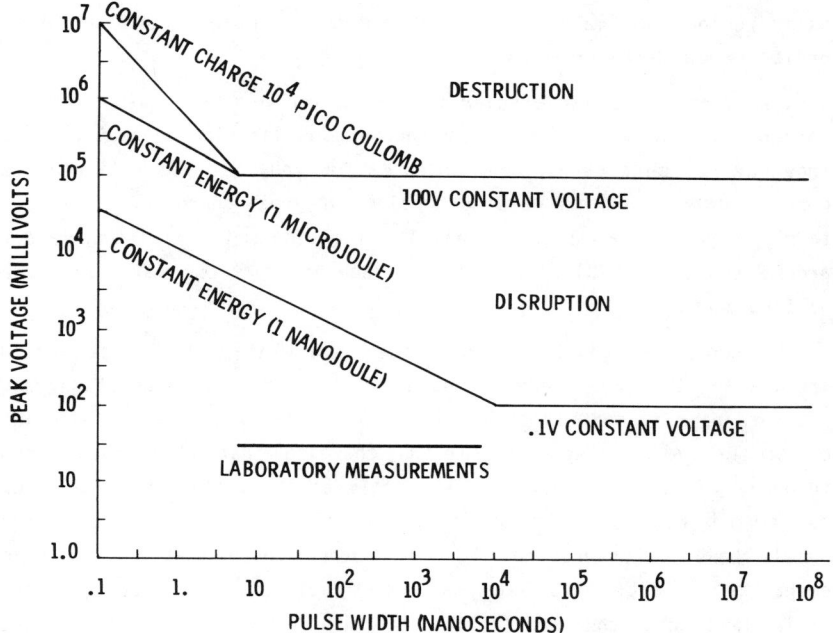

Fig. 10. Typical IC Response to Internal Discharges (1983)

Now, I hope with these two examples you've gotten an insight into some of the things you need to be aware of when you are looking at a radiation environment. The radiation environment can be included in the data base, but you need to understand how to use it and what it's going to do to you.

What does the future hold for radiation concerns for the Shuttle? Well, that rests on two things. The technology will change and the environment will change. The environment will change not only from this factor of ten or so fluctuation that I talked about, but also because there is a lot of thought now being given to going into polar orbit. If you were to do this, you would be flying right through the polar region, and many things that are very simple in an equatorial orbit are not in polar orbit.

If there is a large solar flare and you are in a polar orbit-- come down. You almost certainly should consider single event upsets. They may not be a catastrophic problem for you, but you will need to consider them. You may need to consider internal charging. It's certainly a good idea to ground all the floating metal that you've got around if you possibly can. And surface interactions will of course be important.

I hope I've given you some ideas on what a data base should include and the care needed in using it. You need to understand what the data base is and how to use it. I hope that with the two examples of single event upset and internal charging that I've quickly run through, you can see that just knowing what the charged particle flux is doesn't necessarily mean you've solved all your problems. You really have to understand it. The other thing that I'd like to emphasize is that this data base must pull together not only the environment but the sensitivity of the instrument, whether that instrument is a four-legged monkey or an IC. This means that you need to understand the technology and how these various environments are going to interact with technology.

The research described in this paper was carried out, in part, by the Jet Propulsion Laboratory, California Institute of Technology, under contract with the National Aeronautics and Space Administration.

USER CONCERNS
ENVIRONMENTAL FACTORS AFFECTING SCIENTIFIC USERS OF THE SPACE SHUTTLE

Professor Laurence R. Young
Man-Vehicle Laboratory
Massachusetts Institute of Technology

Local environmental factors which influence the experimenter onboard the Space Shuttle can be classified as follows:

A: disturbances created by other experiments which affect one's measurements.

B: restrictions on one's experiment in order to avoid disturbing others.

C: accurate and timely information concerning the actual environmental experiment during the mission to isolate artifactual measurements.

All of these factors are present, to varying degrees, whether one is performing an experiment on the middeck, inside the pressurized Spacelab module or on an external pallet. The first of these becomes a major design factor to the user from the initial stages of his proposal through the analysis of his data. Realistic estimates of the local disturbances to be encountered determine whether or not a researcher must isolate his experiment and form some protective "micro-environment" around it, or whether it will suffice to measure the disturbance and account for it in data analysis. For example, sufficiently fine specifications on electomagnetic interference, temperature control or particulate matter can reduce the complexity and expense of an experiment by avoiding the need for additional shielding, thermal control and filtering. It is important for the prospective Space Shuttle scientist to know what environmental ranges can be realistically requested and adhered to. As cumulative experimental evidence documents the environmental variables associated with the Shuttle glow and electromagnetic field surrounding it, future experimenters can be guided in the kinds of experiments that are proposed for performance on the Shuttle as opposed to on free flyers. One of the most difficult tasks currently facing new experimenters involves the uncertainties concerning "environmental requirements". A requirement which may be difficult to meet runs the risk of disqualifying an experiment, whereas negotiation concerning the actual specification level and consideration of alternate ways of isolating one's experiment might yield more attractive solutions. The user must be furnished with a Shuttle environment handbook which presents a realistic and contemporary estimate of the expected disturbances.

The converse to the above concern involves the second issue, that of limitations on disturbances which are imposed upon the individual experimenter. Although it may seem reasonable and simple to impose the same level of restrictions upon all experiments and missions, these may, in many cases, involve excessive weight and cost when the specific experiment mix does not need tight limitations on a given variable. For example, in an all-life-sciences laboratory, the specifications on permitted electromagnetic interference might be substantially relaxed, whereas those on outgassing might remain tight. The cost of testing to meet each additional environmental limitation requirement is a substantial portion of space experiment development, and such requirements should only be imposed where they are justified for a particular mission. In all cases, of course, Shuttle systems operation and astronaut safety would set an upper level on disturbance limitations.

The final user concern is one of measuring and documenting, in a timely, accurate manner, the actual environmental conditions experienced during the flight. It hardly is reasonable to require rapid reporting and initial analysis of results by experimenters when potential disturbances associated with environmental variables are not easily accounted for. The STS should routinely provided experimenters with a rapid indication of the "as flown profile" of accelerations, temperature, and external environmental variables effecting viewing and monitoring instruments.

The Space Transportation System, including Spacelab, offers to the space scientist unparallelled opportunities for innovative research. Unlike research in one's own ground laboratory however, facilities must be shared with other experimenters, and all must share the environment of the Space Shuttle itself. By making the expected environment well known and controlled, such restrictions can be minimized in terms of their effect on the scientific value of an experiment.

CHAPTER 2

SUBPANEL REPORTS

ELECTROMAGNETIC INTERFERENCE

William Cutler
Aerospace Corporation

The Electromagnetic Interference Subpanel defined several specific tasks which we would attempt to resolve throughout the week. These tasks evolved from the guidelines presented to each panel chairman and included:

o Evaluation of the available data
o Determine how to use the available data
o Identify data deficiencies
o Identify future data to be obtained
o Determine means for resolving deficient data
o Prepare data for inclusion in the data base

During the meeting, some of the decisons we made allowed us to proceed a little faster than we thought. The data we have obtained thus far provides few guidelines on how to establish an environmental data base for this project. All of the electromagnetic data available have been obtained from measurements taken in the Shuttle Avionics Integration Laboratory (SAIL) at JSC. Some in-flight measurements were taken on the Plasma Diagnostics Package (PDP) on STS-3 and others are planned for the reflight of the PDP on STS-24 (Spacelab II), but these are limited to measurements of S and Ku Band radiated levels and confirmation of some magnetic field and broadband unintentional radiation levels. We went through this available information and determined that the data basically support the levels identified in the latest revision of the Shuttle Orbiter/Cargo Standard Interface Document JSC 07700, Volume XIV, Attachment 1 (ICD 2-19001). This document identifies the maximum electromagnetic environment which can be experienced in and around the Orbiter cargo bay. These levels have been established by JSC as payload compliance requirements to provide a maximum launch availability and to minimize flight costs. It should be noted that no effort to characterize the minimum electromagnetic environment has been identified or contemplated to date. Most DoD payloads are satisfied with the characterization of the electromagnetic environment identified to date. They can provide a payload design which is compatible with this environment and meet their mission objectives. The subpanel recognizes that some people may not be completely happy with this solution; however, the only real way out of that problem today is for a payload manager to buy his own dedicated flight and "manage" the environment as required. It must also be recognized that much of the environment involves essential Orbiter operations, and complete elimination is not practical.

The recognition that the only realistic electromagnetic environment data available were already available in ICD 2-19001 simplified the issue of what should appear in the data base. Although a significant amount of data was presented at the working group meeting and could be used in the data base, it would have to be continuously reviewed and updated. We decided that the purposes of both this working group and JSC would be best served by referencing Sections 7.0 (Electrical Power) and 10.7 (Electromagnetic Compatibility) of ICD 2-19001 in the Natural and Induced Environment Data Base. When additional information becomes available, this document will be updated almost as rapidly as could be accomplished if the data base included all of the information. This would also eliminate any discrepancies between the data base and ICD 2-19001.

At present, the data are based in standard EMI measurement capabilities between 14KHz and 20GHz. The deficiencies in the data base were identified as follows:

   o Data between 1Hz and 14KHz
   o Data above 20GHz
   o S-Band radiated levels outside
     the cargo bay
   o Minimum electromagnetic levels

Due to equipment measurement limitations, it is extremely difficult to take measurements in the frequency ranges listed as deficiencies. In addition, no one has thus far requested that information. However, some of the experimenters in the audience were talking about these low frequencies, and we do identify them as deficiencies. As previously discussed, we have available only maximum electromagnetic environmental levels. That severely limits the experiment capability for operation at levels below that. A realistic estimate of the lowest Orbiter operating electromagnetic environment has never been made. I believe it is up to experimenters to negotiate their low-level ICD requirements with JSC on an individual basis. Another deficiency involves definition of the radiated S-Band environment outside the cargo bay. It has been identified inside the bay but not outside*, and this environment is high enough to effect any experiment which protrudes outside the cargo bay.

Future data will be obtained on S-Band and Ku Band radiated fields outside the cargo bay on the STS-24 Spacelab II mission. These data should be included in the data base as well as JSC estimates on these field levels which should be available now. The DoD has contracted with the University of Iowa to make these measurements. The past data on STS-3 were quite successful even though we had a number of problems pinpointing the accuracy. I think we will do much better on STS-21. The mission involves free flight for about seven orbits, so detailed spatial measurements can be made.

As far as eliminating these discrepancies, a complete elimination would require a large scale, in-flight measurement program which has not been identified to date. Goddard Space Flight Center has identified a method for eliminating many of these deficiencies over a period of time. This method recognizes that many planned scientific experiments will take data which would be helpful to this data base e.g., ERBS magnetometers and SPARTAN magnetic field measurements. These

---

*Editor's Note: Regarding the S-Band antenna patterns outside of the cargo bay: As of November, 1984, an interface revision notice (IRN) JSC-058 has been generated by NASA/JSC to document this environment in ICD 2-19001.

program capabilities could be categorized and the data obtained used in the data base. This effort would also involve significant planning and coordination efforts with the individual experimenters to obtain useable data. It would take considerable long term effort for completion of this task, but some positive gain would be achieved in identification of the cargo bay electromagnetic environment. Details of this capability have been included in the working group minutes; however, no funding presently exists for implementation of this effort. As far as the data discrepancy involving lack of S-Band field intensity data outside the cargo bay is concerned, I believe that JSC already has the data available, and the DoD recommends it be included in ICD 2-19001. These data will then be available as a part of this data base.

Preparation of the data for inclusion in the data base is extremely simplified by the use of ICD 2-19001 Sections 7.0 and 10.7 as the data base elements. All experimenters normally have this information readily available to them. We have also included a section which provides generalized guidelines for contractors who have never done an Electromagnetic Compatibility Study. Included are guidelines on:

- o EMC System Analysis
- o ElectroExplosive Device (EED) Analysis
- o Interface Analysis
- o EMC Testing Requirements

More information on Electromagnetic Interference can be found in Appendix B.

ORBITER MOTION

Roger Chassay
Marshall Space Flight Center
National Aeronautics and Space Administration

The Orbiter Motion Subpanel met to look at the odd assortment of low-g data that is now available. Informal presentations were made as follows. Details are available from each presenter.

I. Daly, Kevin C., C.S. Draper Lab.
   555 Technology Square, Cambridge, MA 02139
   (617) 258-2573
"STS DYNAMIC ENVIRONMENT DURING ORBITAL OPERATIONS"
   This presentation included background information on the Orbiter characteristics, the on-orbit dynamics of the Orbiter and its susceptibility to payload operations, and the flight dynamics instrumentation.

II. Vandervoort, Richard J., Honeywell, Inc.
    13350 U.S. Highway 19, S., Clearwater, FL 33546
    (813) 530-9672
"ADAPTING TO THE SHUTTLE ENVIRONMENT"
   This presentation suggests treating Orbiter motion as an expected disturbance input to a control system designed to isolate the experiment. A low cost, yet effective approach to modeling the control system design was presented, using the Pinhole Occulter Facility as an example.

III. Knabe, Walter E., Senior Physicist, ERNO
     Hunefeldstrasse 1, Bremen D-2800, W. Germany
     0421-539, ext 4446

"THE IN-ORBIT MICROGRAVITY ENVIRONMENT AND TECHNOLOGICAL AVENUES FOR ITS IMPROVEMENT"

This presentation covered predicted and measured levels of acceleration in manned spacecraft, with a listing of sources and acceleration levels and frequencies caused by each. Several possible concepts for improved acceleration sensors were also presented.

"LOW-GRAVITY ENVIRONMENT IN SPACELAB"

This paper was published in ACTA Astronautica, Vol. 9, No. 4, pps. 187-198, 1982. It covers engineering model acceleration measurements which indicate a few narrow band accelerations above 10^{-4}g, assuming restrictions on crew motions.

IV. Chassay, Roger P., Manager, Experiment Carriers Office
Marshall Space Flight Center, AL 35812
(205) 453-1870

"ORBITER ACCELERATIONS"

This presentation covers a description of a low-g accelerometer system and its measurements during STS-7. Also covered were samples of low-g data from STS-3 in the Orbiter middeck and from STS-9 in Spacelab 1.

V. Bailey, Wayne, Teledyne Brown Engineering
300 Sparkman Drive, Huntsville, AL 35807
(205) 837-8221

"PAYLOAD ISOLATION AND STABILIZATION BY A SUSPENDED EXPERIMENT MOUNT (SEM)"

A thorough treatise was presented on an isolation system approach for large payloads in the Orbiter bay; concepts for other options to suspend or isolate payloads were also presented.

VI. Bakken, Gordon B., Manager, System Engineering, Wyle Laboratories
7800 Governors Drive W., Huntsville, AL 35807
(205) 837-4411 ext. 346

"REQUIREMENTS FOR MICROGRAVITY SCIENCE EXPERIMENT APPARATUS"

This presentation covered considerations which are necessary in accommodating science requirements in microgravity experiments.

VII. Malmejac, Yves, Centre D'Etudes Nucleaires
   Department de Metallurgie
   F-38041 Grenoble-Cedex FRANCE
   76-97-4111 ext 3308
"SOME CONSIDERATIONS ON THE MICROGRAVITY LEVELS NEEDED BY
VARIOUS MATERIALS SCIENCE INVESTIGATORS"

This presentation focused on the sensitivity of certain types of microgravity experiments to extremely small accelerations.

SUMMARY AND CONCLUSIONS

Oddly enough, even though the low-g data are very limited, what data exist are still in too bulky a form to be of much use. For example, a set of low-g data from only one portion of a Shuttle flight is a four foot stack of paper. In our panel today, we heard one suggestion that I believe makes a lot of sense: Once we've obtained the bulky low-g data, it will probably be desirable to establish a minimum acceleration threshold, below which no one would be interested, and then print out or plot the data above this threshold. If that later does not satisfy certain users, the data could always be reaccessed to establish a different threshold for data printout. We can also consider data compression to make this bulky data more usable, but I've heard a lot of people talk about that, yet I've never been able to get any of our data specialists to help me with it in a practical way. As I've said, we do have some low-g data from various sources, but we're not quite sure what to do with it, because the data just do not give us a nice clean acceleration level. What we find is something akin to typical vibration data with an enormous amount of indicated acceleration variations. We don't know whether these accelerations are crew-induced, RCS-induced, or simply anomalies in the acclerometer, out-of-calibration of the instrument, or a signal-to-noise ratio of 1.0.

So far the people who have taken these data are preoccupied with other high priority work and have not had the time to analyze the data in depth. They may not even have the capability to do it, since many payload people are not well versed in Orbiter systems and operations, which can be the source of much of the low-g accelerations. So, what we basically have consists simply of raw data. And even though some of it is well over a year old, is not really being analyzed and it's sitting on the shelf.

One of the tasks this subpanel will try to do now is find out what data are missing. We've already discovered that many things are missing, because this is a relatively new area. We have had pointing requirements for instruments, but we haven't yet defined what kind of minimum motion is required to accommodate those instruments on Shuttle. One of the main things which is missing is wide band data. We have narrow band data that in some cases do not even detect RCS firings because the data are integrated over a full one second period. Both positive and negative spikes can integrate out to be a zero value, or at least one that doesn't prominently stand out when integrated over a full second time period. I'm in the process of procuring improved 20 Hertz accelerometers for a materials science payload that will be flying sometime next year. (Mike Sander referred to that yesterday in his talk.) The 20 Hertz band will satisfy most of the materials science users, although we know of at least one that it does not even come close to satisfying. But as a class, the instrument pointing users need something substantially above 20 Hertz. This subpanel will recommend that there be wide band data provided for low-g accelerations and orbit motion up to the 50 Hertz range; we think there ought to be a definite plan for that, and we believe there may be some requirements at even higher bandwidths. We also need angular acceleration data in addition to the translational data. We know we have some data, but they are narrow band data. However, even with only the scattered bits and pieces of data that we have, when they're all woven together, this may provide a useful data base for a preliminary baseline.

Another activity we think we need is to thoroughly map and characterize the typical Orbiter accelerations. Right now it's being done by individuals working in isolation from each other. JSC has provided some middeck accelerometers. The Marshall Space Flight Center has put some accelerometers into Spacelab-3 for which I am now responsible. Recently we put some accelerometers at one location in the Orbiter bay. We randomly go out and purchase accelerometers and we randomly plot that data, or try to plot it, and typically we then stick these raw data on a shelf somewhere. There's no comprehensive, systematic plan for the use of those data, nor to map and characterize the Orbiter with respect to low-g accelerations. So, we will definitely recommend rectifying that situation.

We also need to identify the minimum sensitivity levels. Many of our users, as I indicated, really don't yet know to what levels of motion they're sensitive. There are others who know it very well, and they're worried about the prospect of perhaps not being able to get their requirements met; for example, the Solar Optical Telescope people.

As far as the pointing instrument people and the microgravity experiment people working together, we really haven't scratched the surface as to how we interact properly, since these two disciplines previously have been independent of each other. We've both just been accepting whatever Orbiter motion occurs, more so than specifying what we can tolerate.

Unfortunately, several members of our subpanel will be unable to stay at this workshop any longer for various reasons. (I think perhaps due to the fact that most of these people were here participating on relatively short notice and couldn't arrange to be away for the whole week.) So, we're going to have a brief meeting after this session tonight to try to decide what to do and to work around the fact that we're going to have absenteeism the rest of the week by several of the key people on the subpanel.

I think that summarizes where we are. We feel the group is active and successful thus far. We have a lot of questions about what to do in the future. I suspect that with some degree of good luck and some optimism we can probably come up with a critical mass. As yet I'm not sure we have that, but we do have the essential ingredients to get it.

More information on orbiter motion can be found in Appendix D.

THE PARTICULATE ENVIRONMENT

J. Barengoltz
Jet Propulsion Laboratory
California Institute of Technology

The Particulate Environment Subpanel met in a planned session August 7 and again in an ad hoc session August 8 to complete a busy agenda. The topics discussed were the current status of the data, the current status of analysis and modeling, planned future activities, and future needs in the particulate environment. Nine speakers provided informal presentations covering the first three agenda items. The subjects and the authors were:

KSC In-Line Ground Facilities, D. W. Bartleson
   (LSOC)
KSC Processing Facilities, B. Wenkstern (MDTSCO)
IECM Cascade Impactor Data, D. Wallace
   (Telonic/Berkeley)
IECM PSA and Camera Photometer Data
   from STS-9 (SL-1), E. Miller (MSFC)
Space Debris, J. Park (GSFC)
Optical Environment/Particulates, J. Weinberg
   (U. Fla.)
STS Particulate Modeling, J. Barengoltz (JPL)
SL-2 Infrared Telescope, G. Fazio (Harvard) and
   F. Witteborn (ARC)
Particle Analysis Camera System (PACS),
   J. Wise (AFGL)

The discussion of recent data on STS particulates included reports on ground facilities and flight results. Bartleson's and Wenkstern's information on the various STS ground facilities represents an important body of data covering all of the STS flights. This historical record clearly demonstrates the improvement in cleanliness achieved by facility modifications and changes in procedures. One might also infer a cleaner STS cargo bay, but this is complicated by the unknown contribution of the various payloads.

Wallace presented the first data on the elemental composition of particles collected by the IECM cascade impactor during launch. Launch data are vital because they may affect both attached and deployed payloads. Composition is useful for payload threat assessment, source identification, and model testing.

Miller presented data on the IECM Passive Sample Array (PSA) and the Camera Photometer from its latest flight, STS-9 (Spacelab 1). Findings from earlier flights were a flurry of particulates early in the mission in the vicinity of the Orbiter, while on orbit that decreases later except for short periods correlated to water dumps. However, the present data from the Camera Photometer do not show this time behavior. This issue is crucial for instruments which make observations using the Orbiter as a platform. The discussion may be summarized that at least part of the ambiguity results from experimental bias (i.e., particles are difficult to sight optically). The PSA data on net fallout on selected flat surfaces had in the past reasonably correlated with ground processing cleanliness. On STS-9 the data probably reflected the increase expected from a large complex payload such as Spacelab 1.

Park showed some preliminary data from work in progress on space debris. This important particulate environment refers to particles put in orbit by previous STS flights, deployment devices and other solid propellant rocket motors.

Under modeling and analysis, Weinberg reported on his analysis of his optical observations on STS-3 in terms of an inferred particulate environment. Barengoltz discussed the status of his modeling efforts and reported some success in predicting the Camera/Photometer observations. The model is completely untested for small particles and fast particles because of the limitations and experimental bias of present data.

Two planned instruments that will improve our knowledge of particles in the field-of-view of an Orbiter-based experiment were reported. Wise described PACS, which features an integral strobe light source. Among the system's exciting capabilities, the strobe will permit the detection of particles under most ambient lighting conditions. This capability will eliminate one of the largest experimental biases of the present data. Fazio reported on the infrared telescope to be flown on Spacelab 2 and its capability for detecting very small particles. Since infrared telescopes may be the most sensitive instrument to particles in the field-of-view, the data obtained ought to constitute the benchmark. This instrument will also provide data on particle speed and distance from the Orbiter.

Although they were given in another session, the presentations by Lyle Bareiss (MMA) on the Shuttle/Spacelab Contamination Environment and Effects Handbook and the Shuttle Payload Bay Cleanliness Ground Test Program, and Barengoltz (JPL) on the Interim Operational Contamination Monitor (IOCM), are relevant to future plans for the particulate environment. The handbook being prepared for MSFC will complement several sections of the STS Environment database, including the particulate environment. The test program includes a laboratory simulation of the release of realistic particles from actual bay materials during launch. On its flights, the IOCM (sponsored by USAFSD) will obtain time-dependent particle deposition data throughout an entire STS mission.

The future needs in the STS particulate environment were discussed in an _ad_ _hoc_ session. Lack of the cargo bay and payload measurements, (and the time dependence) prevent characterization of the pre-launch environment. A more representative format for fallout data than MIL STD 1246A is desirable, especially for very large particles. The flight data require measurements with less experimental bias or at least a better determination of the bias. This improvement would yield needed data on large particles released during launch and small, fast particles in the vicinity of the Orbiter while on orbit. Time and mission phase-dependences should be obtained, also. A specific suggestion for more television coverage was made.

In parallel with the data needs, models must be greatly improved and tested against the new data. The models should be capable of bookkeeping the particles and of predicting time dependences. Only such models can make the critical predictions necessary for a particular payload's assessment.

The meetings of the Particulate Environment Subpanel and interested inividuals provided a useful interchange of data and information. There were few surprises in the recent data, and no surprises at all in the discussion of what's needed. However the scope of planned future activities , which address many of the identified deficiencies in our knowledge of the STS particulate environment, was especially noteworthy.

The research described in this paper was carried out, in part, by the Jet Propulsion Laboratory, California Institute of Technology, under contract with the National Aeronautics and Space Administration.

LOADS AND LOW FREQUENCY DYNAMICS SUBPANEL

J. A. Garba
Jet Propulsion Laboratory
California Institute of Technology

INTRODUCTION

The definition of the loads or low frequency vibration environment is required for the design and the qualification testing of the primary payload structures. Primary structure is defined as the main load carrying payload structure which is designed by loads analysis. The payload responses to events such as engine ignitions and shutdowns, liftoff and landing, maximum aerodynamic pressure encounters, staging, and vehicle maneuvers produce the design loads for primary structure. In general, the frequency regime which is of concern for primary structure is up to 50 Hz, and the basic methodology for obtaining design loads is deterministic rather than statistical. The results of payload responses obtained due to the application of a family of forcing functions are sometimes statistically evaluated to obtain expected values. Some events such as maximum aerodynamic pressure also lend themselves to a statistical analysis.

The loads environment is, in general, dependent on the dynamic characteristics of the coupled payload/launch vehicle system and the characteristics, magnitude, and frequency content of the external forcing functions. The objectives of flight response measurements in the low frequency or loads regime, differ from the objectives of the vibroacoustic regime in that the latter aims to define the flight environment statistically while the former requires instrumentation for the verification of methodology.

The design engineer is faced with determining design load factors early in the design process. Since these factors depend on the dynamic characteristics of the combined payload/launch vehicle system they are not readily available for most payloads. The procedure is to make conservative estimates for these load factors and then refine these estimates throughout the design process, either by additional analyses or by flight data from similar payloads.

An effective design data base thus consists of a compilation of realistic load factors for various types of payloads and carriers as well as flight data for the verification of the design data base. The issues in providing the Shuttle users with reliable design tools were the subject of a workshop on STS payloads environmental data held in June 1983, sponsored by the Air Force Space Division, and also a NASA sponsored workshop on Shuttle payload dynamics and load prediction held in January 1984.

SUBPANEL MEMBERS

The subpanel members are:

Ken Hinkle--NASA/GSFC
Alan Thirkettle--ESA/ERT
Tony Sanders--MMC/DEN
John Garba--JPL

JSC, MSFC, and Aerospace Corporation were not represented in this subpanel. Invitees who were not able to attend were Alden C. Mackey, NASA/JSC; Edwin G. Ricks, NASA/MSFC; H. A. Superfine, ESA/ESTEC; Samuel L. Venneri, NASA HQ/RTM; and Robert Wagner, Aerospace Corporation. Several attendees who made significant contributions to the subpanel are:

R. Wlochowicz--NRCC
Ben Bier--MMC
Bill Henricks--LMSC
W. Reynolds--Aerospace Corporation
Bob Blount--NASA/JSC

OBJECTIVES

The overall objectives of the subpanel were to (1) establish a focal point for dynamics and loads data to aid in the design of Shuttle experiments, (2) recommend a development and implementation plan for the loads data, and (3) recommend a format for the information data base. The specific, self-imposed objectives of the subpanel are to :

1. Identify Required Information--User Concerns
2. Summarize Existing Data
    --Reliability
    --Traceability
    --Usefulness/Deficiencies
3. Develop Plan for Continuous Data Collection
4. Establish Method for Incorporating Existing Data into Data Base
5. Identify Technology Issues
6. Identify Required Measurements
7. Propose Plan for Implementation
    --Data Acquisition
    --Data Utilization
    --Architecture for Data Base

USER CONCERNS

Before addressing user concerns, the panel addressed the question, "Who is a user?" It was concluded that in reality all the panel members are users in the sense that they all represented payload organizations in contrast to launch vehicle developers. The panel categorized the main users as the Department of Defense (DoD), the NASA centers, commercial, and academic. The panel further decided that the design data base should primarily serve the commercial and academic users. The reason was that the DoD and NASA users are generally part of large organizations supported by structural dynamic engineers knowledgeable in Shuttle payload dynamics.

The panel received inputs from several users with differing requirements. One of the users required design data for various carriers in order to evaluate possible options for using these carriers. These data were not available, and the user did not know where to get such data. Another user desired to obtain flight measured deflection time histories for solar panel type structures, such that he could extrapolate to another design. At issue here was the superposition of the transient and low frequency acoustic environments using flight data. As a result, the panel decided that there should be two different types of data bases, the first being a design data base containing information concerning load factors and design approaches, and the second, a flight data base which would contain raw and reduced flight data. They further decided that while both data bases are essentially of equal importance, the majority of the users would use only the design data base. Moreover, the flight data base would not be made available to the user; it would only be made available to the structural dynamicists who would then use this data to update the design data base.

Based on the user inputs, the panel discussed the feasibility of such a design data base (Table I). It was concluded that a design data base is required, although there were serious concerns about the responsibility and mechanism of updating, maintaining, and controlling the data base. Furthermore, the maintenance of such a data base would require resources on a continuous basis.

Table I - Feasibility of Data Base

- JUSTIFICATION

   CENTRALIZED DATA BASE IS REQUIRED FOR USERS TO ACCESS DESIGN DATA

- CONCERNS

   CONFIGURATION CONTROL

   - RESPONSIBILITY
   - MECHANISM
   - WORKFORCE

SUMMARY OF EXISTING DATA

The panel did have difficulty in summarizing the existing design data since some of the organizations knowledgeable in these data were not represented. The panel identified potential carrier options and the responsible organizations for these options (Table II). The best documented and most accessible design data available is contained in the Spacelab Payload Accommodation Handbook (SPAH),(Ref. 1). A typical format of the data that are applicable to loads and low frequency dynamics is shown in Table III. Similar data for other carriers are most probably available and are hopefully documented, but this could not be substantiated.

As to the flight data needed to verify the design data base, these are available mainly from STS-1 through STS-5 from the Shuttle Development Flight Instrumentation (DFI) and the NASA Dynamic, Acoustic, and Thermal Environments (DATE) program. These data are limited to lightweight cargoes and have found only very limited applications in the verification of design loads (Ref. 1).

DATA DEFICIENCIES

The panel had difficulty in assessing the design data deficiencies since the extent of the available data was not known. For the known documented data, namely the data from SPAH, the main deficiency was that these data have not been verified by flight measurements. Spacelab-1 flight data are being reduced by NASA/MSFC, but the extent of the use of these data for the updating of the SPAH was not known. The panel unanimously agreed that the Spacelab-1 flight data should be thoroughly analyzed and compared to pre-flight predictions, and that the load factors in the SPAH be reviewed as a result of the comparison.

As to the flight data, the panel identified several deficiencies and limitations (Table IV). The panel felt that the existing flight data had not been thoroughly exploited to verify the design data base. Furthermore, the existing flight data had deficiencies such as: (1) inadequate frequency response. Only a very few measurements were

Table II - Carrier Options

Carrier	Agency
SPACELAB	
MPESS	
IPS	➤ MSFC
APC	
ORBITER LOCKERS	
ORBITER MID-DECK	➤ JSC
CENTAUR	
IUS	➤ AF
MULTIPURPOSE SYSTEMS	GSFC / MSFC / JSC
PAM-D	➤ MDAC
LeRC	

Table III - Spacelab Data Availability

ENVIRONMENT LOCATION	QUASI-STATIC ACCELERATION						RANDOM VIB.		ACOUSTIC NOISE LEVEL (db)	SHOCK
	Lift-Off			Max Q			g^2/Hz (Hz), G_{RMS}			
	X	Y	Z	X	Y	Z	0<M<10 Kg	10 Kg<M<220 Kg		
Module										
· Rack	x	x	x	x	x	x			X_{db} Internal to Module (137 db)	x
· Floor	x	x	x	x	x	x	x	x		
· Overhead	x	x	x	x	x	x	x	x		
· End Cone	x	x	x	x	x	x	x	x		
· Airlock	x	x	x	x	x					
Pallet (inc. IPS)										
· Hard Points	x	x	x	x	x	x			Y_{db} On Pallet (142 db)	x
· Panels	x	x	x	x	x	x	x	x		
· Frames	x	x	x	x	x	x	x	x		
· (Platforms)	x	x	x	x	x	x				

Guidelines and Conditions for Use of Tables:
1. Safety Factors
2. Center of Gravity Limitation
3. Environment Combination Rules
4. Mounting Method (Determinant, etc.)

Table IV - Limitations of Existing Data

- **ACCURACY**
    - DC ACCELEROMETERS ± 5.6%
    - AC ACCELEROMETERS ± 9.1%

- **FREQUENCY RESPONSE**
    - DFI   15 Hz CUTOFF - 100 SPS
    - DATE   INSUFFICIENT DC MEASUREMENTS

- **FIDELITY OF PAYLOAD MODELS**

- **MEASUREMENT LOCATIONS**

- **DATA BASE**
    - OV-102
    - ETR
    - LIGHTWEIGHT CARGO
    - NO CROSSWIND LANDING

made in the desired frequency range, 0 to 50 Hz. Most data measured so far were in the 1.5 to 20 Hz range, which is inadequate for the definition of the low frequency environment; (2) inadequate payload dynamic modeling. Some of the flight data were obtained from payloads that were not adequately verified by test and hence those data are not suitable for methodology verification; (3) lack of statistical data base. The measurements made to date represent insufficient samples and also insufficient repetitive locations for a statistical data base. Furthermore, data have been acquired for the OV-102 vehicle only, for lightweight cargoes only, and no data have been measured on load alleviated payloads; (4) only valid for limited classes of payloads, Shuttle configurations, and launch sites flown to date. Many other variations will be flown which cannot be reliably predicted with current data.

REQUIRED DATA BASE

The panel defined the elements of the data base as (1) a directory for the potential user to contact the responsible agencies to obtain additional data as required, (2) a design data base containing design load factors for all potential carriers, such as those used in Table II and, (3) a flight data base to be used to verify and update the design data base. Only the first two elements of the data base would be made available to the user; the third category of data base would be restricted to the structural dynamicists.

INCORPORATION OF EXISTING DATA INTO DATA BASE

The panel recommended that the data contained in the SPAH be used as a guideline for developing a data base. Similar data for other carriers have to be compiled if available; if not, they will have to be developed. This data base is to be verified by flight data from Spacelab-1 and other flight data. The panel recommended that the directory and the design base be incorporated into the data base as soon as funding is made available. It should be established as soon as possible if SPAH type data are available. If so, these. data should be made part of the data base; if not, tasks should be initiated to develop such data. The approach for designing the data base should be as follows:
1. Use Spacelab Payload Accommodation
   Handbook (SPAH) as a Guideline
2. Develop Similar Data for Carriers
   --MPESS
   --APC
   --Orbiter Mid-deck, etc.
3. Update Data Base Periodically
4. Develop Guidelines for Use of Data Base

All the existing design data should be validated by refined analyses, if available, and by flight data. This requires the establishment of a flight data base and the continuous updating of the design data base using the flight data.

ADDITIONAL REQUIREMENTS

Additional measurements are required to overcome deficiencies in the data base. The panel agreed that future flights should be instrumented to overcome the data deficiencies listed above. Above all, the flight data acquisition should be carefully planned and the data should be analyzed, not just reduced as was done with the DATE data. Analysis should consist of correlation with analytical predictions and comparison with design load factors.

TECHNOLOGY ISSUES

The panel identified several technology issues which have to be resolved before an effective design data base can be achieved. A foremost concern is the improvement of the analytical methods to enhance the correlation of the analytical predictions with flight measurements. To achieve this, measurements are required to determine the forcing functions. Such measurements should be acquired, both on the launch vehicle and on the payloads. These measurements must have a nominally flat frequency response to 50 Hz and be phase-correlated to ±5 degrees or better. If such measurements are recorded on different flight recorders, a system-to-system correlation must be assured to within ±5 degrees. The data should be measured on test verified payloads or cargo elements. The panel recommended that such measurements be made in conjuncion with the appropriate pre-flight and post-flight analyses. That is, the data must be utilized for data base enhancement.

The panel recommended that the following methodology questions be resolved before an effective data base can be implemented: (1) trunnion friction effects on payload load factors, (2) transmissibility effects at cargo element interfaces are to be determined, and (3) methods be developed for the combination of low frequency and high frequency environments. The panel also recommended that the discrepancy between the analytical predictions and the flight data in the 10 to 50 Hz range be resolved as soon as possible. The panel recognized that the development of simplified loads methods are required to support an effective design data base. Ultimately the minimum requirements for a comprehensive data base must be established. It must be determined how many flights are to be instrumented and how the changing launch vehicle configurations are to be handled in the data base.

Furthermore, the panel agreed that a minimum set of operational instrumentation should be defined and flown. The panel recognized that even commercial aircraft carry a flight recorder. Such data on Shuttle flights would be extremely valuable, not only to establish the validity of the existing data base but also to determine anomalies and to assess launch vehicle fatigue life.

PLAN FOR DATA ACQUISITION

The panel recommended that the plan for flight data acquisition for the verification of the design data base exploit all future flight data. The following Shuttle payloads are candidates for such future flight data analyses: SPARTAN (NASA/GSFC), ERBE (NASA/GSFC/Ball Aerospace), TDRS/IUS (NASA/GSFC/TRW/Boeing), the first flight from the Western Test Range (AFSD/MMA), the Galileo/ Centaur mission (JPL/NASA/ LeRC), ISPM/Centaur (ESA/JSC/LeRC), the Space Telescope (NASA/MSFC/ LMSC), the Space Technology Experiment Program (STEP) sponsored by NASA/OAST (NASA/LaRC/JPL), and various GAS bridge experiments (NASA/ GSFC). All of the above programs will collect flight data. Such data should be fully exploited to enhance the design data base.

PLAN FOR DATA UTILIZATION

The panel recommended that a plan for the utilization of flight data be implemented as shown below (Table V). First on the list of such a task should be the flight data analysis of Spacelab-1. This will result in the validation of the SPAH design data to be installed in the data base. Other tasks should include the systematic utilization of the future flight data for the upgrading of the flight data base.

CONCLUSION

The panel recognized that the discipline of loads and low frequency dynamics does not lend itself to the measurement of the environment. Design data must first be estimated conservatively and then these data must be validated by refined analyses and flight data. The conclusions were summarized as follows:

1. Focus in on the design data base
2. Data base is feasible but requires substantial effort
3. Current flight data deficiencies must be overcome
4. Flight data base is required as soon as possible
5. Data base must be kept current

Table V - Development Tasks

- **FLIGHT DATA ACQUISITION**

    - ORBITER RESPONSE DATA
    - PAYLOAD RESPONSE DATA

- **FLIGHT DATA ANALYSIS**

    - VERIFICATION OF METHODOLOGY
    - UPDATE DESIGN DATA BASE

- **PACING TASKS**

    - SPACELAB 1 - VALIDATION AND ANALYSIS OF FLIGHT DATA
    - IMPLEMENTATION OF FLIGHT DATA ACQUISITION

Furthermore, the panel determined that:
1. Future plans can only be implemented to the extent of available resources.
2. A data base for loads design data is feasible and would prove very helpful for Shuttle payload designers.
3. The focus should be a design data base for the academic and commercial user.

4. The design data base should be supplemented by a flight data base which would be used to validate the design data base.
5. There are several technology issues which must be resolved before the data base can become effective.
6. The SPAH design data should be used as a starting point for the design data base. Similar data for other carriers should be compiled, if available, or developed.
7. As a starting point for the resolution of some of these issues the Spacelab-1 data should be analyzed to validate the SPAH design data.
8. The design data base should be kept current using the flight data base.

ACKNOWLEDGEMENT

The research described in this paper was carried out, in part, by the Jet Propulsion Laboratory, California Institute of Technology, under contract with the National Aeronautics and Space Administration.

REFERENCE

1. ESA Document SLP/2104, Issue 1, Revision 7. European Space Agency Spacelab Accommodation Handbook, June 15, 1982.

THERMAL AND HUMIDITY ENVIRONMENT

James Clawson
Rockwell International

The Thermal and Humidity subpanel had an ambitious agenda for today. We did not get completely done with it. First, I would like you to be aware of the members of my sub-panel: Dave Russell from Rockwell, Houston; Jim Fu from JPL; Don Bartleson from Lockheed at the Cape; Brent Winkstern from McDonnell Douglas at the Cape; and Phil Tulkoff from Goddard. I want to express my appreciation to all of them. We have gotten down to where we only have four items left for an ad hoc committee tomorrow, and for what it's worth, they're all fun stuff; it's not the deep stuff that relates to the data base. I think we accomplished our goal today in defining what all we want to do regarding the planned data base. I've just got three things to put up here, and I have no intention of going through them in detail.

Something that was put together earlier is an outline (Figure 1) of what we think are the general things that will go into the data base. In essence, we have divided the Thermal and Humidity activity into three phases. The first phase is the pre-launch and payload integration activity which takes place at the Shuttle launch site. The next phase includes the thermal and humidity environment on the launch pad and during spacecraft ascent. The final phase, considered to be the most important, includes all the on-orbit activity.

I'm just showing you these figures, so you get an idea of the gobs and gobs of data obtained on orbit. We did not have a representative from Marshall relative to the Spacelab. It's unfortunate, but we understand why. They're still in the middle of their data correlation, and it may well work out for the best. I'll explain why in a minute. Lots of data have been taken relative to empty payload bay environments, integrated payload bay environments,

THERMAL AND HUMIDITY SUB-PANEL OUTLINE

TOPIC:	PROPOSED SOURCE
LAUNCH SITE ACCOMMODATIONS	
OPF	KSC
VAB	KSC
OTHER CHECKOUT FACILITIES	KSC
VAFB FACILITIES	AIR FORCE
PRELAUNCH PAD CONDITIONS	
PURGE SYSTEM CAPABILITIES	JANNEY, JSC
PURGE TEMPERATURES	JANNEY, JSC
PURGE HUMIDITY	JANNEY, JSC
PURGE TIMELINE	JANNEY, JSC
PURGE/PAYLOADS INTERACTION	RUSSELL, RI/HOU
ORBITER PRELAUNCH EXTERNAL ENVIRONMENTS	
TEMPERATURES	BROWN, JSC
HUMIDITY	BROWN, JSC
WINDS	BROWN, JSC
ORBITER RESPONSE TO PRELAUNCH ENVIRONMENTS/PURGE	RUSSELL, RI/HOU
ASCENT PHASE CONDITION	
ORBITER/PAYLOAD BAY TEMPERATURES	RUSSELL, RI/HOU
ASCENT PRESSURE DECAY	JENNEY, JSC

Fig. 1. Thermal and Humidity Subpanel Outline

THERMAL AND HUMIDITY SUB-PANEL OUTLINE (CONT.)

TOPIC:	PROPOSED SOURCE
ON-ORBIT CONDITIONS	
EXTERNAL ENVIRONMENTS	
SOLAR FLUX	CLAWSON, RI/HOU
ALBEDO (EARTH REFLECTED SOLAR)	BROWN, JSC
PLANETARY INFRARED RADIATION	BROWN, JSC
ORBITER TEMPERATURES	CLAWSON, RI/HOU
ORBITER THERMAL CONSTRAINTS	CLAWSON, RI/HOU
PAYLOAD BAY OPTICAL PROPERTIES	RUSSELL, RI/HOU
EMPTY PAYLOAD BAY TEMPERATURES	RUSSELL, RI/HOU
INTEGRATED PAYLOAD EFFECTS	RUSSELL, RI/HOU
SOLAR TRAPPING	RUSSELL, RI/HOU
IR INTERACTION	RUSSELL, RI/HOU
BLOCKAGE EFFECTS	RUSSELL, RI/HOU
PAYLOAD BAY PRESSURES	JACOBS, JSC
CONTAMINATION EFFECTS ON OPTICS	JACOBS, JSC
SPACELAB ENVIRONMENTS	
PRESSURIZED MODULE	MSFC
PALLETS	MSFC
CREW CABIN ENVIRONMENTS	STINSON, RI/HOU
PAYLOAD INTERACTION	STINSON, RI/HOU
CABIN ARS CAPABILITY	STINSON, RI/HOU
ATCS/PAYLOAD CAPABILITIES	STINSON, RI/HOU
PAYLOAD CARRIERS	
TYPICAL ENVIRONMENTS	RUSSELL, RI/HOU
TYPICAL THERMAL/ATTITUDE CONSTR.	BROWN, JSC
TYPICAL PAYLOAD THERMAL CONSTRAINTS	BROWN, JSC
OXYGEN INTERACTION EFFECTS ON OPTICS	LEGER, JSC
RCS THRUSTER PLUME HEATING ENVIRONMENTS	JSC

Fig. 1. continued

and crew cabin environment. This latter is important because people certainly work in the crew cabin. I want to express my appreciation to the people who showed up from Ames and told us, correctly, that we needed to increase our emphasis on the cabin environment, which we will do. The rest of the proposed outline (Figure 2) relates to entry, post-landing, and ferry flight environments, and some descriptions of thermal math models. Again, this is just to give you an idea of what we plan to have in the data base.

We had about 20 people show up this morning at our formal meeting, and I really appreciated it. I was surprised at the number of people from the Cape, and I really appreciated that.

One of the things that came out of our meeting was that we needed to emphasize a little bit more an area that I felt was simple: the closely controlled activities, like the purges and what is done to control the cabin environment. I'm a passive thermal control guy so ±10 degrees is beautiful to me. I now realize that there are lots of people in the payload world who really are concerned about plus or minus one or two degrees inside the cabin. That's not my world, but we will emphasize what goes on in terms of purge and time lines inside the cabin, inside the Orbiter. We learned of a new system that we didn't know anything about thanks to some constant pressure from a gentleman from Canada. This is a dedicated payload purge system--I'll give it that name, I don't know if that's the right name--in which certain payloads can receive either nitrogen or helium purge gas, very clean stuff. The users have to provide their own gas supply, and a few other things, but it will help them during prelaunch. It's a very clean environment, and dry--something new none of us knew about.

To digress for a moment, one of the challenges we had as a panel was to answer the questions, "What's the status of the data? Are they any good or not?" I'm going to summarize the answers to these. There are two categories of data that came out of the Orbiter program. There is operational instrumentation data, OI data, and developmental flight instrumentation, or DFI, data. Right now, we feel that the OI data, which are always there, are in excellent shape, and it's very good data, and the data base is sound and clean. The DFI data are available

THERMAL AND HUMIDITY SUB-PANEL OUTLINE (CONT.)

TOPIC:	PROPOSED SOURCE
ENTRY CONDITIONS	
ORBITER THERMAL RESPONSE	CLAWSON, RI/HOU
PAYLOAD BAY THERMAL RESPONSE	RUSSELL, RI/HOU
PRESSURE PROFILE	JANNEY, JSC
HUMIDITY PROFILE	JANNEY, JSC
POST LANDING CONDITIONS	
ENVIRONMENTS	
THERMAL	RUSSELL, RI/HOU
HUMIDITY	BROWN, JSC
WINDS	BROWN, JSC
PURGE SYSTEM CAPABILITIES	
TEMPERATURES	JANNEY, JSC
HUMIDITY	JANNEY, JSC
TIMELINE	JANNEY, JSC
ORBITER THERMAL RESPONSE POST LANDING	RUSSELL, RI/HOU
FERRY FLIGHT CONDITIONS	
ALTITUDE/PRESSURE PROFILES	JANNEY, JSC
TEMPERATURES	JANNEY, JSC
HUMIDITY	JANNEY, JSC
ORBITER INSTRUMENTATION	CLAWSON, RI/HOU
MATHEMATICAL MODELS	
390 NODE MODEL	RUSSELL, RI/HOU
SOTS MODEL	RUSSELL, RI/HOU
CABIN MODELS	STINSON/RI/HOU
PAYLOAD MODEL REQUIREMENTS	RUSSELL, RI/HOU

Fig. 2. Thermal and Humidity Subpanel Outline

in at least two different modes. One mode is the one that most of us are used to using, and it's a big, complicated story as to what it is, but the problem is that it only samples data every 10 minutes, and as a result of this and some other sampling problems, it misses things. It will give you apparent false peaks and valleys that aren't really true readings. However, in the unanimous opinion of the panel members, these data were accepted as sufficient for model correlation, and I will touch on model correlation in a minute. We feel like any engineer or scientist will always want additional data. Our position, rather simply, is that the Orbiter program has acquired all the data that it plans to gather, and we all understood that. But there are experiments and payloads to be flown that could, if they are capable and the experimenters desire it, measure something about their thermal environment for themselves, and make these data available for inclusion in the thermal data base. There are systems--the Air Force has one, and there are probably others--which might require regular measurements of temperature on their carrier. We would like to get those data when they becomes available and include them in the data base.

Dr. John Park came in briefly and gave us a presentation on some experiments on how white coatings are affected by the on-orbit oxygen reaction. This was enlightening, to say the least. Most of us in the thermal control world have been aware for years of what white coatings do and don't do, and it's never good, but we really appreciate what John had to tell us about some of the ideas of silicone overcoats, etc.

At the very last minute, sort of "a la A Team" and Hannibal Smith--"I love it when a plan comes together"--it came together at the last minute. We were sitting there trying to summarize our work, and we suddenly realized what needs to go into the data base. There are gobs of data; they would fill up the room, almost. Rather than mess with the details of the myriad flight data, what we propose to put in the data base are things from our models. I will need to explain. The flight data were good enough to correlate the models. There are many models, thermal models around the world, but the important ones are payload integration models. There's something called the 390-node model and there's a more simplified one. Neither of these models is easy for users. They're difficult for a user to get hold of. If you can get hold of it it's tough to use. But, they are excellent models. They correlate probably within the order of 10 degrees Fahrenheit with flight data. That's really good for on-orbit stuff. As a result, what we propose to do for the data base, even though it's as yet unfunded, is to expand the number of cases that we analyze with these models, both empty bay and with payload combinations, and get that kind of data into the data base. It will take some time and effort to do this. But something that we've got to figure out is how we go about it from here. So, rather than putting flight data directly into the data base--that's not going to help anybody--what we want to do is put integrated model results into the data base so an experimenter will know what to expect when you've got several different payloads stacked together. We're convinced that's the way to go, and we've got excellent models to do it with.

There are several issues that concern us. We are all concerned that a user might inadvertently misinterpret and misuse a number that's put in the data base. He could just take it and run with it. It's not just a concern in the thermal area. The gentleman on Loads and Low Frequency Dynamics expressed the same concern. I don't know yet how we're going to solve this problem. We're going to try somehow, and it won't happen immediately, to key the user to think about the consequences of what he sees in the data base. You know, to cause the lightbulb to go "Click", which causes the user to think, "Gee, I've got to think about that."

One of the other major concerns in the thermal world, and I don't know how to solve this one either, is trying to guess the mission assignment attitude time line that a user's going to get two years from now, when he has no idea what mission he's going to be flying on, what payload he's going to be integrated with. I don't know how to solve that problem. For the big users, the big Class A Galileo payloads, even some of the other smaller things, they pretty well know they're going to get integrated in a certain manner. Thermal analysis for these payloads will be fairly easy. But the poor little guy with something just "this big" that he wants to hang on a GAS can, or on the longeron, or on one of these other carriers that we've talked about, has no idea what mission he's going on. It's tough know how to solve that one. Maybe, with some ideas that we came up with right after our meeting, we can develop, if we build the data base right and the guy knows approximately where he's going in the payload bay, we can develop a first cut of an "equivalent sink" temperature, and that's something else we don't need to talk about now.

The last issue we identified (and I have already alluded to it) is that there are some things that we need to do, like run the models to develop the database since they are indeed the best source, the correlated models, and continue to support the database development (it's currently unfunded). I'm sure other sub-panels can identify with this problem.

Well, that pretty well covers where we stand today. Are there any questions in the thermal area? There will be, just if somebody's interested, four little presentations tomorrow afternoon. They're fun stuff. It's not deep.

Thank you very much.

Moderator: Thank you, Jim. I can tell you from a user of the cabin temperatures that I wish I knew a lot more about it. I knew it was cold the first couple of days and was warm the rest of the mission. I also know that this lovely billion-and-a-half-dollar Shuttle doesn't have an automatic thermostat in it, and we ended up pulling up the floor panels and actually manually tweaking the diverter valves, and doing it once every couple of days to try and keep the temperature within, like you said, ±10 degrees or so. I'm talking about the Orbiter, the mid-deck, and especially the flight deck. Every time you roll around and get the direct thermal input through those windows, it gets very warm up in the flight deck in a hurry, and it swings on a 90-minute cycle. The mid-deck is a little more stable, but it responds to the solar input quite rapidly. Of course, we had a special case with the Spacelab onboard, and we had the payload heat exchangers on and took a lot of the cooling capability away from the mid-deck, so it's probably not quite so bad if you have just some cargo in the payload bay other than the Spacelab.

More information on the Thermal and Humidity Environment can be found in Appendix C.

SURFACE INTERACTIONS

Henry Garrett
Jet Propulsion Lab
California Institute of Technology

A spacecraft, the Shuttle or any other vehicle in space, does interact with its environment. In fact, that's the major reason we are here--because of those interactions. The specific interests that the surface interactions panel had, however, are unique. Most of these areas would not have been present three or four years ago, and some of them have come about in the last few years. Of these, we are primarily concerned with six types of surface interactions:

      Shuttle Surface Glow
      Spacecraft Surface Charging
      Spacecraft Surface Erosion
      High Voltage Surface Interactions
      Shuttle Wake/Ram Plasma Variations
      VxB Induced Currents

The first characteristic that I would like to stress about the Shuttle is that it is in a very complex geophysical environment, perhaps the most complex one that we have in the solar system. The reason for that is because we not only have to contend with the neutral atmosphere, which is still very substantial at these altitudes and varies greatly with geomagnetic activity, but we also have a very dense ionospheric population. Figure 1 shows views looking down on the polar cap of the earth and the electron densities in units of particles per cubic centimeter. It also lists the temperature of those particles in thousands of degrees. There are two characteristics which you should take note of: first, the very dense electron population, and second, the very cold comparative temperatures of these plasmas. We use the term comparatively cold because the rest of the magnetosphere temperatures typically run thousands of electron volts. In most cases, these temperatures are lower than an electron volt, 10,000 degrees, which is very, very cold for plasmas in space. This phenomenon has a very real effect on the type of environment that we are dealing with.

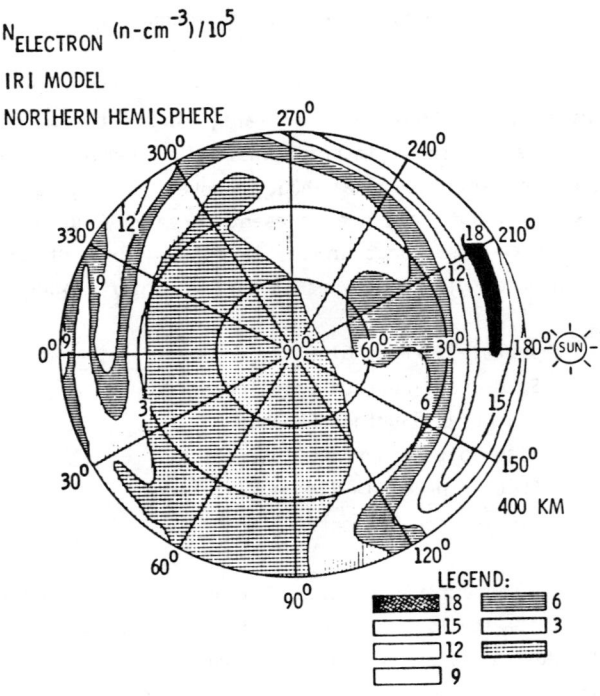

Fig. 1. Polar View of the Electron Environment at 400 km for December as Predicted by the IRI Model (sunspot number R is 100). Electron Density at 400 km as Predicted by the IRI Model

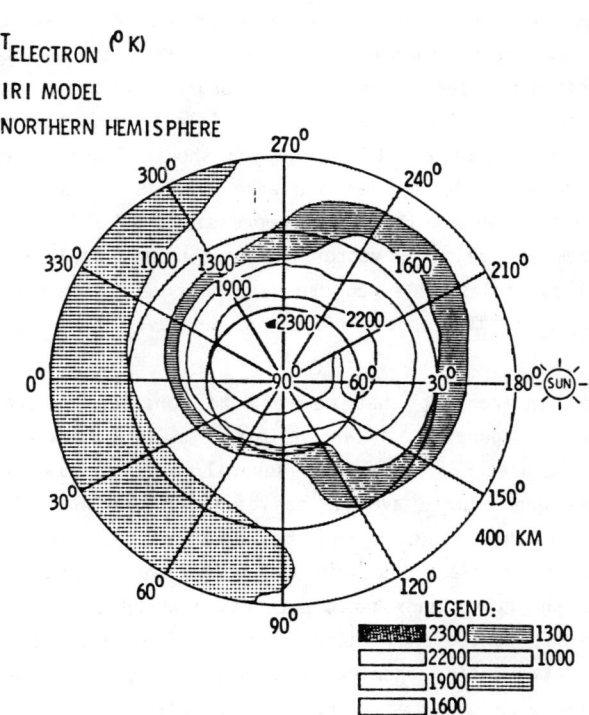

Fig. 1. continued

Another aspect that must be taken into account is the rapid motion of the Shuttle. Even at low altitudes, it is moving at approximately seven to eight kilometers a second, whereas the atmosphere and ionosphere are nearly stationary, moving with the earth's rotation rate. This results in a supersonic passage through this environment and, hence, a very pronounced wake phenomenon. Figure 2 shows some examples of wakes for very small satellites at various altitudes. Zero degrees or 360 degrees is the velocity vector direction, the direction in which the vehicle is moving, the so-called ram direction. There are two words you should learn if you are not used to this terminology: ram and wake. Ram is the direction of the velocity vector, where there is compression. Wake is the direction opposite from the velocity vector where we have a depletion in plasma and neutral particles. As you can see, this depletion can reach well over a factor of two orders of magnitude, even for a small spacecraft.

Another environment that we need to consider is the so-called auroral zone. Figures 3, 4, and 5, based on data from the Air Force Geophysics Laboratory, show the auroral/polar environment. They indicate the approximate average positions of the so- called auroral oval which in this case, is for the Northern Hemisphere. There are intense fluxes of electrons over the polar caps. Their energies, in contrast to the cold ionospheric population which they penetrate, are on the order of tens of Kev in some cases. Thus we are dealing with a very strange population, dominated by the cold plasma which leads to the so-called wake phenomenon and then penetrated by the very intense auroral zone particles (over the polar caps). Thus far, the Shuttle has not spent much time in those regions. I believe that 57 degrees magnetic is the highest that we have flown manned flights. In the auroral zone, you're skirting the edge of those latitudes, but typically, it's up around 62 degrees. Even in the edges of the auroral zone, you certainly do not enter the polar cap regions.

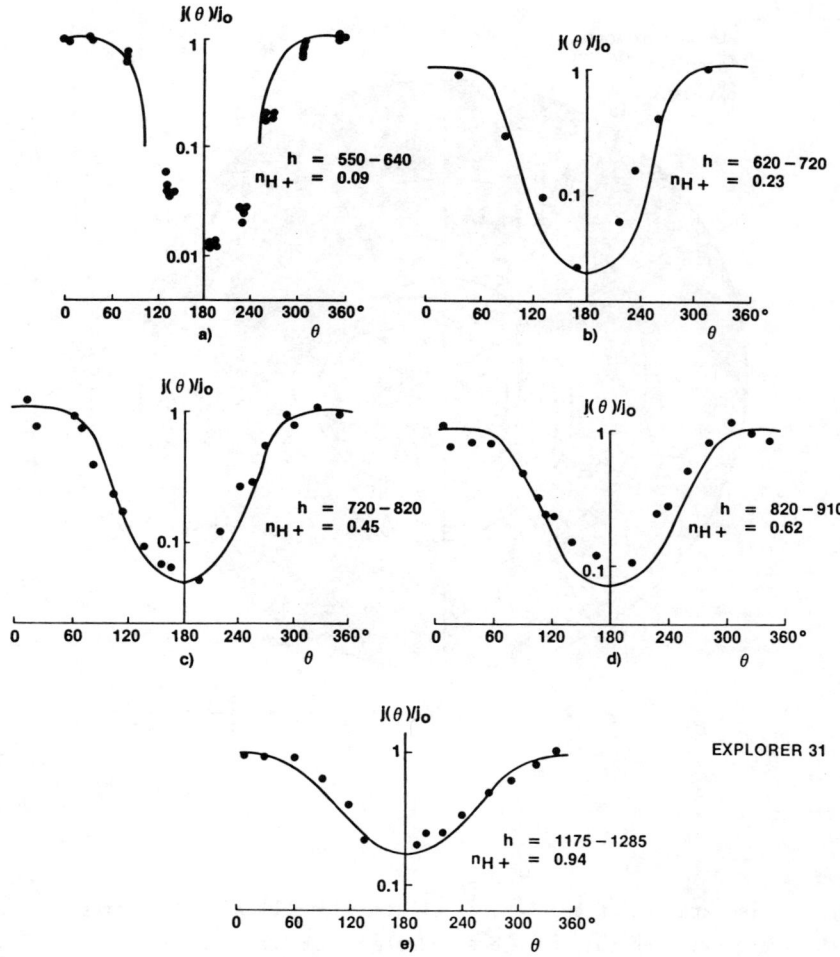

NORMALIZED ELECTRON CURRENT DENSITY
vs ANGLE, ALTITUDE AND PERCENT H+

Fig. 2. Plasma Wakes

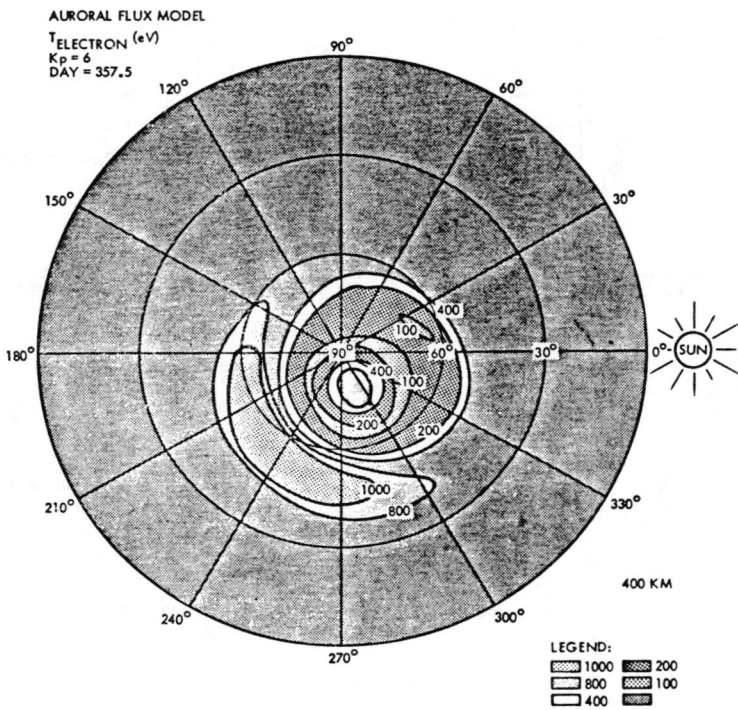

Fig. 3. Temperature of the Auroral Electron Flux at 400 km. Units are eV. The data have been adapted from the auroral flux model.

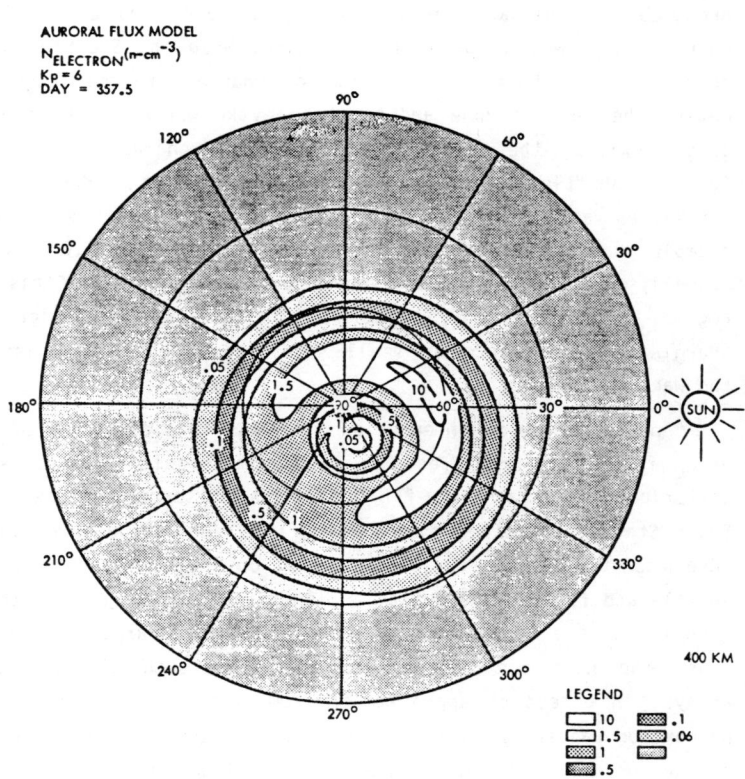

Fig. 4. Density of the Auroral Electrons at 400 km. Units are number per cubic centimeter

The principal phenomenon that we are addressing here is that the Shuttle is a large object, much larger than we have typically flown in these regions in the past. It creates a wake. In that wake, presumably on passage through an auroral arc, you could elicit the phenomenon known as spacecraft charging. When the Shuttle is in this cold plasma environment, it does not charge very much, but when it enters the auroral zone and creates a wake where the cold plasma no longer exists, the auroral particles could charge a body separated from the Shuttle, such as an astronaut. There is evidence from other satellites at somewhat higher altitudes, the DMSP satellites for example, suggesting that potentials on the order of 1,000 volts have actually been observed on vehicles passing through this orbital regime. So, as of now, one of the problems which we might face is charging of astronauts or small satellites electrically isolated in the wake of the Shuttle.

Another type of interaction that was considered in our working group is that of the effects on high-voltage solar arrays. Aside from batteries, solar arrays will be the major source of power for the Space Station and other systems that need very large power levels. We have recently had a four-day workshop on power systems both for the Shuttle and for the Space Station, and it became very clear that solar arrays are here to stay and will be the major source of power for at least the next one to two decades. For such high-voltage solar arrays, in excess of approximately 200 volts, we find that we get a phenomenon called snapover for positively biased arrays. We have cover glasses on most solar cells that protect them from the plasma environment. Between them, however, are wires called "interconnects." These interconnects are often exposed. When we bias one of these solar arrays, you can see the potential change at the interconnects as shown in Figure 5. You can see the slight blips in the positive voltage at the surface on the right in Figure 5.

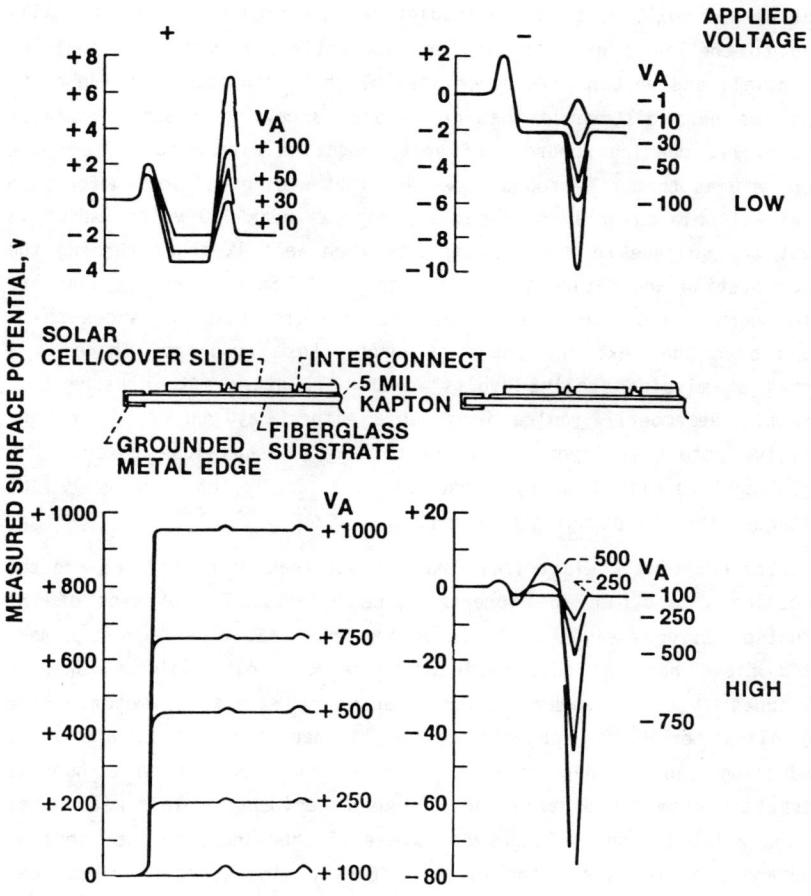

Fig. 5. Surface Potential Profiles for Biased Solar Cells

As we crank up the voltage, however, we find that that structure disappears and we actually see the potential of the array across the entire surface. We have an explanation for that. It deals with what is referred to as the secondary emission process. The net result is that even a small hole in the insulation on a highly biased array will lead to the whole array having that potential, a very high positive potential, and we can have power loss by that technique. On the other hand, we have discovered that most solar arrays will actually float negatively, meaning hundreds of volts negative with respect to space when we bias them. We found, however, that arcing can set in for such negatively biased array surfaces at voltages near 200 volts, which is about the voltage we are dealing with when we talk about running the Space Station and flying the large deployable Shuttle arrays that are being considered. In fact, there are several Shuttle missions to be flown over the next few years to study these problems. There is a series of missions called "Volts" that will attempt to measure this effect. Remember: you're in trouble either way you go. For high positive potentials, you can attract electrons and lose power. For very negative-biased arrays, you can get arcing at relatively low voltages, thereby damaging your array.

The end of the electrical interactions that we considered are the so-called V x B, or the Lorentz force effects. The physics of the situation is very simple. A conducting body that moves across a magnetic field has an induced electric field. That field at Shuttle altitudes is on the order of three tenths of a volt per meter. There are plans for flying objects called "tethers." Tethers are long, conducting cables that could be run as much as ten to a hundred kilometers from the Shuttle for various experiments. This means that we are talking about voltages in excess of thousands, perhaps tens of thousands, of volts for the electric field. The types of electrical effects that are induced on the ends of those cables and the grounding of the cables to the space environment both pose unique problems that need to be considered.

The earth's magnetic field is another environmental effect that leads to complexity in a surface interaction. Figure 6 shows the earth's magnetic field viewed from above, in terms of contours of equal intensity. Figure 7 is the induced electric field for a Shuttle in a polar orbit at 400 kilometers. You can see that the induced electric field varies fairly rapidly from roughly three-tenths of a volt per meter down to less than a tenth of a volt per meter. For a tether experiment over the poles you can get into quite complex V x B effects that need to be considered.

Another aspect of the surface interactions that our panel considered is the so-called "Shuttle glow" phenomenon. The Shuttle glow phenomenon extends roughly 15 centimeters from the surface of the Shuttle. It is generally slightly subvisual but it does become intensified when the thrusters fire. It is a very pronounced phenomenon that was not really anticipated. Although there were a few papers prior to the actual launch of the Shuttle that mentioned a glow phenomenon and a recent Geophysics Research Letter volume dedicated to those data, we still do not really understand the phenomenon. We do, however, have some hypotheses concerning the causes of this phenomenon. For example, we believe that it is more of a chemical rather than a plasma interaction.

Closely related to the Shuttle glow phenomenon, we believe--again we are not certain--is the so-called oxygen erosion problem. Figure 8 shows a sample of the Galileo thermal blanket material. Briefly, the material had a Kapton substrate with a partially conducting paint on top of the surface. On the left are the pre-flight surfaces. On the right are about four days of Shuttle flight data. You can see the erosion down to the Kapton beneath the paint. Due to the intense charged particle environment at Jupiter that we expected to encounter, we were very concerned with having a conducting spacecraft. The fact that this happened on Shuttle has led to a reassessment of the coatings on Galileo. As you can see, it has a major impact on the Shuttle as a transport system from our standpoint.

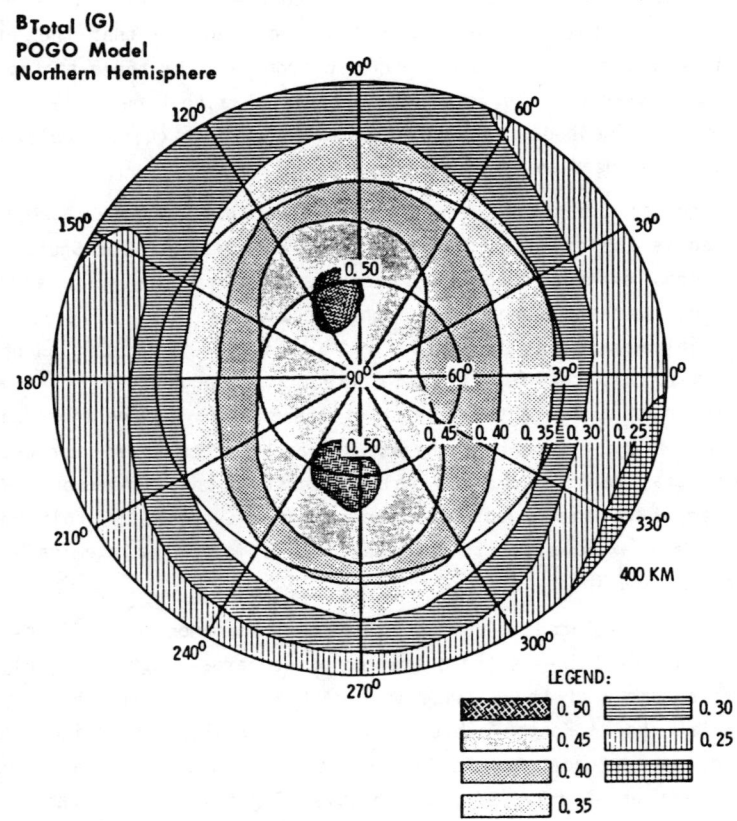

Fig. 6. Total Magnetic Field at 400 km for POGO Magnetic Field Model. Units are G.

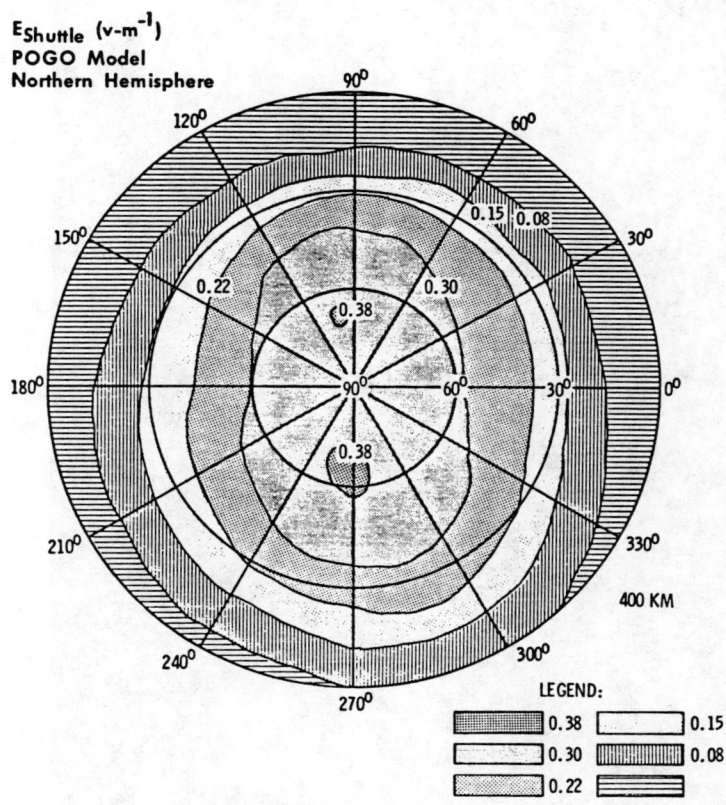

Fig. 7. Absolute Value of the V x B electric field induced on a body in a 90° inclination orbit for the POGO model. Units are volts per meter.

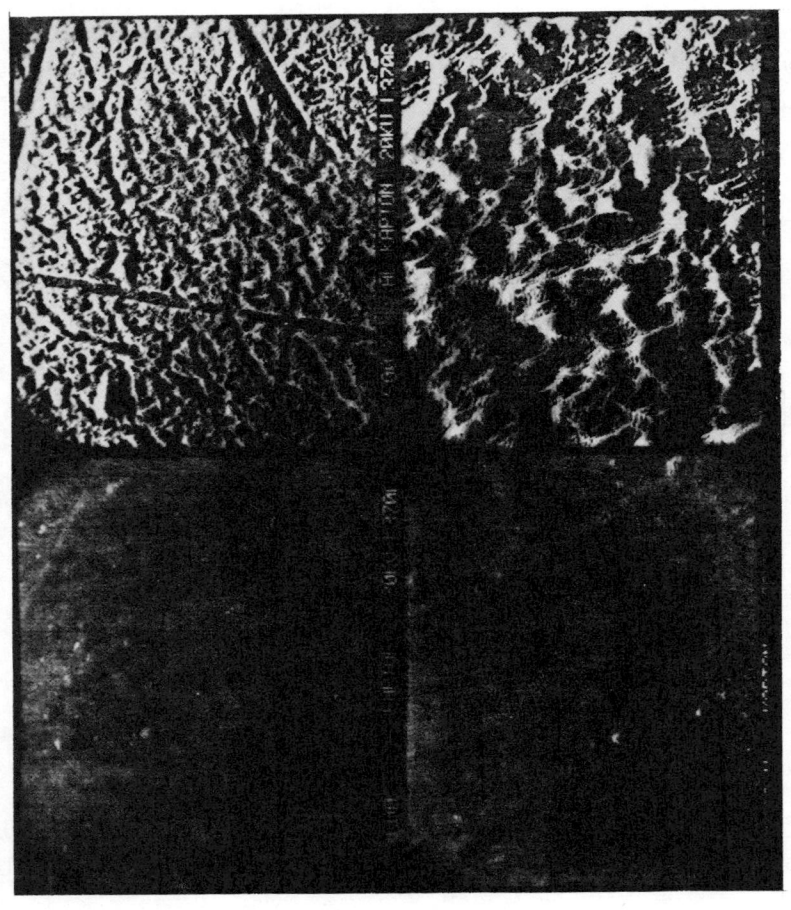

Fig. 8. STS-3/CMP Thermal Blanket

So, in a nutshell, you have the six interactions that we looked at. To repeat, we looked at the surface charging, the high-voltage effects, V x B, the ram/wake effects, the surface glow phenomenon, and finally, the surface erosion phenomenon. We had a very successful scientific interchange culminating in a conference call across the country to discuss some of the latest results.

I would like to conclude by discussing the way in which we intend to address our responsibility as a subpanel to this meeting. One person was assigned to each of the six topics to provide data on the interaction in a fixed format. We will first describe each of the various types of interaction for the layman. Next, we are going to list the references and the key players. This will enable users to go to the computer data base, find out what the phenomenon is, why they should be concerned with it, whom to talk to, and where the information is. We will also list what information exists, what format it is in, and where it is available. For six such diverse interactions, you can imagine that the types of data that exist are quite different. A lot of the data exists only in the form of models, for example, wake models. We have wake models that were discussed here that we believe give a fair representation of what the wake looks like, but we do not have many data at this point.

The next part of our data base will list what the ideal world be like. That is, if we had all of our information for these spacecraft interactions, what would those data look like? You subtract this from the preceeding and you find out what we have to do. This is what's missing in the form of data and models. Next we will outline what experiments need to be flown to fill in the missing data and models. And finally, for the engineering community, we hope to present solutions to the problems. We've assigned people to carry out those tasks. Within about the next two months, we will have this data base in existence for your use. In fact, much of the work is already being put together now. That summarizes the surface interaction area; a more detailed account of surface interactions has been prepared for the NASA data base, and is given here in Appendix A.

Question: How serious of a problem is blanket degradation/ optical degradation?

First, as discussed at this conference, many of the blankets and things of that nature that go into space have been contaminated on the ground, while many of the tests and measurements that we make are in idealized laboratory conditions. So there is always a dichotomy between what we think the blanket really does and what we observe in the laboratory. Based on the conversations we've had here, it is now clear that the blankets erode rapidly and that optical surfaces are easily damaged at the lower shuttle altitude. Two things happen: one is, in some cases, the material properties actually improve. People at the Air Force Weapons Laboratory tell us that some of their mirrors, in certain frequency ranges, actually improved their reflectivity after being exposed to this environment, whereas for most materials, they're actually structurally damaged and degraded, as you've seen. The point I'm trying to lead up to is that it depends strictly on the type of material and the conditions under which it was exposed. People are talking about Kapton getting a dull look to it, or even in some cases, about getting yellow deposits on the material. Clearly the thermal properties are degrading in such cases.

Question: Do we have any information on long term (10 or more years) exposure?

Yes, we have basically two sources of that kind of information. First of all, some of the most interesting data are from the Solar Max mission. They actually brought back some of the thermal blankets and some of the surface samples. The preliminary results were briefly discussed in our own working session. That's approximately four years of exposure. The LDEF (Long Duration Exposure Facility), when it comes back, will have about a year of exposure of these materials, whereas with the Space Shuttle, you're talking basically of seven-day missions. But even in the seven-day scale, we now have tables that Lubert Leger has put together from his work and the work of his co-investigators. There were 47 samples in one flight, 40 in another flight. So you have almost 100 different samples.

The problem is that you have to look on a material-by-material basis because we don't understand the reactions well enough at this point to tell you exactly what's going to happen. If you happen to have one of the materials that has been flown, we can tell you what happens, but if if was a new material, you might not really know.

I would like to thank the committee for their excellent job.

The research described in this paper was carried out, in part, by the Jet Propulsion Laboratory, California Institute of Technology, under contract with the National Aeronautics and Space Agency.

VIBROACOUSTIC ENVIRONMENT

Don Wong
Aerospace Corporation

1. Technology Issues

The Vibroacoustic Environment Subpanel addressed several issues including:

LIFTOFF ACOUSTICS
    o Forward 1/3 of the Cargo Bay Acoustic Environment
    o Radial Acoustic Gradient
    o Payload Effort on Cargo Bay Environment

TRANSONIC ACOUSTICS
    o Vent Noise Not Adequately Characterized

STRUCTURE-BORNE VIBRATIONS (ORBITER-INDUCED)
    o Ground Test and Flight Test Do Not Correlate
    o Interface Vibration Spec (IRN-048)

VLS ACOUSTIC VERIFICATION

COMBINED LOADING

Liftoff generates the highest overall acoustic noise environment in the payload bay. However, the liftoff acoustic environment has not been adequately defined for payloads from the existing flight data. The acoustic environment in the forward one-third of the cargo bay has not been measured at all except on the cargo bay forward bulkhead. Cargo bay acoustic data indicate an aft-to-forward increasing noise level gradient, with the forward bulkhead measurement being one of the highest recorded. The gradient is contrary to pre-flight expectations and requires verification. Adequate flight data are also lacking to characterize the acoustic environment in the proximity of the orbiter doors and sidewalls. The doors are the dominant sound transmission path into the cargo bay, and this region appears to have significantly more intense sound pressure levels than the other interior regions of the bay. Perhaps the greatest concern is centered on the effects of large payloads on the environment. Model data and limited flight data show that large payloads have a significant effect on local acoustic environments due to acoustic modes set up between large payloads surfaces and orbiter cargo bay surfaces. This effect may be compounded when a payload is near the doors.

The payload bay acoustic environment during transonic flight is characterized by discrete noise emanating from pressure equalization vents. This narrow band vent noise, which occurs in approximately the 280Hz to 340Hz frequency range, exceeds the spectral levels occuring during liftoff. The vent acoustic environment is not adequately defined. The trend of the limited data available from microphones near the aft vents is inconsistent, and the forward vent environment is unknown. The effect of large payloads on the localized acoustic environment by the vents is also unknown. Test simulation methodology for this discrete acoustic tone environment has not been developed and verified.

Differences between the KSC and VLS facilities will also affect the payload bay environment. These launch site differences include the launch pad design, VLS thrust augmentation, and terrain differences. A data base is needed to define the effects of orbiter-to-orbiter differences and KSC/VLS differences on the payload bay vibration and acoustic environments.

The relative contributions to the payload loads of the structure-borne vibration induced by acoustic excitation of the orbiter payload bay structures, and the vibration generated by launch vehicle transient events, are not well understood. The relative contributions vary during the launch phase, and considerable controversy exists as how to combine the effects to obtain resultant loads. Flight data acquired to date have been of limited use for this problem due to difficulty in separating the resulting effects as functions of the excitation source.

2. Proposed Approaches to Technical Issues

For each of the technology issues, the subpanel outlined ways to improve upon these as follows:

LIFTOFF ACOUSTICS
    o Develop Data Base
    o Improve Prediction Techniques
    o Update Criteria (ICD-2-19001)

TRANSONIC ACOUSTICS
    o Develop Data Base
    o Characterize Vent Noise
       - Study Wind Tunnel Test Feasibility
       - Conduct Test/Develop Analytical Models
       - Verify Results with Flight Data
       - Update Criteria (ICD-2-19001)

    o Develop Ground Test Simulation Methods

STRUCTURE-BORNE VIBRATION (ORBITER - INDUCED)
    o Develop Data Base
    o Flight and Ground Test Vibration Correlation
    o Improve Component Vibration Criteria/Methodology
       - VAPEPS Computer Program
    o Clarify IRN JSC-048

VLS ACOUSTIC VERIFICATION
    o Require Minimum of 7 Mic's for 4 Flights

COMBINED LOADING
    o Assess Existing Flight Data
    o Perform Source Identification Study
    o Methodology Study

The key to improving the definition of the liftoff acoustic environment is the acquisition of additional flight data. Opportunities for acoustic measurements of the forward one-third of the cargo bay should become available in the future as add-ons of instrumentation to payloads scheduled to be flown in that location of the bay. Obtaining adequate measurements of the acoustic radial gradients near the cargo bay doors will require early integration with planned Shuttle payloads. A boom structure should be integrated with a pallet payload to provide a mounting structure for microphones at various distances from the cargo bay doors. It will be difficult to totally define the effects of large payloads on the local acoustic environment with flight measurements, because of the relatively limited number of large payload flights and the variety of payload configurations that must be evaluated. Analytical techniques for predicting large payload effects, such as PACES (Payload Acoustic Environment for Shuttle), should be developed or improved and validated with future flight data. The degree of simulation of STS cargo bay liftoff acoustic environments achieved by conventional reverberant chamber acoustic tests should be evaluated by correlating flight acoustic and payload response measurements with payload ground test results. If conventional acoustic tests are non-conservative, as some data seem to indicate, or if the tests are overly conservative, methods for improving the flight environment simulation should be investigated.

As with the liftoff acoustic environment, additional flight data are required to improve the definition of the transonic vent noise environment. Although some improved definition of this environment can be obtained by add-on instrumentation to developed payloads, it is recommended that early integration with a pallet payload be accomplished to implement the installation of an array of microphones around a vent, or vents, to characterize the noise as a function of direction and distance. Flight data should be supplemented with ground experimental and analytical studies. The feasibility of performing a wind tunnel test on a vent should be investigated to more thoroughly characterize the noise source. If such a test is feasible, the results can be employed to develop analytical models of the noise source. The analytical models would be verified using flight data. Once the vent noise is adequately characterized, potential ground test simulation methods should be studied to determine which method provides an adequate test of payloads without being overly conservative.

Adequate characterization of the orbiter-induced, structure-borne vibration environment requires that existing and future flight and ground test data be consolidated and evaluated. A data base is needed on a variety of payload configurations from both STS flight measurements and ground test measurements. Flight measurements are required both at payload attach points and payload response points. Phase correlation information for the vibration measurements and the cargo bay acoustic measurements would facilitate understanding of the data trends. Vibration correlation of the flight data with ground test data should be performed, and improved payload component vibration prediction methodology be developed, to account for payload configuration variations.

A data base management and prediction system named VAPEPS (<u>Vi</u>bro<u>a</u>coustic <u>P</u>ayload <u>E</u>nvironment <u>P</u>rediction <u>Sy</u>stem) has been developed for the vibration and acoustic data obtained from Space Shuttle and Expendable Launch Vehicle payload components. VAPEPS accepts vibroacoustic/structural data and constructs a data base that can be used by the aerospace community to establish the environment of new payload components. Additionally, VAPEPS provides a full range of software routines that can be used to manipulate and perform computational operations on data. The computer program is jointly sponsored by NASA and Air Force Space Division.

A data base is required for characterization of the effects of orbiter-to-orbiter differences or significant modifications to orbiters and differences between the KSC and VLS facilities on the payload bay vibration and acoustic environments. Flight instrumentation is also required for anomaly investigation. The recommended minimum orbiter payload bay instrumentation to define differences in the VLS environments is seven microphones and nine high frequency accelerometers monitored for four flights. Recommended minimum requirements for each orbiter, or major modifications to an orbiter, are four microphones and six accelerometers for four flights. Recommended minimum orbiter payload bay instrumentation, for continuous monitoring to support anomaly investigations, is two microphones and three high frequency accelerometers. This orbiter instrumentation may supplement but should not replace payload instrumentation intended to resolve specific technology issues.

The first step required towards a combined loading methodology to improve vibroacoustic and transient environments is to identify the sources of payload loads for each dynamically significant STS event. These sources are primarily mechanically transmitted transient vibration, mechanically transmitted acoustic vibration, and direct acoustic excitation of the payload. Existing flight data should be assessed to identify sources. However, preliminary evaluations indicate that quantitative values cannot be determined for the relative contributions of the sources to the total loads, due partly to the lack of phase correlation information for the data. Additional flight measurements should be proposed and implemented. All proposed measurements should allow for obtaining phase correlation information. A well documented analysis should be available for payloads selected for measurements. A proposal for an instrumented payload experiment, consisting of simple structures that have been thoroughly analyzed and test verified, should be considered. Once the source characteristics of the payload loads have been adequately defined, a study should be perfomed of existing and proposed combined loading methodologies. The most promising methodologies can then be refined on the basis of the STS payloads source characteristics.

3. Recommendations

The Subpanel's recommendation can be summarized as follows:

THREE REQUIRED LEVELS FOR DATA BASE
- o Executive Data Management
  - Description of Orbiter/Carriers and Environment
  - Reference List and Directory

- o User Data Management
  - Orbiter/Carrier User Handbooks
  - Technical Discipline Review Panel

- o Technology Data Management
  - Establish Center for VAPEPS Vibroacoustic Data Management
    * Update Future Flight and Ground Data
    * Provide Consultation for Users
    * Data Accessible by Outside Terminals
    *Inform Users of New Changes
    * Data Review Panel
  - Validate VAPEPS Modeling Techniques
  - VAPEPS Program Maintenance/Improvement
  - Increase VAPEPS Data Base with Existing Ground Acoustic Tests

ESTABLISH LONG TERM DATA ACQUISITION PLANS
- o Coordination of all Acquisition Planning
- o Coordination of all Data Reduction and Evaluation
- o Data Recording Systems
  - OASIS (7 Flights)
  - MADS (Expanded Capability Required)
  - ARS (Advanced Solid-State Recording Systems)
    * Development (Ongoing)
    * Future Development Funding Required

The Vibroacoustic Panel recommended that future data base requirements consist of three levels of data management. The top level data management consists of the description of the various carriers (i.e. IUS, PAM-D, Centaur, etc.) and their design environments. The second level data management is to provide quick design environment information to the user. This information consists mainly of the vibroacoustic/shock interface requirements for the various carriers. The third level data management is the raw data that are obtained during each flight and the ground test data for each payload. It is recommended that the required vibroacoustic/shock data base management system utilize the VAPEPS program.

A center needs to be established for VAPEPS vibroacoustic data base management to optimize utilization of the code. The center would convene a panel of experts to review all future ground and flight data submitted to confirm its validity, to input and maintain the data base, to provide consultation to users, and to inform users of new changes to the code and data base. The data would be accessible by outside terminals. The VAPEPS modeling technique must be validated for various payload configurations, and needs sustained maintenance and improvements to the code implemented as new requirements are established. Also, the VAPEPS data base should be expanded with existing ground acoustic test data.

Finally, for long term data acquisition plans, it is recommended that the payload/carrier community coordinate all data measurement plans with the VAPEPS data base management center (JPL is the recommended data base management center). All data reduction and evaluation should be also coordinated with the data base management center.

Data recording systems for the future are limited to the OASIS (maximum seven flights) and the MADS. The MADS system is not presently available for payload users. A new recording system is needed for long term data acquisition. The Air Force Space Division has taken the initiative to contract for a prototype recorder called ARS (Advanced Solid-State Recording System) using state-of-the-art magnetic bubble memory devices with four megabit capacity. The ARS is expected to be complete by June 1985.

GASEOUS BACKGROUND

J. Barengoltz
Jet Propulsion Laboratory
California Institute of Technology

The Gaseous Background Subpanel met in a planned session and in an ad hoc session. One of the Subpanel's first actions was to change its name to the Molecular Contamination Environment Subpanel to reflect STS user interest in both background and deposited molecules, and to explicitly include the large body of useful data on molecular deposition. The agenda of the meetings comprised recent data, the status of analysis and modeling, planned future activities, and future needs. Eight speakers provided informal presentations on recent data, planned future activities, and future needs. The subjects and the authors were:

KSC Ground Facilities, B. Wenkstern (MDTSCO)
IECM TOCM, D. McKeown (Faraday)
IECM OEM and M/S, E. Miller (MSFC)
STS-4 M/S, W. Swider (AFGL)
Solar Max Venting, J.Triolo (GSFC)
Shuttle Payload Bay Cleanliness Ground
    Test Program, L. Bareiss (MMA)
IOCM, J. Barengoltz (JPL)
German Infrared Laboratory (GIRL),
    O. Lemke (MPIA)

The discussion of recent data on STS molecular contamination included reports on ground facilities and flight results. Wenkstern presented data on non-volatile residue (NVR) for the KSC ground facilities which show that a 1 $mg/ft.^2$ - month requirement is easily met. These data were obtained by rinsing stainless steel witness plates with carbon tetrachloride and performing infrared spectroscopy. This method is sensitive to hydrocarbons, but the residue is difficult to analyze for composition. Wenkstern noted that the time dependence of the deposition is also an open issue. He also reported that they are considering the procedure in use at CCAFS: Solution of the sample in a 1,1,1-trichloroethane/ethanol solution, solvent evaporation, and direct weighing of the residue. This procedure's advantages are the direct weight and the potential for chemical analysis of the residue.

Two papers on the recent flight (STS-9, Spacelab 1) of the Induced Environment Contamination Monitor (IECM) were presented. McKeown reported on a forward-looking temperature controlled quartz crystal microbalance (TOCM) located aft of Spacelab 1 and most of the instrument pallet. He noted the apparent time constants are small; i.e., at $0°C$ collection temperature, 5 $\mu g/cm^2$ collected in the first 16.7 hours and the deposition only increased to a total of 8 to 12 $\mu g/cm^2$ after 100 hours. Bay exposure to the sun was observed to enhance deposition rates, presumably due to increased outgassing of materials at higher temperatures. Miller presented data from the Optical Effects Module (OEM) and the mass spectrometer. A small real-time degradation in gold surface reflectivity due to contamination was noted. Surprisingly, an attempt to characterize the oxygen erosion effect by transmission measurements on carbon and osmium thin film samples in the OEM failed, in that the observed change in transmission was virtually nil. Miller suggested that perhaps this finding resulted from a contamination effect. The mass spectrometer measured a water vapor column density (return flux) of 1 to 2 x 10^{12} cm^{-2}. This is an upper limit because part of the water vapor may be direct from the payloads, given the cargo configuration. The total flux of species with molecular weight above 50 was approximately one percent of the water vapor flux or less, including a known freon leak.

Swider reviewed the AFGL mass spectrometer data from the STS-4 flight. The largest contaminant, water vapor, was observed to correlate with surface temperatures except during water dumps and thruster firings. Helium was a major contaminant, with concentrations approaching the water vapor value. Thruster firings caused pressure pulses of 1 to 2×10^{-6} torr which decayed to ambient within a few seconds of engine cut-off. The major exhaust products were identified as N_2, H_2O and H_2. Important interactions between the gas (neutral) background and the plasma were also noted. (A paper by R. Narcisi et al., "The Gaseous and Plasma Environment Around Space Shuttle", AIAA-83-2659, October 31-November 2, 1983 is available.)

Triolo described the observed contamination deposits associated with venting orifices on the recently recovered Solar Maximum Satellite. Although not Shuttle contamination, this effect, from a very clean spacecraft, highlights the importance of payloads as a source of contamination in the Shuttle cargo bay.

The joint ad hoc session of the Surface Interactions, Particulate Environment and Molecular Contamination Environment Subpanels in telecommunication with Marsha Torr (Utah State) should be noted here. Dr. Torr kindly discussed her recent data from the Imaging Spectrometric Observatory, "A Preliminary Spectroscopic Assessment of the Spacelab 1/Shuttle Optical Environment.

There were no presentations on completed work in analysis and modeling. However, Bareiss reported on the planned Shuttle Payload Bay Cleanliness Ground Test Program. This program is concerned primarily with particulates, but should provide important data on the effect on non-volatile residues of candidate cargo bay cleaning procedures.

Barengoltz described the Interim Operational Contamination Monitor (IOCM). The planned flights of this instrument will obtain molecular contaminant deposition data at multiple locations in the Shuttle cargo bay. The temperature controlled quartz crystal microbalance data will augment the body of data of the IECM, for certain selected Shuttle payloads.

Bareiss' presentation on the "Shuttle/Spacelab Contamination Environment and Effects Handbook", given in a plenary session, should also be noted as a planned future activity in this area.

The future needs in the STS molecular contamination environment were reviewed appropriately by Lemke in his talk on the German Infrared Laboratory (GIRL). This cooled infrared telescope is sensitive to gas (and particles) in the field of view and the deposition of molecular contaminants on the cooled surfaces. Lemke noted the quantitative requirements which can be established by the goal that the instrument be limited, not by contamination, but rather by the natural background and the system noise equivalent power. For GIRL and many other instruments, the time history of the environment is necessary to plan the protective shuttering opening. More specific to GIRL and infrared detection systems, the temperature and the short-term stability in concentration of the gas background are needed. A stable concentration would permit background correction by means of differential measurements.

The future needs in the STS molecular contamination are more data in a data base useful to both experimentalists and to contamination technologists. Routine data from relatively simple engineering instruments are necessary to compile a statistical base for the prediction of later flights. The hard-to-obtain data required for sensitive scientific instruments should be obtained by their predecessors. This means that the experimentalists must be encouraged to extract environmental and effects data as an additional objective, by means of appropriate support. Finally, further development in the modeling of the Shuttle with realistic payloads is needed.

The meetings provided a useful dialogue on the STS molecular contamination environment. The participation of STS scientific users was especially helpful. The extraction of molecular contamination data from scientific instruments, as discussed at the meetings, represents a significant potential improvement in our knowledge of this environment.

The research described in this paper was carried out, in part, by the Jet Propulsion Laboratory, California Institute of Technology, under contract with the National Aeronautics and Space Administration.

MICROBIAL AND TOXIC CONTAMINANTS

Bonnie Dalton
Ames Research Center
National Aeronautics and Space Administration

The panel met throughout the week and shared issues in our discipline interests pertaining to microbiological and toxic contamination. Our primary goal was to establish a set of data responsive to the user community needs and to provide those data in a format suitable for inclusion into the electronic data base.

The primary concern, with respect to contamination, is the crew members. For each of us, the term crew member has a relatively different meaning. To personnel from Johnson Space Center (JSC), a crew member is thought of in the classical sense, i.e., those individuals manning the Orbiter and performing the experiments aboard the Shuttle Transportation System (STS). To the users and experiment developers, crew members include: those growing by photosynthesis, all forms walking on four or more legs (the non-human elements), and the humans entering mid-deck or Spacelab areas. Finally, personnel from Kennedy Space Center (KSC) must respond to all groups passing through their facilities prior to launch. Regardless of our primary interests, the issue of microbial and toxic contamination is a relevant issue to us and to the potential proposers of future experiments, since living systems are affected by contaminants both before and during flight.

The means of containing such contamination resides in procedures implemented and equipment used during pre-flight support and flight activities. These will be discussed briefly later.

Because of the man interface, concern over toxic contaminants has existed in all Space flights prior to Shuttle. A potential threat to humans exists in closed environments and during environmental stress conditions at zero gravity. Today, STS users are therefore required to verify that no toxic contaminants exist in their equipment by adherence to established materials lists and by passing their equipment through out-gassing tests at the National Aeronautics and Space Administration (NASA) White Sands or Marshall Space Center test facilities. The criteria for these tests are found in a NASA document-NHB 8060.1B. The arbitration for permissible loads of toxic contaminants has resided in the Structures and Thermal Division at JSC and with Mission management.

Will Ripstein, a former member of that division and consultant to the organization, presented our group with insights on the classification of toxic contaminants and the permissible levels as seen in Table I.

TABLE I

MAJOR TOXIC EFFECTS BY CHEMICAL COMPOUNDS

Toxicity Category	Chemical Examples
Irritants	Aldehydes, ketones, esters
Asphyxiants	Carbon monoxide, carbon dioxide, methane
Central Nervous System Depressants	Aliphatic hydrocarbons, alcohols
Systematic Poisons	Chlorinated hydrocarbons, aromatic hydrocarbons
Carcinogens	Benzene, arsenic compounds

Note that the category of toxicity is based on human physiological responses. The JSC toxicologists, as indicated previously, began their work directing their interests primarily toward protecting the crew members; it was for their benefit that permissible levels were instituted. Those of us now interested in use of the STS for experiments involving plants and animals are also concerned in containment of toxic levels. Table II represents a list of applicable documents addressing the STS toxicology elements.

TABLE II

APPLICABLE DOCUMENTS

Document	Title
NHB 8060.1B	Flammability, Odor, and Offgassing Requirements and Test Procedures for Materials in Environments That Support Combustion
SE-R-0006A	JSC Requirements for Materials and Processes
JSCM 8080	Manned Spacecraft Criteria and Standards
SE-S-0073C	Space Shuttle Fluid Procurement and Use Control
JA-016A	Spacelab Payload Project Office
JSC 17858	Flight Crew Equipment List for STS Operations
JSC 16768	Flight Crew Equipment Landing Dispositioning Manual
JSC/NSI #82-987	Shuttle Preflight Atmospheric Analysis
JSC/NSI #82-985	Shuttle Atmospheric Analysis of Inflight Samples

Table III, excerpted from Appendix D, NHB 8060.1B represents the Maximum Allowable Concentrations (SMACS) for organic constituents. Included are alcohols, aldehydes, aromatic hydrocarbons, esters, ethers, chlorofluorocarbons, fluorocarbons, and a host of others. Inorganic acids, mercaptans, and nitrogenous compounds are also indi-

TABLE III

APPENDIX D - SPACECRAFT MAXIMUM ALLOWABLE CONCENTRATIONS (SMACs)
OF
ATMOSPHERIC CONTAMINANTS IN MANNED SPACECRAFT AND USAGE GUIDELINES

I. Maximum Allowable Concentrations (SMACs) of Contaminants for Missions up to 7 Days*

	Mol. Wt.	SMACs 7-Day ppm	(mg/M^3)
Alcohols			
allyl alcohol (2-propanol)	58.08	0.5	(1)
n-amyl alcohol (1-pentanol)	88.15	35	(126)
isobutyl alcohol (2-methyl-1-propanol)	74.12	40	(121)
n-butyl alcohol (1-butanol)	74.12	40	(121)
sec-butyl alcohol (2-butanol)	74.12	40	(121)
tert-butyl alcohol (2-methyl-2-propanol)	74.12	40	(121)
cyclohexanol	100.2	30	(123)
ethyl alcohol (ethanol)	46.07	50	(94)
ethylene glycol (1,2-ethanediol)	62.07	50	(127)
2-hexyl alcohol (2-hexanol)	102.2	40	(167)
methyl alcohol (methanol)	32.04	40	(52.4)
octyl alcohol (1-octanol)	130.2	40	(213)
phenol	94.11	2	(7.7)
n-propyl alcohol (1-propanol)	60.09	40	(98.3)
isopropyl alcohol (2-propanol)	60.09	40	(98.3)
Aldehydes			
acetaldehyde (ethanal)	44.05	30	(54)
acrolein (propenal)	56.6	0.05	(0.11)
benzaldehyde (benzenecarbonal)	106.1	40	(173)
butyraldehyde (butanal)	72.10	40	(118)
crotonaldehyde (trans-2-butenal)	70.09	0.6	(1.7)
formaldehyde (methanal)	30.03	0.1	(0.12)
furfural (2-furancarbonal)	96.08	2	(7.9)
propionaldehyde (propanal)	58.08	40	(95.0)
valerdalehyde (pentanal)	86.13	30	(106)
Aromatic hydrocarbons			
benzene	78.11	0.1	(0.32)
cumene (isopropylbenzene)	120.2	15	(73.7)
decalin (decahydronaphthalene)	138.2	2	(11)
ethylbenzene	106.2	20	(86.8)
1,2-ethylmethylbenzene (1-ethyl-2-methylbenzene	120.2	5	(25)

* For missions longer than 7 days consult the NASA Toxicologist for SMAC values.

cated. (Because of its length, one sheet only of this table will be represented in the Proceedings.)

Table IV represents data available at this time (August, 1984) on contaminants detected in the Shuttle Orbiter atmospheric samples. Contamination appears to progressively decrease with each mission, especially if we compare STS Missions 1 and 2 with 41C. No concentration levels are indicated in this table. Note that most trace gas contaminants were present in the spacecraft atmosphere at concentrations less than 0.1 milligrams per cubic meter (or less than 0.1 parts per million).

Table V indicates the levels of contaminants found and the SMAC limits for comparison. Freon 113, methane and ethanol remain predominant throughout the missions. This may reflect cleaning solvents used on the equipment. This concludes the toxicology data; all of this material will be in the data base.

Microbiological contamination and containment has become very relevant in the era of the reusable spacecraft. Reusability also provides the ability for accumulative buildup of microorganisms from flight to flight and creates a very real health concern. Data from earlier Space flights have shown that cross contamination does occur among crew members. A comprehensive monitoring program was established to identify potential problems. This has also extended to the non-human elements being introduced into the STS.

Though we are unable to provide a listing of all relative documents in this Summary Session which are applicable to the subject of microbial contamination, we do intend to have them available in the Conference data base. This will include not only the elements of human crew microbiology, but documents pertaining to microbial specifications for animal and plant selection and maintenance, surveillance of ground support facilities and personnel, and surveillance of the late access Pad equipment and personnel.

Dr. Duane Pierson, of the Biomedical Laboratories Branch at JSC presented data obtained from some of the STS flights to date. Table VI indicates the host of bacteria and fungi isolated thus far by surface sampling of the Spacecraft.

TABLE IV

CONTAMINANTS FOUND IN SHUTTLE ORBITER ATMOSPHERIC SAMPLES*

Compound Identity	1	2	3	4	5	6	7	8	9	SL1	41B	41C
Acetic Acid, n-Butyl Ester	X							X				
Acetic Acid, 2-Ethoxyethyl Ester	X											
Acetic Acid, Ethyl Ester						X	X					
C_4-Alkene						X						
Benzaldehyde		X										
Benzene	X	X	X			X	X			X	X	X
Bromotrifluoromethane (Halon 1301)		X	X							X	X	X
1-Butanal (Butyraldehyde)	X	X	X									
1-Butanol	X	X								X	X	
2-Butanone	X	X	X			X	X			X	X	
Butene						X						
n-Butylbenzene	X											
Carbon Dioxide							X					
Carbon Disulfide		X										
Carbon Monoxide	X	X	X			X	X	X	X	X		X	X
Cyclohexane		X								X		
Decamethyltetrasiloxane										X		
Decane		X										
Dichlorodifluoromethane				X								
1,1-Dichloroethane		X										
Dichloromethane	X	X	X			X	X	X		X	X	X	X
1,2-Dimethylbenzene	X	X	X				X			X	X	
1,3-Dimethylbenzene	X	X	X			X	X	X		X	X	
1,4-Dimethylbenzene	X						X			X	X	
1,1-Dimethylethanol	X											
Ethanal (Acetaldehyde)	X	X	X			X	X	X		X	X		X
Ethanol	X	X	X	X	X	X	X			X	X	X	X
Ethylbenzene	X	X	X				X			X	X	
2-Ethylhexanal		X										
Ethyl-2-Propenyl Ether							X					
1-Heptanal	X											
Heptane		X	X									
2-Heptanone	X											
3-Heptanone	X											
Hexamethylcyclopentane		X										
Hexamethylcyclotrisiloxane	X							X		X		
1-Hexanal	X									X		
n-Hexane	X	X										
Hydrogen										X		
Indan		X										
C_7-Ketone						X						
Limonene											X	
Methane	X	X	X	X	X		X	X	X	X	X	X
Methanol	X	X										
2-Methyl-1,3-Butadiene	X											
Methylcyclopentane	X	X				X						
Methylethylcyclopentane			X									

* Most trace gas contaminants were present in the spacecraft atmoshpere at concentrations of less then 0.1 milligrams per cubic meter (or less than 0.1 parts per million).

TABLE IV (cont'd)

CONTAMINANTS FOUND IN SHUTTLE ORBITER ATMOSPHERIC SAMPLES*

Compound Identity	1	2	3	4	5	6	7	8	9	SL1	41B	41C
6-Methyl-2-Heptanone		X										
2-Methylpentane		X										
2-Methyl-1-Propanol		X								X		
2-Methyl-2-Propanol		X	X			X						
4-Methyl-2-Propantanone	X		X									
Napthalene		X										
Nonane		X										
Octamethylcyclotetraisiloxane											X	X
Octamethylcyclotrisiloxane						X						
Octamethylsilicone											X	
Octane		X										
1-Pentanal	X	X	X									
Pentane			X	X					X			
1-Propanal	X	X	X			X		X	X			
1-Propanol						X						
2-Propanol (Isopropyl Alcohol)	X		X		X	X		X	X	X	X	X
2-Propanone (Acetone)	X	X	X	X	X	X	X	X	X	X	X	X
Prophylbenzene	X	X										
Silicone, Molecular Weight 452										X		
Silicone, Molecular Weight 532												X
Silicone, Molecular Weight 236								X				
Siloxane				X								
Toluene	X	X	X			X	X	X	X	X		
1,1,1-Trichloroethane	X	X	X			X	X	X	X	X		
Trichloroethene		X	X				X	X				
Trichlorofluoromethane (Freon 11)	X	X	X				X		X	X		
1,1,2-Trichloro-1,2,2-Trifluoro-ethane (Freon 113)	X	X	X	X	X	X	X	X	X	X	X	X
Trimethylsilanol		X								X		X
C_7-Aliphatic Hydrocarbons		X										
C_8-Aliphatic Hydrocarbons		X	X									
C_9-Aliphatic Hydrocarbons		X	X									
C_{10}-Aliphatic Hydrocarbons	X	X	X									
C_{11}-Aliphatic Hydrocarbons	X	X	X									
C_{12}-Aliphatic Hydrocarbons	X	X	X									
C_{13}-Aliphatic Hydrocarbons		X										
C_{14}-Aliphatic Hydrocarbons		X										
C_8-Alkane			X		X							
C_9-Alkane			X		X							
C_{10}-Alkane			X									
C_{11}-Alkane			X									
C_{12}-Alkane			X									
C_{13}-Alkane							X				X	
C_8-Olefinic Hydrocarbon		X										
C_3-Substituted Benzene	X	X										
C_4-Substituted Benzene	X	X										

TABLE V
ATMOSPHERIC CONTAMINANTS IN EXCESS OF 1MG/M³
MAXIMUM CONCENTRATIONS IN MG/M³ (PPM) FROM AIR SAMPLES TAKEN DURING EACH MISSION

CHEMICAL COMPOUNDS	STS MISSIONS												SMAC LIMITS
	1	2	3	4	5	6	7	8	9	SL-1	41-B	41-C	
Bromotrifluoromethane (Halon 1301)	5,697 (0.749)	16,292 (2,674)	2,318 (0.383)	10,214 (2.332)		17,281 (2.254)			5,045 (0.658)	77,331 (12.778)		8,060 (1.323)	608.8 (100)
1,1,2-Trichloro-1,2,2-Trifluoroethane (Freon 113)		13,091 (1.707)	2,124 (0.277)							4,316 (0.658)			383 (50)
Trichlorofluoromethane (Freon 11)			1,709 (0.304)										561.8 (100)
Methane	18,384 (28.10)		4,933 (7.54)	89,351 (135.54)	75,108 (114.774)		15,760 (24.089)	19,044 (29.108)	66,781 (102.074)	68,041 (104.280)	83.56 (102.2)	52,931 (89.904)	171 (2700)
Carbon Monoxide	1,019 (0.890)		2,610 (2.28)		1,837 (1,921)	1,773 (1.549)							28.5 (25)
Ethanol		4,724 (2.626)	2,272 (1.208)			3,854 (2.044)	1,916 (1.019)		14,940 (7,943)	10,758 (5,719)	3,090 (1.643)	2,917 (1.551)	94 (50)
2-Propanol						1,287 (0.525)			8,784 (3.580)	34,580 (14.095)			98.3 (40)
2-Methyl-2-Propanol		2,247 (0.742)											121. (40) 75.3 (20)
Toluene		63,877 (16.960)											75.3 (20)
Butane				1,563 (0.683)	1,322 (0.351)								458 (200)
2-Butanone (Methylethyl Ketone)										1,256 (0.426)			59 (20)
Octamethylsiloxane										1,125 (0.093)			114 (12.5)
2-Methyl Pentane		2,625 (0.746)											360 (100)
Hexane		1,362 (0.387)											176 (50)
Methylcyclopentane		1,091 (0.553)											172.3 (60)
1,3-Dimethylbenzene	1,383 (0.319)												86.8 (20)

TABLE VI

MICROORGANISMS ISOLATED FROM
SPACECRAFT SURFACE SAMPLES

BACTERIA	FUNGI
*Acinetobacter calcoaceticus	Alternaria species
*Acinetobacter calcoaceticus var. lwoffi	*Asperigillus aculeatus
Bacillus species	*Asperigillus amstelodami
Corynebacterium species	*Asperigillus caespitosus
Enterobacter agglomerans	*Asperigillus candidus
*Enterobacter cloacae	*Asperigillus chevalieri
Escherichia coli	*Asperigillus flavus var columnaris
Micrococcus luteus	*Asperigillus glaucus group
Micrococcus species	*Asperigillus janus
*Pseudomonas fluorescens	*Asperigillus niger
*Pseudomanas putida	*Asperigillus niger group
Pseudomonas species	*Asperigillus penicilloides
*Staphylococcus aureus	*Asperigillus phoenicis
	*Asperigillus repens
	*Asperigillus restrictus
	*Asperigillus ruber
	*Asperigillus sydowi
	*Asperigillus terreus
	*Asperigillus ustus
	*Asperigillus versicolor
	*Candida krusei
	*Candida parapsilosis
	*Curvularia lunata
	*Curvularia senegalensis
	*Drechslera hawaiiensis
	*Geotrichum candidum
	*Penicillium species
	*Rhodotorula rubra
	*Scopulariopsis brevicaulis
	Scopulariopsis species
	*Trichosporon beigelii

*Potential Pathogens

Though no crew member has become ill because of these particular organisms, it is to be noted that the majority are classified as potential pathogens. Table VII illustrates a mission by mission quantitation of bacteria and fungi.

TABLE VII

SPACECRAFT SURFACE SAMPLES

QUANTITATION*

SHUTTLE MISSION	BACTERIAL			FUNGAL		
	F-30	F-2	L+0	F-30	F-2	L+0
1	2	135	538	<1	12	7
2	4	1895	74	4	<1	<1
3	42	36	3004	13	4	41
4	58	199	237	19	18	18
5	60	106	2057	11	8	6
6	136	222	63	5	48	95
7	127	3574	NS	8	9	NS
8	24,908	172	2657	29	30	9
9	171	63	1989	3	<1	124
11	108	184	1461	3	4	7
13	65	66	NS	1	4	NS

*CFU/25 cm^2

Though numbers appear to increase from 30 days prior to launch to L + 0, Mission 8 exhibited a reversal characteristic; perhaps this was due to the activities aboard that particular Mission. Figures 1 and 2 indicate airborne bacteria on the mid-deck and flight deck, respectively.

Figures 3 and 4 indicate fungi located in the same areas during Missions 1 through 11.

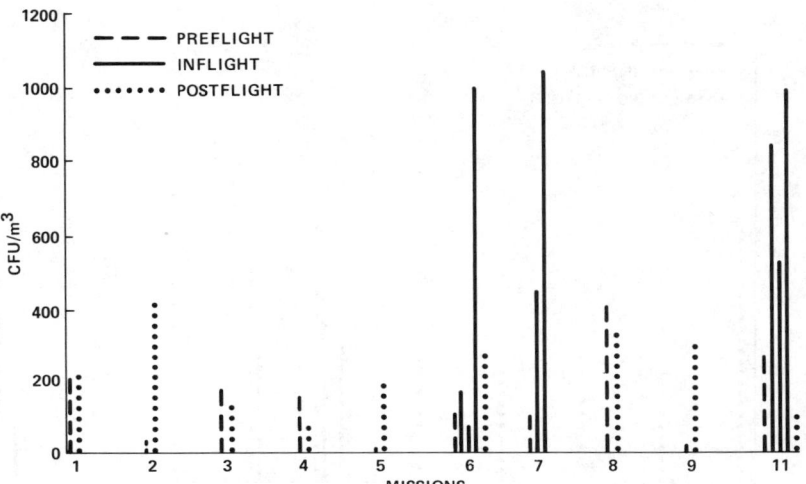

Fig. 1. Airborne Bacteria--Mid Deck

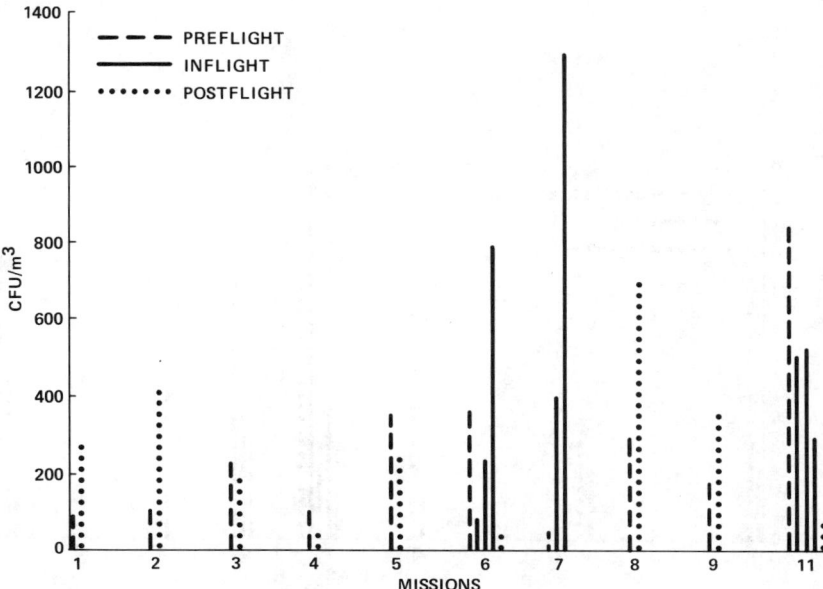

Fig. 2. Airborne Bacteria--Flight Deck

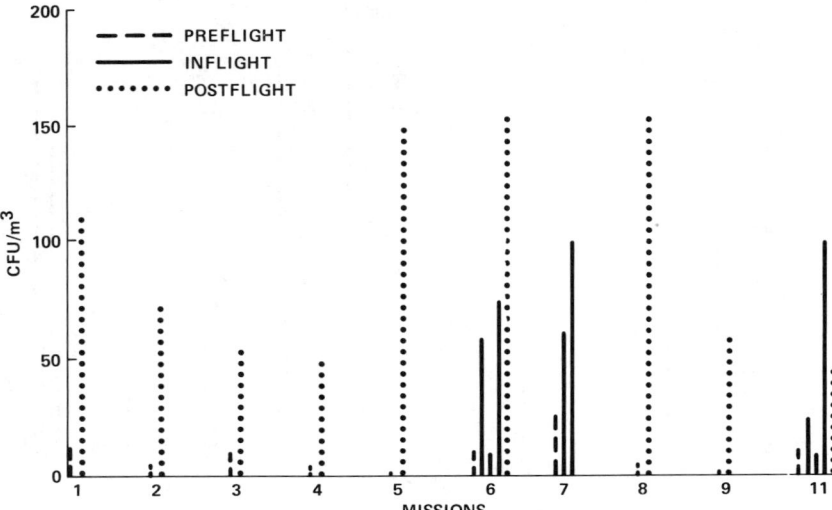

Fig. 3. Airborne Fungi--Mid Deck

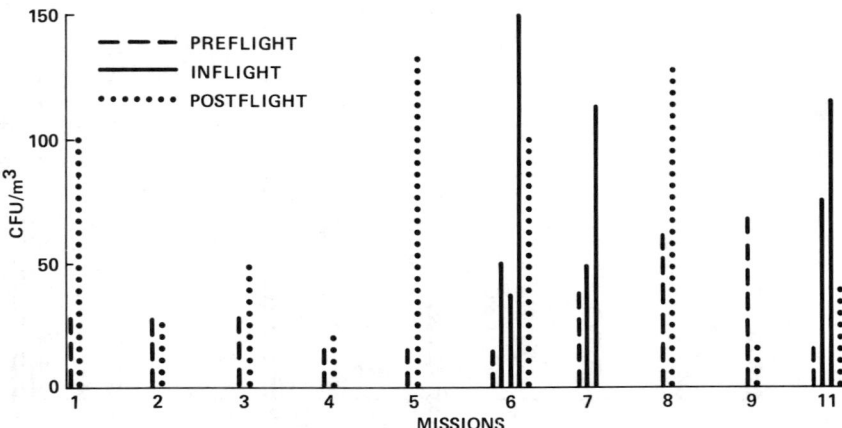

Fig. 4. Airborne Fungi--Flight Deck

Crew members are very thoroughly screened to determine their microbial flora prior to, during, and post flight. The following viral agents were reported among STS-5 through -8 crew members: Hepatitis A and B, rubella, rubeola, mumps, varicella zoster, cytomegaly, Epstein-Barr, adenoviruses, enteroviruses, herpes simplex, and seven types of respiratory virus. Do not be alarmed! Such gardens of microflora are typical of the general population. Tables VIII and IX illustrate micro-organisms isolated from specific crew members.

TABLE VIII

MEDICALLY IMPORTANT MICROORGANISMS ISOLATED FROM STS-5 CREWMEN

CREWMAN		F-10	F-2	L+0
			SAMPLE PERIOD	
A	Nose	Staphylococcus epidermidis P.O. Enterobacter aerogenes	S. epidermidis P.O. S. aureus 96 E. cloacae E. aerogenes	S. aureus P.O.(NT) E. aerogenes
	Throat	β-hemolytic Streptococcus P.O. S. epidermidis E. aerogenes	β-hemolytic Streptococcus P.O. E. cloacae Candida parapsilosis Rhodotorula rubra	β-hemolytic Streptococcus P.O. E. aerogenes
	Ear	Micrococcus sp. P.O.	S. epidermidis P.O. C. parapsilosis	S. epidermidis P.O.
B	Nose	S. epideridis P.O. β-hemolytic Streptococcus S. aureus 53 E. aerogenes	Corynebacterium sp. P.O. S. aureus 53	β-hemolytic Sreptococcus P.O. S. aureus 53 E. aerogenes
	Throat	β-hemolytic Streptococcus P.O. E. cloacae C. albicans C. parapsilosis	β-hemolytic Streptococcus P.O. E. cloacae E. aerogenes	β-hemolytic Streptococcus P.O. C. albicans
	Ear	S. epideridis P.O. Aspergillus ustus	S. epidermidis P.O.	Micrococcus sp. P.O. S. aureus 53

TABLE IX

MEDICALLY IMPORTANT MICROORGANISMS ISOLATED FROM STS-6 CREWMEN

CREWMAN		F-10	F-3	L+0
A	Nose	Staphylococcus aureus P.O. 94/96	Micrococcus sp. P.O. Staphylococcus aureus 94/96 Drechslera hawaiiensis	S. epideridis P.O. Staphylococcus aureus 94/96 Serratia marcescens
	Throat	Streptococcus, alpha P.O. Candida albicans	Streptococcus, alpha P.O. Candida albicans	Streptococcus, alpha P.O.
	Ear	S. epidermidis P.O.	Micrococcus sp. P.O.	S. epidermidis P.O. Serratia marcescens
B	Nose	S. epidermidis P.O.	S. epidermidis P.O. Staphylococcus aureus Nontypable Escherichia coli	Staphylococcus aureus Nontypable P.O.
	Throat	S. epidermidis P.O.	Streptococcus, alpha P.O. Rhodotorula rubra Candida parapsilosis	
	Ear	S. epidermidis P.O. Escherichia coli Staphylococcus aureus Nontypable	S. epidermidis P.O. Staphylococcus aureus Nontypable Aspergillus niger	Serratia marcescens P.O. Rhodotorula rubra

As stated, these are common; under stress of microgravity, some may prove harmful and thus serve as an explanation for any potential crew health problems. This remains an unknown. Though these data reflect only STS-5 and -6 isolates, the vigilance is constant for all missions--and, in the same manner, wil be maintained on those "other" crew members referenced earlier.

Over two years have been spent establishing the criteria for non-human crew microbiology. Table X indicates those organisms which must not be present in rats.

TABLE X

SPF CRITERIA FOR RATS

MICROORGANISM	CULTURE SITE/MATERIAL OR IDENTIFICATION TEST
BACTERIA:	
Streptobacillus moniliformis	Oral
Spirillum minus	Oral
Streptococcus pneumoniae	Oral, Nasal
*Streptoccocus, beta hemolytic	Oral, Nasal
Bacillus piliformis	Liver (Invoke with Cortisone)
Corynebacterium kutscheri	Fecal, Oral
Salmonella sp.	Fecal
Pasteurella pneumotropica	Oral, Nasal
Leptospira sp.	Urine
Klebsiella pneumoniae	Fecal, Oral, Nasal
Klebsiella oxytoca	Fecal, Oral, Nasal
Campylobacter sp.	Fecal
Pseudomonas aeruginosa	Oral, Fecal
MYCOPLSMAS:	
Mycoplsma pulmonis	Blood, Nasal Aspirant
Mycoplasma arthritidis	Blood, Nasal Aspirant
VIRUSES:	
Lymphocytic choriomeningitis virus	Blood (Serology)
Rat parvoviruses	Blood (Serology)
Rat coronavirus	Blood (Serology)
Sialodacryadenitis virus	Blood (Serology)
Sendai virus	Blood (Serology)
FUNGI:	
All Dermatophytes	Skin
Ecto parasites	Skin, Hair
Endo parasites	Feces, Caecal contents

* — (Group A)

Table XI indicates those which must be absent from squirrel monkeys.

TABLE XI

SPF CRITERIA FOR SQUIRREL MONKEYS

MICROORGANISM	CULTURE SITE/MATERIAL OR IDENTIFICATION TEST
BACTERIA:	
Shigella sp.	Fecal
Salmonella sp.	Fecal
Streptococcus pneumoniae	Oral, Fecal
Klebsiella pneumoniae	Oral, Fecal
Mycobacterium tuberculosis	Skin Test, X-Ray
Pasteurella multocida	Nasal, Fecal
Campylobacter sp.	Fecal
Leptospira sp.	Urine
Streptococcus, Beta Hemolytic (Group A)	Oral, Nasal
VIRUSES:	
Lymphocytic choriomeningitis virus	Blood (Serology)
Herpes tamarinus	Blood (Serology)
ENDOPARASITES:	
Trichomonas	Oral
Acanthocephalans	Feces
Strongyloides	Feces
Entamoeba histolytica	Feces
Hemoprotozoa	Blood
FUNGI:	
All Dermatophytes	Skin

This listing is referenced as the specific pathogen free (SPF) list. Extensive testing is done prior to obtaining animals and during their maintenance preflight to assure that all animals are SPF. Testing is not only conducted on the animals, but on all associated personnel at the vendor site, at an investigator's laboratories, at Ames Research Center (the prime developer on non-human experiments) and at KSC, the final stop before launch. The data base will reference the SPF listing and documents establishing procedures for SPF activities. Issues of animal selection and maintenance are important to the user since the question of supply source may influence the experimenter's pre-flight baseline tests. Animals obtained from different breeders have inherently different characteristics although breeders adhere to the SPF criteria and can provide the same species. In other words, a laboratory rat (rattus norvegicus) from Simonson Laboratories may exhibit differing growth rates from a similar rat obtained from Camm or Taconic Laboratories. Not only the microbiological, but also the physiological subtleties become relevant to the experiment proposer /user in establishing his pre-flight data base. Such differences may negate his flight data should he unknowingly obtain animals from a laboratory different from those established by NASA.

The last item in the data base will be a listing of equipment available for containment of non-human species and activities with those species. Units have been constructed to protect humans from animal microbiology and, in turn, to protect plants and animals from the crew microbiology. The animal hotels are called Research Animal Holding Facilities (RAHFs) and may be used for rats (figure 5) and for squirrel monkeys (figure 6).

A unit is under development for maintenance of frogs in the Spacelab environment. Other items available for working with the specimens under containment conditions include a General Purpose Work Bench (GPWB) illustrated in figure 7.

The unit provides containment of fixatives and solvents and also allows interaction of animal specimens and tissues with a microscope designed for zero gravity use. We feel that knowledge of such existing equipment provides the potential user with further guidelines in determining the feasibility of his proposed experiment.

In totality, we assesed toxic and microbial contaminant both in flight and in ground operations.

Toxic Contaminants Panel met August 16 to address current microbial contamination issues relative to Spacelab 3 (SL-3). Issues specifically addressed were: flight animal procurement and adherence to specific pathogen free (SPF) criteria, animal facility maintenance

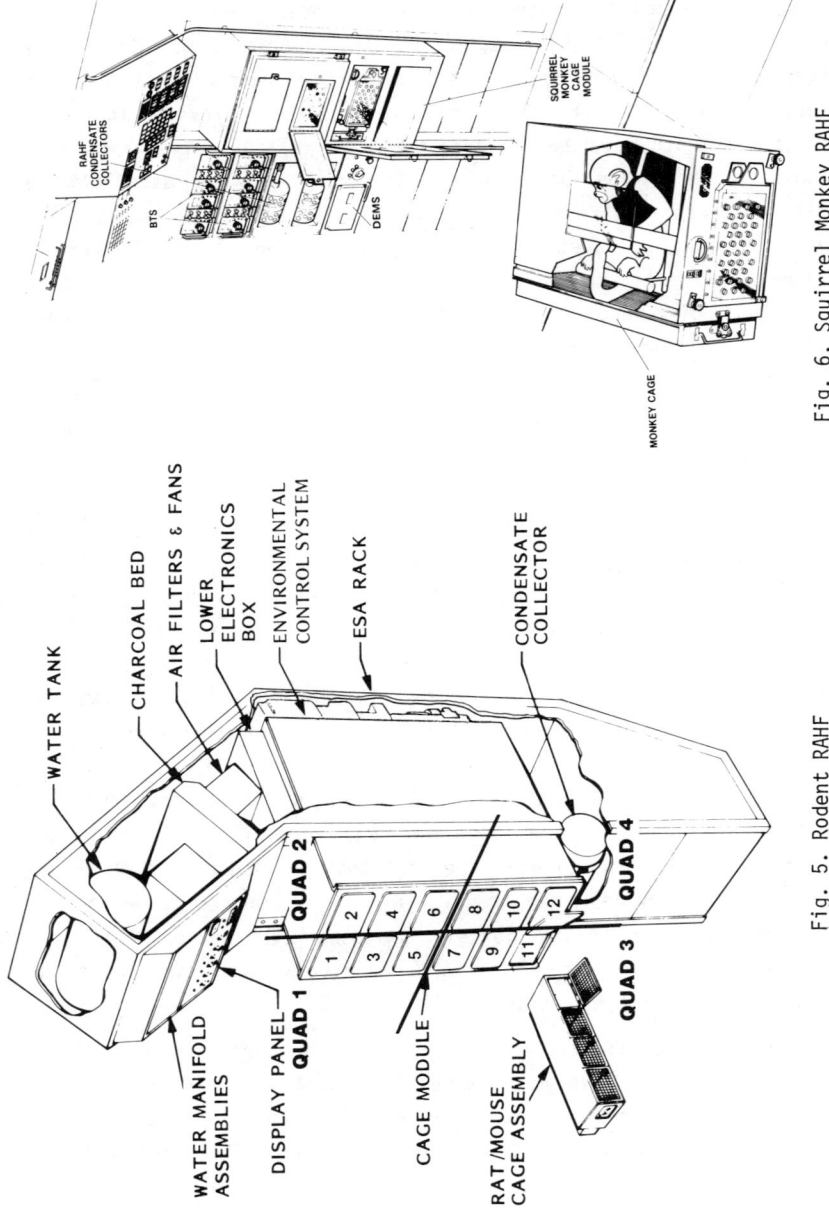

Fig. 6. Squirrel Monkey RAHF

Fig. 5. Rodent RAHF

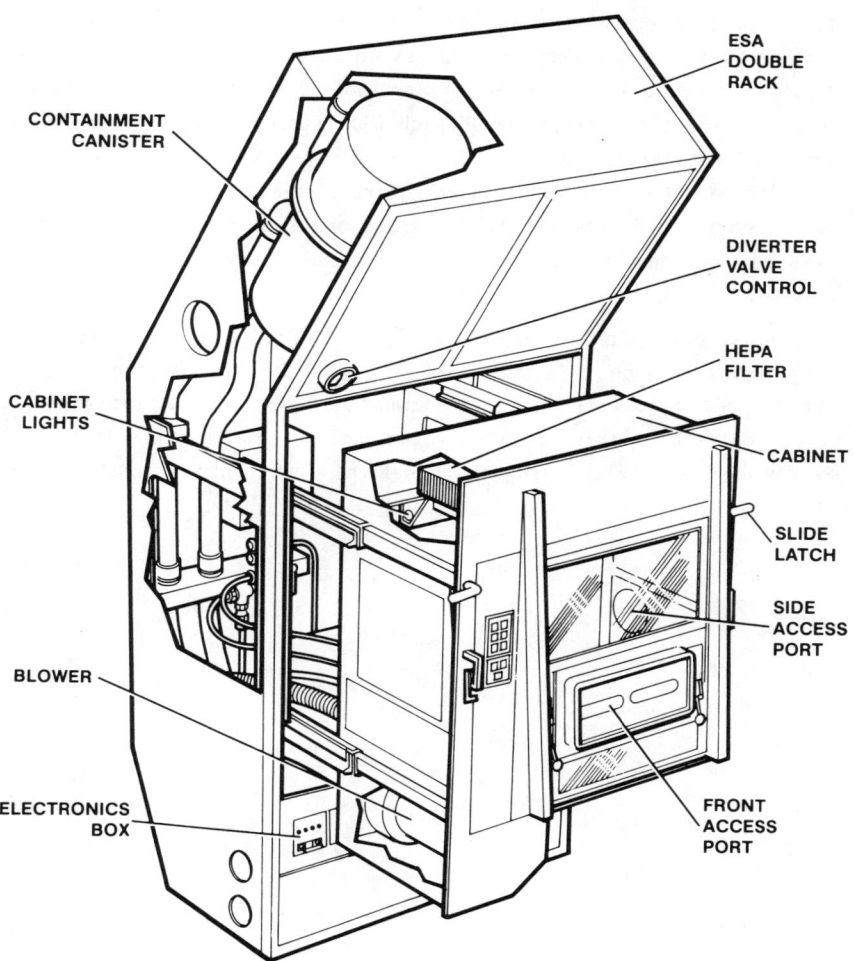

Fig. 7. General Purpose Workstation

during support operations at the Hangar L facility at Kennedy Space Center (KSC), and personnel procedures to assure SPF conditions during mission operations. The SL-3 Microbiology Plan is the document evolving from these discussions which will serve as primary reference during all activities at the KSC. This document will be referenced in the data base to alert future users of requirements for Spacelabs. As data are received from the SL-3 mission and support operations, they will be fed into the data base. Proposed entry is 60 days post mission.

Human crew microbiology for flights following STS-6 along with toxic contamination data following 41C will be reviewed for availability and processed into the data file over the coming months. Target data for entries up to SL-3 is March, 1985. Mission data will be reviewed yearly each August to determine new entry into the data file.

CHAPTER 3

FUTURE DIRECTIONS

LOADS AND LOW FREQUENCY DYNAMICS

J. A. Garba
Jet Propulsion Laboratory
California Institute of Technology

The panel decided that the subjects to be covered in this summary session would be: (1) the approach to the design data base, (2) the development tasks required to achieve such a data base, and (3) a summary of the panel findings and recommendations.

The panel decided that the emphasis should be on the design data base. We recognized that an effective design data base cannot be implemented without a flight data base. Steps towards such a data base are outlined below. The panel recommends that the Spacelab Payload Accommodation Handbook (SPAH) be used as a guideline for the data base. Although these data are limited to one particular type of carrier, it is a complete and thorough data base. One of its shortcomings is the lack of verification by flight data. The panel strongly recommended that the SPAH data be verified and updated using Spacelab #1 and Spacelab #2 flight data as soon as possible.

Approach to Design Data Base for Loads and Low Frequency Dynamics

o Use Spacelab Payload Accommodation Handbook (SPAH) as a guideline
o Develop similar data for Carriers

- MPESS
- APC
- Orbiter Middeck, Etc.

o Update Data Base Periodically

o Develop Guidelines for use of Data Base

Furthermore, the panel members recommended that the availability of similar data for other carriers be investigated. If such data are available, their applicability and accuracy should be checked and then these data should be incorporated in the design data base. All data should be updated periodically using flight data. The panel recognized that this will take resources. We recommend that guidelines for the use of the data base be developed. These are required to avoid misinterpretation and misuse of the data base.

Next, let me turn to the development tasks which are required to implement a design data base (see list below). The panel emphasized that, in the loads and low frequency dynamics discipline, the approach is to develop methodology for the design of payloads and then verify the methodology using flight data. To accomplish this, flight data are required - specifically Orbiter and payload response data for the verification of the design data base. We agreed that the flight data must be analyzed, not just reduced, to verify the methodology and to update the design data base. The pacing tasks here are the Spacelab #1 data analysis and the implementation of future flight data acquisition. The Spacelab data should be thoroughly analyzed and compared to the design data contained in the SPAH. The SPAH should be updated using Spacelab #1 and Spacelab #2 data. A plan for future flight data acquisition should include the analysis of these data, and such analyses should be directed towards enhancing the data base.

Development Tasks for Loads and Low Frequency Dynamics

- Flight Data Acquisition
    - Orbiter Response Data
    - Payload Response Data
- Flight Data Analysis
    - Verification of Methodology
    - Update Design Data Base
- Pacing Tasks
    - Spacelab 1--Validation and Analysis of Flight Data
    - Implementation of Flight Data Acquisition

To summarize, the panel concluded that the focus of future activities should be on the design data base. The panel realized that this will require a substantial effort since there are some technology issues. The current flight data deficiencies must be overcome by acquiring flight data and then updating the design data. Finally, the design data base must be kept current as new data become available.

Summary, Loads and Low Frequency Dynamics

- o Focus is on the Design Data Base
- o Data Base is feasible but requires substantial effort
- o Current flight data deficiencies must be overcome
- o Flight Data Base is required as soon as possible
- o Data Base must be kept current

I would like to again take this opportunity to acknowledge the cooperation of the panel members.

VIBROACOUSTICS

Don Wong
Aerospace Corporation

Measurements obtained to date have been extremely valuable in defining the vibroacoustic environments of the cargo bay. However, there are still significant uncertainties in the definition of these environments. Future data requirements better define the vibroacoustic environments as follows: a) The acoustic environment in the forward one-third of the cargo bay has not been measured at all except on the cargo bay forward bulkhead. During liftoff, the cargo bay acoustic data indicate an aft to forward gradient, with the forward bulkhead measurement being one of the highest recorded; b) The cargo bay doors are the dominant sound transmission path into the cargo bay, and the acoustic environment near the doors has not been defined; c) The acoustic environment during transonic flight, which is characterized as a high level discrete noise emanating from the pressure equalization vents, is not adequately defined; d) Structure-borne vibration transmitted into payloads from the Orbiter is not adequately characterized; e) The effects of a large payload configuration on the cargo bay acoustic environment are not well understood; and f) the effects of the difference between the KSC and VLS pad and the Orbiter configuration need to be defined.

The basic element required for the resolution of many of the Shuttle payload bay vibroacoustic environment technology issues is the development of a vibroacoustic data base. This is to be acquired from STS flight data and payload ground test data. The data base and prediction methodologies utilizing the data will need to be maintained and updated for optimal use. Funding is required from both NASA and Air Force to acquire the necessary data, and to analyze and organize the data into a useable format. It is recommended that the required vibroacoustic data base management system utilize the VAPEPS (Vibroacoustic Payload Environment Prediction System) program. A center needs to be established to manage VAPEPS and the vibroacoustic data base, and to optimize utilization of the computer code. The center would chair a panel to review all future ground and flight data, confirm their validity, input data and maintain the data base, provide consultation to users, and inform users of new changes to the code and data base. The data would be accessible by outside terminals. The long term plan is to expand the responsibilities of the data base management center to include coordination of all Shuttle payload bay flight environmental measurements.

ELECTROMAGNETIC INTERFERENCE

William Cutler
Aerospace Corporation

The Electromagnetic Interference Subpanel has concluded that adequate information is available in the Shuttle Orbiter/Cargo Standard Interface Document JSC 07700, Volume XIV, Attachment I (ICD 2-19001) Sections 7.0 and 10.7 to provide a complete data base for Engineering Design. A significant discrepancy exists between this data base and that required to provide electromagnetic environment levels for design of scientific experiments. Requirements for such a scientific data base must be identified and funded by NASA Headquarters in order to completely resolve the issues involved. This subpanel formulated the following future plans and conclusions:

    o Evaluation of Available Cargo Bay EMI Data Base

        * Engineering EMI Data

            - complete definition
            - use ICD 2-19001 in data base

        * Scientific EMI Data

            -incomplete definition

    o Future Data to be Obtained

        * Engineering Data

            - S-Band Radiation
            - Ku-Band Radiation

        * Scientific EMI Data

            - Radiated Levels
            - Magnetic Field Levels

o Conclusions

   * Orbiter EMI Environment Sufficiently
     Identified for Engineering Use

     - No additional data required

   * Orbiter EMI Environment not Sufficiently
     Defined to be Considered as a National
     Resource for the Scientific Community

   * NASA Headquarters Develop Requirements for
     Eliminating Deficiency

     - Provide funding

PARTICULATE ENVIRONMENT

J. Barengoltz
Jet Propulsion Laboratory
California Institute of Technology

A complete prediction of the Shuttle particulate environment for a specific payload would require ground facility data, cargo bay data, data on other cargo elements, and flight data. For controlled conditions, the body of data must correspond to the expected conditions for the specific payload. Realistically, the body of data must also have adequate completeness to statistically bound the uncontrolled variations. Time-dependent data are also required, because the various sources of particulates and transport mechanisms change during payload processing and flight. Finally, validated predictive models are the only way to supplement gaps in the data in the time or spatial domain, to interpret the data in a standardized format, and to extrapolate to conditions previously unmeasured.

The status of knowledge of the Shuttle particulate environment is that there are good data on the ground facilities and limited, but valuable, flight data. For the cargo bay and cargo elements, quantitative data to characterize operationally defined cleanliness standards are needed. Certain routine (i.e., frequent) flight measurements must be accomplished to obtain adequate statistics. Existing data are also deficient for the launch period and for small particles in the on-orbit period. The extraction of particulate environment data from the observations of science instruments with other objectives should be financially supported.

Modeling is in a rudimentary state, especially with regard to particulate sources. Existing models must be tested with available and new data. Better models should be developed. The models and the data should be used to derive the fundamental particulate environment in standard formats for intercomparison among different measurement techniques.

Activities already planned for the future can accomplish much of what needs to be done. Important efforts include the quantitation of the "visibly clean" levels of cargo bay cleanliness, the additional flight of the Induced Environment Contamination Monitor (IECM), and the planned flights of the Interim Operational Contamination Monitor (IOCM). Necessary data would be obtained by the Particle Analysis Camera System (PACS) on-orbit and by at least one of the concepts for collecting particles released during launch. Especially noteworthy are the plans to analyze the data to be obtained by the infrared telescope on Spacelab 2, and for particulates in the field of view. However, the funded tasks must continue to be funded, and support must be provided to the proposed activities.

The research described in this paper was carried out, in part, by the Jet Propulsion Laboratory, California Institute of Technology, under contract with the National Aeronautics and Space Administration.

MOLECULAR CONTAMINATION ENVIRONMENT

J. Barengoltz
Jet Propulsion Laboratory
California Institute of Technology

A Shuttle-launched payload requires knowledge of the molecular contamination environment for the assessment of effects on its performance. The description of the environment must be sufficient to predict both the deposition on critical payload surfaces and the column density in the field-of-view of instruments. Ideally, contaminant species data are also available. To some extent species data can be provided by direct effects from appropriate analogous measurements. Effects data also serve to alert new Shuttle users to consider contamination sensitivity assessments.

Time-dependent deposition rates are needed from ground operations through the end of the mission. Both the sources and the deposition depend strongly on temperature and background pressure. Column densities are only relevant to the observational period on-orbit but are intimately related to the previous deposition history.

A completely general specification of the Shuttle molecular contamination on the basis of data alone is extremely difficult. The variability introduced by other payloads, an important source of molecular contamination in a mixed cargo, renders it impossible. Payloads must only meet, at most, outgassing requirements; none must identify their outgassing species. A major dependence on modeling is inescapable.

The status of our knowledge of the contamination environment may be summarized as good deposition data in ground facilities, good data on-orbit for molecular deposition, volume densities and species, and limited effects data. The time dependence of deposition and the identification of contaminants in the ground facilities are open questions. The flight data are not complete enough to span the variability due to the payloads themselves, solar exposure, plasma conditions, and Shuttle thruster firings. Column densities have not been measured directly, but are inferred from volume densities. With one exception, all available effects data suffer from the uncertainty inherent in a retrospective study and a lack of sensitivity. The Optical Effects Module of the Induced Environment Contamination Monitor (IECM) may also be incapable of as sensitive a response as certain scientific instruments that may fly on Shuttle.

Operational contamination models are available. The Shuttle/ Payload Contamination Evaluation (SPACE) program and the Contamination Analysis Program (CAP) are both capable of flight analysis. Both could be improved, however, by better source data. Models of effects are particularly deficient.

Activities planned for the future will only accomplish part of what remains to be done. The change under consideration at the KSC ground facilities for the non-volatile residue (NVR) measurement of witness plates should be implemented and the NVR analyzed for species. This analysis should be performed at all ground facilities. The deposition time history, which is not under consideration, should be investigated. Additional necessary flight data will be obtained by the final flight of the IECM and the flights of the Interim Operational Contamination Monitor (IOCM). However, at this writing, the crucial measurements may only be made by certain scientific instruments. This is an unfortunate circumstance, as the information may be obtained indirectly in the form of data and performance degradation.

The research described in this paper was carried out, in part, by the Jet Propulsion Laboratory, California Institute of Technology, under contract with the National Aeronautics and Space Administration.

MICROBIAL AND TOXIC CONTAMINANTS

Bonnie Dalton
Ames Research Center
National Aeronautics and Space Administration

Members of the Microbial and Toxic Contaminants Panel met August 16 to address current microbial contamination issues relative to Spacelab 3 (SL-3). Issues specifically addressed were: Flight animal procurement and adherence to specific pathogen free (SPF) criteria, animal facility maintenance during support operations at the Hangar L facility at Kennedy Space Center (KSC), and personnel procedures to assure SPF conditions during mission operations. The SL-3 Microbiology Plan is the document evolving from these discussions which will serve as primary reference during all activities at the KSC. This document will be referenced in the data base to alert future users of requirements for Spacelabs. As data are received from the SL-3 mission and support operations, this will be fed into the data base. Proposed entry is 60 days post mission.

Human crew microbiology for flights following STS-6 along with toxic contamination data following 41C will be reviewed for availability and processed into the data file over the coming months. Target data for entries up to SL-3 is March 1985. Mission data will be reviewed yearly each August to determine new entry into the data file.

SUMMARY COMMENTS BY INEP CHAIRMAN

Gerald W. Sharp
Eyring Research Center Institute
Provo, Utah

I would like to begin by expressing my personal thanks first to those subpanel chairmen (and their stand-ins) who have worked so hard to make this Workshop so successful, and second to all of you who have so actively participated in the work of these past four days. I believe that we have accomplished what we intended to accomplish when we planned and organized this Workshop. All of you can take credit for that. We are well aware that our activity here has not been carried out under the most ideal circumstances, but you have all been good sports and made the best of it.

It is really important to understand that this Workshop is not an end in itself but we have tried to use it as a means to an end. There is yet much to be done before we get to the end product of the activity which we began this week here at New England College. When we talk of work yet to be done, one of the first things we think about is money. Where is the money going to come from to do the work? To answer that, we have to understand that money may be needed for us to carry out the mechanics of our assignment as subpanel chairmen and, what seems to concern some of you even more, money is needed to fund measurements and data analysis to have what you consider to be an adequate data base for the future experimenters to utilize. If funds are needed for any of you subpanel chairmen to carry out your assignment, then you should appeal directly to Dr. Michael Lauriente at the Goddard Space Flight Center. He receives funding from Mike Sander of NASA's Office of Space Science and Applications for the administration of this activity.

I shall now address the matter of funding for measurements and/or data analysis activities. For some time now, the science and applications funding organizations and their experiments have expected the Space Transportation System organization to make measurements and analyze data to characterize the environment of the Space Shuttle. I think it is fair to say that the STS folks feel they have done their job as far as this is concerned. We may disagree with them, but that is not likely to change things. So, if there are still significant measurements that need to be made in order to satisfy some or all of the user community, it is probably going to fall on the shoulders of the Office of Science and Applications at NASA Headquarters or some other Air Force office to support such measurements. In particular, if some experiment requires knowledge of some environmental parameter for its success, then the experimenter had better find a way to include that measurement in with his experiment.

Let's talk for a minute about organization. Each of the Natural and Induced Environment Subpanels has been constituted based both on qualification of the members as well as their availability to attend this Workshop. As we move in to the next phase of our activity, I think it is important for each subpanel chairman to take a good hard look at his subpanel membership. You may need to make some changes and/or additions. I think that you should try to have subpanel members who can be key contributors to the data that will go into the data base. No important segment of data should be overlooked. It will make your job of collecting and reviewing data for suitability much easier if you have arranged for these key people to participate. Feel free to make these needed changes and, if you need some help from me in getting people committed, just let me know and I will be glad to do what I can.

To date we have focused our attention on getting a successful Workshop organized and carried out. We have done this now and it is time to focus on the real end product of this activity, the compilation of a useful Shuttle Environment data base. Please concentrate your time and attention on this particular matter. It is a new experience for most of us and it will take some time and effort to get a product that will do what we would like it to do. Be sure to include appropriate data for the various carriers flown in the Shuttle as well as the data for the Shuttle itself. As you have time and it seems appropriate, you may, as a subpanel, make recommendations for measurements that yet need to be made in order to make the data base more useful to the experimenter.

Again, my thanks to all of you for your willing participation in this Workshop activity, and my special thanks to the Subpanel Chairmen and their substitutes for their superb work these past four days. Go forth and do good work.

SUMMARY OF ESA REACTIONS TO THE WORKSHOP AND CLOSING REMARKS

D. J. Shapland
Directorate of Space Transportation Systems
European Space Agency, Paris

This workshop has proved most valuable for all present and has meant that views could be exchanged between potential users and those experimenters who have already had experiments flown by the Shuttle and who have first-hand knowledge of the environment. There is no doubt that a data bank of information related to the environment of the Space Shuttle and its payloads (of which Spacelab is particularly important) is essential for good experimentation. I believe that a good start has been made in generating and classifying this type of information. It is now our job to see that this initial base is enlarged upon and properly disseminated to future users. To this end I would like to make a few observations that represent the views of all members of ESA contingent at this workshop.

Design Data vs. Information Data

"Design to" data should not be confused with information data. For example, the Spacelab Payload Accomodation Handbook (SPAH) provides the design data for the interfacing of Spacelab experiments and must be followed at all times. The data bank should give information to help in interpreting the results and possibly as an indicator for experiment accomodation.

Experimenter Needs

The specific information required by a potential experimenter must be included in the data bank. Hence, the experimenter himself must state the type of information he needs to ensure meaningful results from his experiment. This implies that the User Subpanels should cover all aspects of Shuttle use.

Data Bank or Face-to-Face

The provision of a data bank should not be substituted for face-to-face contact between experimenters and engineers. The two are complementary and each can provide valuable information.

Get Top People Involved

In view of the effort and money required to fill the gaps in the data bank, the people in charge of the various aspects of STS must be involved.

Data Bank to Include Documentation References

For follow-up actions, the data bank should include good references to the literature and the names of experts in the field.

Seek Means and Actions to Improve Environment

The data bank would provide extremely useful data but does not substitute for an improved environment. Those familiar with the problems should (1) devise experiments to fill the data gaps, and (2) suggest methods (hardware or methodology) for improving the situation.

Where are the Verification Flight Test (VFT) Data?

No VFT data have been presented at this Workshop although the programme was designed to give as much of the information sought after. MSFC should be contacted on this subject and their results fed into the data bank.

Need for Correlation

The experiment data provided must be correlated with the conditions prevailing at the time and carefully screened by experts in the field to ensure relevance to future experiments.

Continued ESA Support

European experimenters need a comprehensive data bank so that their experiments will yield meaningful results. ESA on their behalf will continue to support the Shuttle Environment Plan. At the same time, it is recognized that personnel, time, and money are required to make the system work well.

The provision of data should be a simple process and the experimenter who is new to space research should be given special consideration. A suggested schematic for data flow is as follows:

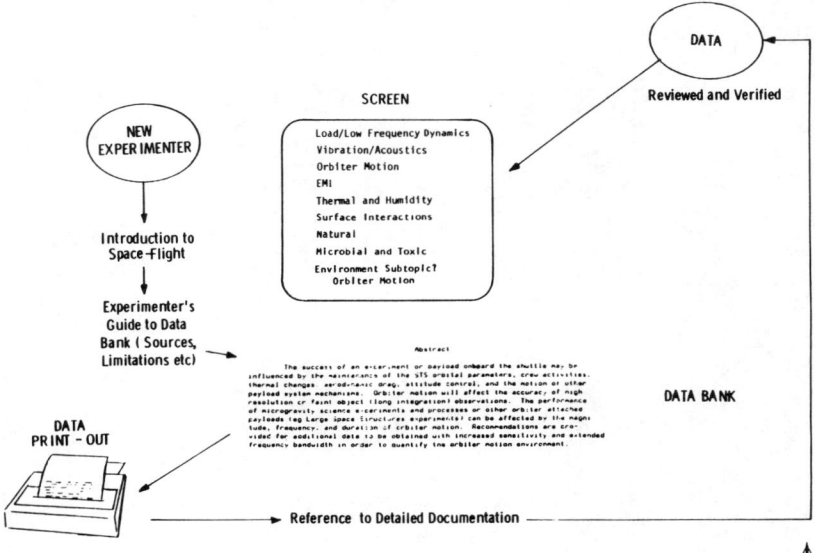

CLOSING REMARKS BY THE GENERAL CHAIRMAN

There is a slight disadvantage to being the last speaker in that most of the important issues have been revealed. Let me say that I echo the summary statements presented by Dai Shapland and the others. Also, I want to express my deep appreciation for the generous support by ESA in providing the international flavor. Their presence enriched the dialogue that took place at this meeting. I look forward to this relationship continuing for another meeting in the future.

The rural atmosphere of this lovely New England campus, devoid of urban distractions, was conducive to the productivity of the meeting. Many times in the planning of this meeting, I was asked "Why five days for a meeting?" From our experience here, we realize that even more time would have been desirable to do justice to the challenging topics at hand.

Gil Moore brought up an important question that I want to address apropos the paper by Lyle Bareiss. Perhaps this issue serves as an example of the need for communications, and reflects the intense pace of technology. When that study was initiated, it was considered appropriate to develop a handbook. This was the first public disclosure of this study and an appropriate place to articulate the subject. It is now NASA's plan to integrate this effort into ENVIRONET. To respond to Gil, he can be reassured that the labors of this meeting will not be in vain and will be integrated with Lyle's results and reviewed by the subpanels. Let me emphasize that this is a completely unified effort.

This Workshop has uncovered considerations not treated in the earlier conference where there was mainly a call for help by the users of the Shuttle to understand the environment in which their instruments are expected to operate. Quantitative data useful for design of later experiments were scarce. The present Workshop made a considerable advance in response to the need for Shuttle environmental information. The status of data was reviewed in the light of subsequent flight experience, and steps were taken to catalog the data into ten main topics and future projections. Microgravity and life sciences were new topics at this meeting.

NASA expects that a similar Workshop will be offered in two years (1986) at which time the Space Station will be a timely topic. This is expected to call for a review of the needs expressed here. A further elaboration of the process of perfecting the comprehensiveness and accessibillity of the Shuttle Environment data could be a principal agenda item for the projected 1986 conference.

Michael Lauriente, General Chairman

APPENDICES

APPENDICES

APPENDIX A

SURFACE INTERACTIONS SUBPANEL: DATA BASE INPUT

Dr. Henry Berry Garrett, Chairman
Jet Propulsion Laboratory (144-218)
4800 Oak Grove Dr.
Pasadena, CA 91109
Telephone: (818) 354-2644, (FTS) 792-2644

INTRODUCTION TO SURFACE INTERACTION DATA BASE

Far from being a benign vacuum, the space environment at Shuttle altitudes can seriously affect spacecraft surfaces. Besides contamination and particle damage (covered by other subpanels), several such interactions with spacecraft surfaces have been observed in the last few years. The principal ones of concern to this subpanel are:
1) Spacecraft surface charging
2) VxB induced currents
3) High voltage surface interactions
4) Shuttle wake/ram plasma variations
5) Shuttle surface glow
6) Spacecraft surface erosion

Of these, Shuttle glow, surface erosion, and high voltage surface interactions (resulting in power loss and increased surface arcing) are considered the most serious, as they pose direct threats to Shuttle operations. As for the other interactions, recent evidence from the AF experiments on DMSP indicate that potentials as high as 1 kV have been observed in the auroral zone at low Earth orbit. The implications for Shuttle need to be addressed. Likewise, variations in the Shuttle plasma sheath are tied to interactions with the Shuttle surface. These variations, and the regions of plasma enhancement and depletion that they create, will greatly influence the environment around the Shuttle and therefore must also be accurately characterized. VxB effects are probably the least serious of the interactions to be considered. Even so, given the size of the Shuttle and some of the proposed Shuttle-based systems such as the tether, the VxB electric field may significantly affect operations.

The purpose of the surface interactions section of the Shuttle Environment Data Base is to review the current status of ongoing investigations into these interactions, indicate possible sources of existing data, and recommend necessary experiments to assess the processes involved. Where possible, the known effects of these interactions are catalogued, and potential engineering solutions are presented. Models, both analytic and theoretical, of the interactions are a critical part of the discussion. The work presented has been performed within the framework of the Space Shuttle Experiment and Environment Workshop; namely, the information has been developed so as to make it readily integrable into this computer data base. If the data are not sufficient, the user is referred to the key players listed under each interaction category.

ORGANIZATION OF DATA ON SURFACE INTERACTIONS

TOPICS:

1) Surface Charging (I. Katz and A. Rubin)
2) VxB (I. Katz and W. Raitt)
3) High Voltage (I. Katz)
4) Ram/Wake (A. Rubin, W. Denig, and W. Raitt)
5) Glow (S. Mende, J. Gregory, B. D. Green, and E. Murad)
6) Surface Erosion (L. Leger)

Appendix A

SUBTOPICS:

    A. Description
    B. References
    C. Key Players
    D. Existing Data
    E. Required Data
    F. Possible Experiments
    G. Mitigation

PRINCIPAL TOPICS

1. <u>SURFACE CHARGING</u> (by I. Katz and A. Rubin)

1A. Description

    Spacecraft charging occurs at geosynchronous orbit during magnetic substorms, when dense clouds of kilo-electron volt (keV) plasma envelop spacecraft. These plasma clouds charge up the spacecraft surfaces to potentials ranging up to 19 kilovolts. Surface potentials of many kilovolts lead to arc discharges, with accompanying currents of up to 1000 amps, which radiate rf power into the spacecraft electronics, and have led to false commands, upsets and, in one case, complete loss of a spacecraft.

    At Shuttle Orbiter altitudes of 200 to 500 km, the ionospheric plasma is cold and dense, a state which ordinarily would not support charging. Special conditions must exist to permit spacecraft charging; these special conditions exist in polar orbit, when the spacecraft passes through auroras. Auroras are produced by fluxes of energetic electrons, with energies up to 20 kilo-electronvolts. Ionospheric spacecraft have an orbital velocity of 8 km/sec, so that their directed velocity is much higher than the thermal velocity of ionospheric oxygen ions, although much smaller than the electron thermal velocity. Because of the high directed orbital velocity, flow effects are important.

The Space Shuttle Orbiter in polar Earth orbit sees an environment which differs dramatically from that in equatorial orbit. Near the Arctic circle, the Orbiter may experience intense fluxes of energetic (5 to 10 keV) electrons propagating down along magnetic field lines. Shuman et al. (1) have reported peak electron fluxes at about 800 km of 5.4×10^9 eV $(cm^2\text{-s-sr})^{-1}$ near 0 degree pitch angle in the midst of an inverted-V event where peak energy reached 9.5 keV, corresponding to a hemispheric current of about 50 micro amps/m^2. For worst-case scenarios, even larger fluxes should be considered.

Ionospheric ions pile up on the front surface of the spacecraft, an effect called "ram", and a wake is produced as well. This wake represents the shadowing of ion flow by the spacecraft body. Measurements show that the near wake is depleted of ions, in some cases by many orders of magnitude, relative to the concentration of electrons.

On occasion the auroral electron current exceeds the ram current, so that charging of the entire vehicle can take place. On the other hand, the ion depletion in the wake allows charging to take place at lower auroral electron fluxes. Charging of the aft region of a spacecraft in polar auroral orbit would be the most frequently observed phenomenon. Charging involves a balance between ram current and auroral electron current.

The first indication that spacecraft can charge up to tens of volts in an aurora was in a paper by Knudsen and Sharp written in 1967. The next finding of charging was a report by Sagalyn and Burke in 1977 of charging to greater than -50 volts on Injun 5.

In 1980 Parks and Katz computed the charging of a large spacecraft in low polar orbit. They showed that a spacecraft could charge to potentials of the order of a kilovolt and, further, that the charging depends on the vehicle size, being larger for larger vehicles.

The combination of intense auroral fluxes, rapid charging times, and observed high negative potentials creates a convincing argument for the possibility of large differential potentials on the Orbiter.

Appendix A

The probability that exactly the correct conditions for high-voltage charging occur on a given flight has not been addressed. However, recent data from DMSP, a much smaller spacecraft in an 800-km polar orbit, have shown charging to -800 V on one orbit and several examples of charging greater than -100 V.

In 1982, a Workshop on Natural Charging of Large Space Structures in Near Earth Polar Orbit was held at AFGL, 14-15 September, chaired by Sagalyn and Rubin. At this Workshop, Katz et al. showed results of the AFGL POLAR code, for 3-dimensional computation of auroral charging and wakes of complex large spacecraft.

The POLAR code can model the Shuttle Orbiter with more than 1,000 surface cells, and can compute charging by the aurora on every cell. The Earth's magnetic field is taken into account, and flow of ionospheric plasma, in the vehicle reference frame, is treated. The VxB electric field is included as well. At the present NASA Workshop, Katz showed that the POLAR code describes ion density in the orbiter bay as a function of angle with respect to the flow direction. Data were taken by Murphy et al. of Iowa.

The electric field must be included in the computation in order to explain the observed ion densities in the bay.

At the Spacecraft Environmental Interactions Technology Conference in Colorado Springs in 1983, Cooke, et al. of S-Cubed and Rubin of AFGL showed POLAR Code results on a 3-D Shuttle Orbiter. Charging was computed, together with 3-D potential and density contours.

At this conference, Besse, Rubin and Hardy modeled a charging event found by Hardy on DMSP. The measured electron spectrum was used in a probe theoretical treatment. Burke et al. showed charging data for the 18 January 1983 event on DMSP, and employed a Gurevich wake charging model. More charging events on DMSP have been found by Hardy, Gussenhoven and Yeh of AFGL, with potentials of up to -600 volts (Private communications).

1B. References

Knudsen, W.C. and Sharp, G.W., J. Geophys. Res. 72, 1061-1072 (1967).

Parks, D.E. and I. Katz, "Charging of a Large Object in Low Polar Earth Orbit", in Spacecraft Charging Technology 1980, NASA Conf. Publ. 2182, AFGL-TR-81-0270 (1980).

Proceedings of the Air Force Geophysics Laboratory Workshop on Natural Charging of Large Space Structures in Near Earth Polar Orbit. (14-15 September 1982) Editors: R.C. Sagalyn, D.E. Donatelli, I. Michael (25 Jan 1983).

> Relevant Articles: Whipple, E.C. "An Overview of Charging of Large Space Structures in Polar Orbit", p. 11.
>
> Hardy, D.A., "The Worst Case Charging Environment", p. 141.
>
> Samir, U., "The Interaction of Large Space Structures with the Near- Earth Environment", p. 235.
>
> Rubin, A.G. and A.L. Besse, "Shuttle Orbiter Charging in Polar Earth Orbit", p. 253.
>
> Katz, I., D.L. Cooke, M.J. Mandell, D.E. Parks, J.R. Lilley, J.H. Alexander, A.G. Rubin, "POLAR Code Development", p. 321.

Katz, I., D.L. Cooke, D.E. Parks, M.J. Mandell, A.G. Rubin, "Electrostatic Charging on the Shuttle Wake in Polar Orbit". AIAA Shuttle Environment and Operations Conference, Washington, D.C. (Oct. 31 - Nov. 2, 1983).

Cooke, D. L., I. Katz, M.J. Mandell, J.R. Lilley, Jr., A.G. Rubin, "A Three-Dimensional Calculation of Shuttle Charging in Polar Orbit", Spacecraft Environmental Interactions Technology Conference, Colorado Springs, CO (4-6 October 1983).

Appendix A

Besse, A.L., A.G. Rubin, and D. Hardy, "Charging of the DMSP Spacecraft in the Aurora on 10 January 1983". Ibid. S/C Conference (1983).

Burke, W.J., D.A. Hardy, F.J. Rich, A.G. Rubin, M.F. Tautz, N.A. Saflekos and H.C. Yeh, "Direct Measurements of Severe Spacecraft Charging in the Auroral Ionosphere," Ibid. S/C Conference (1983).

Katz, I. and D.E. Parks, "Space Shuttle Orbiter Charging," J. Spacecraft and Rockets, 20, 22-25 (1983).

Cooke, D.L., I. Katz, M.J. Mandell and J.R. Lilley, Jr., "A Three-Dimensional Calculation of Shuttle Charging in Polar Orbit," presented at Spacecraft Environmental Interactions Technology Conference, U.S. Air Force Academy (4-6 October 1983).

Katz, I., D. C. Cooke, D.E. Parks, and M.J. Mandell, "Polar Orbit Electrostatic Charging of Objects in the Shuttle Wake," presented at Spacecraft Environmental Interactions Technology Conference, U.S. Air Force Academy (4-6 October 1983).

Winckler, J. R., "Application of Artificial Electron Beams to Magnetospheric Research," Rev. of Geophys. and Space Phys., 18, 659 (1980).

Linson, L. M. and K. Papadopoulos, "Review of the Status of Theory and Experiment for Injection of Energetic Electron Beams in Space," Science Applications, Inc. Report SAI-023080-459 LJ (April 1980).

Obayashi, T., "Space Experiments with Particle Accelerators," Science, 225, No. 4658, p. 195 (13 July 1984).

1C. Key players

Dr. William F. Denig
AFGL/PHK
Hanscom AFB
Bedford
MA 01731 USA
(617) 861-3989

Dr. George Inouye
TRW STGM21145
1 Space Park
Redondo Beach
CA 90278 USA
(213) 535-8448

Dr. Carolyn K. Purvis
Lewis National Research Center MS77-4
21000 Brookpark Road
Cleveland
OH 44135 USA
(216) 433-4000

Dr. Al Rubin
AFGL/PHK
Hanscom AFB
Bedford
MA 01731 USA
(617) 861-2933

Mr. N. John Stevens
Space and Communications Group, Hughes
P.O. Box 92919
Los Angeles
CA 90009 USA
(213) 647-4975

Appendix A   283

Dr. Nobie H. Stone
Marshall Space Flight Center
Huntsville
AL   35812   USA

Dr. Ira Katz
S-Cubed
Box 1620
LaJolla
CA   92038   USA
(619) 453-0060

Dr. Karl Knott
Space Science Dept., ESA
Keplerlaan 1
2200 AG Noordwijk
NETHERLANDS
31-1719

Dr. Henry Berry Garrett
Jet Propulsion Laboratory (144-218)
4800 Oak Grove Dr.
Pasadena, CA 91109
(818) 354-2644
(FTS) 792-2644

1D. Existing data

1. Who has data/models?
    POLAR Code: S-cubed, AFGL
2. What format are data/models in?
    POLAR is a large simulation code which is being developed to run on UNIVAC and CDC computers. POLAR models the interaction of a three-dimensional spacecraft with the ambient plasma, including precipitating auroral electrons, magnetic fields and ram/wake effects.

3. Where do the data models reside?

POLAR - S-Cubed - UNIVAC 1100/81, 3398 Carmel Mountain Road, San Diego, CA 92121 POLAR - CYBER 750 - AFGL - Hanscom Air Force Base, MA 01731

4. What is the uncertainty?
There are no data for large structures in polar orbit.
5. What is the spread in the data?
Typical satellite voltages at 800 km are about -1 volt. DMSP charged to approximately -800 volts.

1E. Required data

1. What information should exist?
There should be models that have been successfully compared with space flight data and will predict accurately surface charging on large structures.

2. How should information be accessed?
The models should be accessed _via_ interactive computer networks.

3. Models
POLAR should be an easily accessible tool available to spacecraft and instrument designers over the NASA computer network. The input should be _via_ interactive graphics. Results should be accessible via an interactive searching program with graphics capabilities.

1.F. Possible experiments
There is a dire need for large object charging experiments in polar orbit, e.g., IMPS.

1.G. Mitigation
See NASA Spacecraft Charging Design Guidelines Handbook, by C.K. Purvis, H.B. Garrett, A. Whittlesey, and N.J. Stevens; NASA Lewis publication (1984).

Appendix A

2. V x B CURRENTS (by I. Katz and W. Raitt)

2A. Description

The induced V x B electric field caused by an object moving at orbital speed through the geomagnetic field is 0.2 to 0.3 Volts/meter maximum. Thus, since the induced emf between the points separated by a vector L is

$$E = V \times B \cdot L,$$

the 10's of meter dimensions of the Orbiter mean that we can expect potential differences of up to 20 V to exist between different points on the Orbiter.

The potential difference is not cancelled out when one of the conductors in the circuit is not moving at orbital speed. This is the situation for plasma probes mounted on the Orbiter which return currents through the stationary ionosphere. The probes become biased by a voltage additional to their intended bias by an amount given by the equation above. The value for L is the vector between the probe current collector and the connection to the ionosphere for the return current. Experiments onboard STS-3 have shown this effect is well modeled by assuming the return current is through the rocket engine nozzles.

The recent announcement of opportunity for TSS-1 describes the first flight of an object large enough that the V x B electric potentials will dominate all others on the Shuttle Orbiter. The TSS-1 tether is a 20 kilometer wire that will be extended vertically from the Orbiter. It will generate 5000 to 7000 volts potential difference between the Orbiter and a 1.5 meter subsatellite located on the end of the tether. Interaction with the plasma will be a passive collection of electrons by the subsatellite, the emission of energetic electrons from the Orbiter and the emission of cold plasma from the Orbiter. The aim of this experiment is to draw more than an ampere of current through the tether. This would serve as proof of principle for future applications of electrodynamic tether generator systems for large space structures.

2B. References

Al'pert, Ya.L., A.V. Gurevich and L. P. Pitaevski, Space Physics with Artificial Satellites, Consultants Bureau, New York (1965).

Arnold, D. A. and M. Dobrowolny, "Transmission Line Model of the Interaction of a Long Metal Wire with the Ionosphere," Radio Science, 15, 1149 (1980).

Banks, P. M. and TSS Team, "The Tethered Satellite System, Final Report OM from the Facility Requirements Definition Team," NASA (1980).

Banks, P. M., Williamson, P. R. and Oyama, K-I., Planet Space Sci., 29, 119-147 (1981).

Belcastro, V., P. Vettri and M. Dobrowolny, "Radiation from Long Conducting Tethers Moving in the Near-Earth Environment," Il Nuovo Cimento, 5, 537 (1982).

Booker, H. G., Cold Plasma Waves, Chapter 18, Martinus Nijhoff Publishers, Amsterdam (1983).

Colombo, G., Gaposchkin, E. M., Grossi, M. D. and Weiffenbach, G. L., Smithsonian Astrophysical Observatory, Reports in Geoastronomy 1 (1974).

Csiky, G. A., "Measurements of Some Properties of a Discharge from a Hollow Cathode," NASA Technical Note, NASA-TN-D-4966 (February 1969).

Drell, S. D., Foley, H. M. and Ruderman, M. A., J. Geophys. Res., 70, 3131-3145 (1965).

Dwight, H. B., Tables of Integrals and Other Mathematical Data, The MacMillan Company, New York, Fourth Edition, p. 247 (1961).

Grossi, M. and D. A. Arnold, "Engineering Study of the Electrodynamic Tether as a Spaceborne Generator of Electric Power," Smithsonian Astrophysical Observatory, SAO Technical Report, NASA Contract NAS8-35497 (June 1984).

Hargreaves, J. K., The Upper Atmosphere and Solar Terrestrial Relations, An Introduction to the Aerospace Environment, Van Nostrand Reinhold, New York, p. 60 (1979).

Ichimaru, S., Basic Principles of Plasma Physics: A Statistical Approach, Benjamin/Cummings, London, p. 292 (1973).

Jones, S. G., J. V. Staskus, and D. C. Byers, "Preliminary Results of Sert II Spacecraft Potential Measurements Using Hot Wire Emission Probes," NASA TM-X-2083 (1970).

Kahalas, S. L., "Excitation of Extremely Low Frequency Electro-Magnetic Waves in the Earth-Ionosphere Cavity by High-Altitude Nuclear Detonations," J. Geophys. Res., $\underline{70}$, 3587 (1965).

Katz, I., J. J. Cassidy, M. J. Mandell, D. E. Parks, G. W. Schnuelle, P. R. Stannard and P. G. Steen, "Additional Application of the NASCAP Code. Vol. II, SEPS, Ion Thruster Neutralization and Electrostatic Antenna Model," NASA CR-165350 (February 1981).

Katz, I., D. E. Parks, D. L. Cooke and J. R. Lilley, Jr., "Polarization of Spacecraft Generated Plasma Clouds," submitted for publication in Geophysical Research Letters (July 1984).

Knishnan, M., R. G. John, W. F. von Jaskowsky and K. E. Clark, "Physical Processes in Hollow Cathodes," AIAA Journal, $\underline{15}$, 1217 (1977).

Langmuir, I. and K. Blodgett, "Currents Limited by Space Charge Between Concentric Spheres," Phys. Rev., $\underline{24}$, 49 (1924).

Linson, L. M., "Current Voltage Characteristics of an Electron-Emitting Satellite in the Ionosphere," Jour. Geophys. Res., 74, 2368 (1969).

Mandell, M. J., I. Katz, and D. L. Cooke, "Potentials on Large Spacecraft in LEO," IEEE Trans. Nucl. Sci., NS-29, 1584 (1982).

Mandell, M. J., I. Katz, and G. A. Jongeward, "Computer Simulation of Plasma Electron Collection by PIX-II," presented at AIAA 23rd Aerospace Sciences Meeting, Reno, NV (January 14-17, 1985).

Morse, P. M. and H. Feshbach, Methods of Theoretical Physics, McGraw-Hill Book Company, New York, p. 530 (1953).

Parker, L. W. and B. J. Murphy, "Potential Buildup on an Electron-Emitting Ionospheric Satellite," J. Geophys. Res., 72, 1631 (1967).

Parks, D. E. and I. Katz, "Charging of a Large Object in Low Polar Earth Orbit," USAF/NASA Spacecraft Charging Technology Conference III, U.S. Air Force Academy, CO (November 1980).

Parks, D. E., M. J. Mandell and I. Katz, "Fluid Model of Plasma Outside a Hollow Cathode Neutralizer," J. Spacecraft and Rockets, 19, 354 (1982).

Raitt, W. J., Banks, P. M., and Williamson, P. R., ESA SP-195, pp. 361-367 (1983).

Raitt, W. J., Siskind, D. E., Banks, P. M. and Williamson, P. R., Planet. Space Sci., 32, 457-467 (1984).

Samir, U., P. J. Weldman, F. Rich, H. C. Brinton and R. C. Sagalyn, "About the Parametric Interplay between Ion's Mach Number, Body-Size, and Satellite Potential in Determining the Ion Depletion in the Wake of the S3-2 Satellite," Jour. Geophys. Res., 86, 11161-11166 (1981).

Siegfried, D. E. and P. J. Wilbur, "An Investigation of Mercury Hollow Cathode Phenomena," AIAA/DGLR 13th International Electric Propulsion Conference, San Diego, CA, p. 78, (April 1978).

Stevens, N. J., "Review of Biased Solar Array-Plasma Interaction Studies," NASA TM-82693 (1981).

Thompson, W., <u>An Introduction to Plasma Physics</u>, Addison Wesley, p.108 (1964).

Ward, J. W. and H. J. King, "Mercury Hollow Cathode Plasma Bridge Neutralizers," J. Spacecraft, <u>10</u>, 1161 (1968).

Williamson, P. R. and Banks, P. M., Final Report, NOAA (1976).

Williamson, P. R., Denig, W. F., Banks, P. M., Raitt, W. J., Kawashima, N., Hirao, K., Oyama, K-I, and Sasaki, S., <u>Artificial Beams in Space Plasma Studies</u> (Ed. B. Grandal) Plenum Press, 645-656 (1982).

2C. Key Players
1. Ira Katz, S-Cubed, theory.
2. P. M. Banks, Stanford University, experiments and theory
3. P. R. Williamson, Stanford University, experiments
4. W. J. Raitt, Utah State University, experiments
5. M. Grossi, Harvard, experiments and theory
6. D. Crouch, Martin Marietta Denver, technology
7. G. Manarini, CNR, Frascati, Italy, experiments and theory

2D. Existing data

1. Who has data/models?
    SAO - M. Grossi
    CNI - M. Dobrowolny
    Stanford - P. Banks and R. Williamson
    S-CUBED - D. Parks and I. Katz

2. What format are the data/models in?
    SAO - Computer model of electrical circuitry and dynamics
    S-CUBED - NASCAP/LEO computer model of spacecraft interactive analytical and computer model of plasma contactors

3. Where do the data/models reside?
    SAO
    NASCAP/LEO - S-CUBED
            - NASA/LeRC

4. What is the uncertainty?
    a. The closure path for current in the ionosphere
    b. The effect of magnetic fields on electron collection
    c. The action of electron beams on the Shuttle environment
    d. The efficiency of plasma contactors

5. What is the spread in the data?
    Rocket shots with electron beams collected tens of amperes of electron current at potentials of less than 20 volts. PIX-II in 800-km polar orbit collected 4 milliamperes of electron current at a potential of 1000 volts.

2E. Required data
    A. What information should exist?
    There should be complete predictive models for the electrical operation of tethered spacecraft for all Earth orbits.

    B. How should it be accessed?
    The models should be interactive and accessible on a standard NASA network <u>via</u> a graphics terminal, so that a planner can predict the performance of a system under design.

    C. How should it be formatted? The models should be simulations which have been verified by comparison with flight data. Actual flight data should be available for verification of models under construction, but since the details of the interactions are so dependent on the physical structure of the vehicle, the orbital

Appendix A   291

parameters and the termination circuitry, flight data themselves will
not be completely useful for designs of future systems.

2F. Possible experiments
   1. Data
      A. Current-voltage characteristics for large high voltage
         tethered satellites in the several kilowatt range.
      B. Experiments on the performance of plasma contactors in
         space.
            i. as cathodes
            ii. as anodes
   2. Models
Model development must progress beyond the present "science tool"
level into validated, sophisticated, transferable technology.

3. HIGH VOLTAGE (by I. Katz)

3A. Description
   For the solar power systems proposed for the Space Station and
other large space structures, the major environmental interaction
concern is negative potential arcing. Because electron collection
rates are hundreds of times larger than ion collection rates for equal
surface areas, solar arrays when configured for high output voltages
will attain current balance with much of the array at high negative
potentials with respect to the plasma environment. Numerous experiments both in space and in the laboratory have recorded arcing at
negative potentials as low as 250 volts, and only recently has a
theory of this discharge process been proposed. Biasing the entire
system positive with respect to the plasma is not an acceptable
solution due to the large power losses arising from parasitic currents
flowing in the plasma. This leakage current flowing through the
plasma can be calculated with computer models (NASCAP/LEO and POLAR)
presently being developed and there is successful comparison with
flight and ground test data.

3B. References

Banks, P. M., P. R. Williamson and W. J. Raitt, "Results from the Charged Particle Beam Experiments on the Space Shuttle," AIAA Paper 83-0307 (1983).

Bekey, I., "Big COMSATS for Big Jobs at Low User Costs," Astron. and Aeron., 17m, 42-56 (1979).

Chen, F. F., "Electric Probes," in Plasma Diagnostic Techniques, (R. Huddlestone and S. Leonard, Editors) Pure and Applied Physics, Vol. 21, Academic Press, 113-119 (1965).

Cole, R. W., H. S. Ogawa and J. M. Sellen, Jr., "Operation of Solar Cell Arrays in Dilute Streaming Plasmas," NASA CR-72376, (1968).

Domitz, S. and N. T. Grier, "The Interaction of Spacecraft High Voltage Power Systems with the Space Plasma Environment," Power Electronic Specialists Conference, IEEE, NJ, 62-69 (1974).

Grier, N. T., "Experimental Results on Plasma Interactions with Large Surfaces at High Voltages," NASA TM-81423 (1980).

Grier, N. T., C. Smith and L. M. Johnson, "Plasma Interactions with Solar Arrays at High Voltages," Spacecraft Charging Technology-1980, NASA CP-2182/AFGL-TR-81-0270, AD A114426, 922 (1981).

Herron, B. G., J. R. Bayless and J. D. Worden, "High Voltage Solar Array Technology," AIAA Paper 72-443 (1972).

Johnson, R. D. and C. Holbrow, Editors, Space Settlements, A Design Study, NASA SP-413 (1979).

Katz, I., M. J. Mandell, G. W. Schnuelle, D. E. Parks and P. G. Steen, "Plasma Collection by High Voltage Spacecraft at Low Earth Orbit," J. Spacecraft and Rockets, 18, 79 (1981).

Kennerud, K. L., "High Voltage Solar Array Experiments," NASA CR-121280 (1974).

Knauer, W., J. R. Bayless, G. T. Todd and J. W. Ward, "High Voltage Solar Array Study," NASA CR-72675 (1970).

Mandell, M. J., I. Katz and D. L. Cooke, "Potentials on Large Spacecraft in LEO," IEEE Trans. Nucl. Sci., NS-29m, 1584-1588, (December 1982).

Mandell, M. J., I. Katz, G. A. Jongeward and J. C. Roche, "Computer Simulation of Plasma Collection by PIX-II," to be presented at AIAA 23rd Aerospace Sciences Meeting, Reno, NV (January 14-17, 1985).

McCoy, J. E. and A. Konradi, "Sheath Effects Observed on a 10-meter High Voltage Panel in Simulated Low Earth Orbit Plasmas," Spacecraft Charging Technology-1979, NASA CP-2071/AFGL-TR-79-0012, 315 (1979).

Outlook for Space, NASA SP-386 (1976).

Parker, L. W., E. G. Holeman and J. E. McCoy, "Sheath Shapes: a 3-D Generalization of the Child-Langmuir Sheath Model for Large High-Voltage Space Structures in Dense Plasmas," AFGL-TR-83-0046, Proceedings of the Air Force Geophysics Laboratory Workshop on Natural Charging of Large Space Structures in Near-Earth Polar Orbit, (14-15 September 1982).

Parker, L. W., "Contributions to Satellite Sheath and Wake Modeling," ESA SP-198, Proceedings of the 17th ESLAB Symposium, Noordwijk, The Netherlands, (13-16 September 1983).

Purvis, C. K., N. J. Stevens and F. D. Berkopec, "Interaction of Large, High Power Systems with Operational Orbit Charged-Particle Environments," NASA TMX-73867 (1977).

Shawhan, S. D., R. R. Anderson, N. D'Angelo, L. A. Frank, D. A. Gurnett, G. B. Murphy, H. D. Owens, D. Reasoner, N. Stone, H. Brinton and D. Fortna, "Beam-Plasma Interactions and Orbiter Environment Measurements with PDP on STS-3," AIAA Paper 83-0308 (1983).

Snoddy, W. C., "Space Platforms for Science and Applications," Astron. and Aeron., 19, 28-36 (1981).

Snyder, D. B., "Discharges on a Negatively Biased Solar Array in a Charged Particle Environment," NASA TM-83644 (1983); "Characteristics of Arc Currents on a Negatively Biased Solar Cell Array in a Plasma," NASA TM-83728 (1984).

Stevens, N. J., "Solar Array Experiments on the SPHINX Satellite," NASA TMX-71458 (1973).

Stevens, N. J., F. D. Berkopec, C. K. Purvis, N. T. Grier and J. V. Staskus, "Investigation of High Voltage Spacecraft System Interactions with Plasma Environments," AIAA Paper 78-672 (1978).

Stevens, N., "Interactions Between Spacecraft and the Charged-Particle Environment," Spacecraft Charging Technology-1978, NASA CP-2071/AFGL-TR-79-0082, AD A084626, 268-294 (1979).

Stevens, N. J., "Review of Biased Solar Array-Plasma Interaction Studies," NASA TM-82693 (1981).

Woosley, A. P., O. B. Smith and H. W. Nassen, "Skylab Technology Electrical Power System," ASME Paper 74-129 (1974).

Appendix A

3C. Key players

Dr. Ira Katz
S-Cubed Box 1620
LaJolla CA 92038 USA
(619) 453-0060

Mr. William F. Denig
AFGL/PHK Hanscom AFB
Bedford MA 01731 USA
(617) 861-3989

Dr. George Inouye
TRW STGM21145
1 Space Park
Redondo Beach CA 90278 USA
(213) 535-8448

Dr. Carolyn K. Purvis
Lewis National Research Center MS77-4
21000 Brookpark Road
Cleveland OH 44135 USA
(216) 433-4000

Dr. Al Rubin
AFGL/PHK
Hanscom AFB
Bedford MA 01731 USA
(617) 861-2933

Mr. N. John Stevens
Space and Communications Group, Hughes
P.O. Box 92919
Los Angeles CA 90009 USA
(213) 647-4975

Dr. Henry Berry Garrett
Jet Propulsion Laboratory (144-218)
4800 Oak Grove Drive
Pasadena, CA 91109
(818) 354-2644, (FTS) 792-2644

3D. Existing data

1. Who has data/models?

   a. Data - NASA/LeRC has data from both ground and flight experiments. Contact C. K. Purvis, N. T. Grier, D. B. Snyder.

   b. Models - S-CUBED is developing the NASCAP/LEO code to model high voltage interactions. Contact I. Katz, M. J. Mandell, G. A. Jongeward, S-CUBED, P.O. Box 1620, LaJolla, VCA 92038. Lee Parker is developing CLEPH3D to model space charge limited collection in three dimensions.

2. What format are data/models in?

   NASCAP/LEO is a three-dimensional computer code capable of modeling the potential about, and current collection by, a high voltage, large spacecraft in a short Debye length plasma. It includes good models for the charging of surfaces, including such phenomena as solar cell snapover. CLEPH3D is a self-adjusting, finite element code that has been used to model the space charge-limited sheath surrounding conducting objects.

3. Where do the data/models reside?

   NASCAP/LEO resides on the UNIVAC at S-CUBED (contact I. Katz or M. J. Mandell) and on the UNIVAC and CRAY at NASA/LeRC (contact J. C. Roche). CLEPH3D is the property of Lee W. Parker, Inc., Concord, MA.

4. What is the uncertainty?

   For a positive potential, snapover occurs at higher potential in space than in laboratory. Computer simulations agree with laboratory data. Electron collection by large structures with a magnetic field is not understood. Sheath ionization effects are important for rockets and may be important for Shuttle.

   For negative potentials, arcing threshold is uncertain. Arcing in space has been observed as low as -255 V. Arcing threshold is inversely related to plasma density. We have a promising theory to understand arcing, but it needs theoretical development and experimental testing.

Appendix A

5. What is the spread in the data?

For positive potential, data are needed for Shuttle altitudes. More data are needed at negative potential. For both polarities, we need data for current technology solar cells.

3E. Required data

1. What information should exist?

   a. At positive potentials, we need sufficient flight data for large objects to determine the role of magnetic fields in limiting collection. Also required are data on turbulent sheath interaction which has been proven to enhance current collection at low altitudes.

   b. For negative potentials, we need data on large, modern active arrays to determine arcing thresholds for different orbital parameters.

2. What is therefore required?

Data--Sufficient data are required, especially at negative potentials, to validate theory.

Models--NASCAP/LEO model should incorporate our best understanding of experimental data. Theory of negative arcing should be pursued.

3F. Possible experiments

Large active solar arrays should be flown to study arcing and collection. Environment must be well-characterized. Also, effect of arcing on power system needs further study.

3G. Mitigation

Little has been done here. We need to study filtering, insulation and interconnection materials.

4. RAM/WAKE (by A. Rubin, W. Denig, and W. Raitt)

4A. Description

The existence of a pronounced plasma wake effect has been observed from the early flights of plasma probes on unmanned spacecraft (Samir, et al., 1983; Samir and Wrenn, 1969) and studied theoretically by a number of workers (Gurevich, et al., 1966, 1968, 1969, 1973; Mathews, 1971; Raadu, 1979; Anderson, et al., 1980). Most of this work shows little effect of buildup of plasma density in the ram direction. Measurements with directionally sensitive probes oriented in the ram direction are usually assumed to give the ambient ionospheric plasma density.

Recent plasma measurements made within and near the Shuttle cargo bay indicate that the ram/wake characteristics of this vehicle in Low Earth Orbit (LEO) are distinctly different than for smaller satellites. Specifically, enhanced electron temperatures, large amplitude turbulence, and molecular ion contamination are observed near Shuttle away from the wake (Raitt, et al., 1984; Siskind, et al., 1984) while near the wake and in daylight the electron temperatures and ion characteristics more properly reflect ionospheric conditions (Raitt, et al., 1984). These measurements suggest that some mechanism unique to the Shuttle, i.e., a large space structure of irregular shape, is operating which generates the turbulence and heats the electrons. The molecular ions, possibly CO_2^+, O_2^+, NO^+, or some combination thereof, appear to have been created via the anomalous ionization of a gaseous cloud traveling with the Orbiter. The effects of such processes on the distribution of plasma around the Shuttle and on the charging of a small satellite within the wake have not been addressed.

A space vehicle traveling supersonically--in the ion sense-- through a plasma will create a region of depleted density immediately behind itself. Both neutral and charged particle species are thus affected, and the structure of the wake depends upon the diffusion rate for each population. Although the individual diffusion rates are dependent, in part, on the temperature and mass of the species, there does exist a coupling between the charged particles--an ambipolar

Appendix A   299

effect--which tends to create a region of negative potential (relative to the plasma potential) within the wake. Data from thermal plasma probes on spinning satellites in low Earth orbit (LEO) have been used to characterize the ram/wake asymmetries for an object traveling supersonically (Mach 6-7) through a plasma (see Samir and Stone, 1980, and references therein). These studies clearly indicate that the electron density depletions within the wake are dependent upon the ambient electron temperature and the average ionic mass. For example, the ion thermal current to a probe mounted on a 32-cm boom was used as a measure of the plasma density near the AE-C satellite. These measurements indicate that the density depletion in the wake was reduced two orders of magnitude from the ambient density in a O^+ plasma with T_e = 1000 degree K (Samir, et al., 1979). Consistent with general diffusion concepts, the depletion was not as extreme for a higher ambient electron temperature or a plasma consisting of a higher percentage of the less-massive hydrogen ion. A number of measurements have also indicated that there is an apparent electron temperature enhancement within the wake of orbiting satellites (Samir and Wrenn, 1972; Troy, et al., 1975). Illiano and Storey (1974) have shown that this anomalous measurement is merely an artifact of the plasma probe being within the negative potential well of the satellite wake. Under such conditions the electron energy distribution, as detected by the probe, ceases to be Maxwellian so that temperature measurements are biased and not correct.

Murphy, et al., (1983) and Raitt, et al., (1984) report that under certain conditions the electron density trough within the wake is at least four orders of magnitude below the ambient ionospheric density. These measurements were limited by the sensitivity of the instruments (minimum $10^3/cm^3$) and were made within the cargo bay on STS-3. At distances of 5 to 10 meters outside of the bay the spatial width of this depleted volume is narrower than that which is detected nearer the bay, and the density trough is only 2 to 3 orders of magnitude below the ambient density (Murphy, et al., 1983). It should be noted, however, that the extremes in plasma density occur when the Orbiter is in darkness. Raitt, et al., (1984) have noted that under sunlit conditions, their instruments sometimes do not detect an obvious wake, a condition they suggest may be due to Orbiter outgassing or a deviation of the plasma wake from the geometric wake. The latter explanation is probably correct, i.e., an artifact of the data sampling, since under a different attitude maneuver on STS-3 the ram/wake asymmetry in density is clearly visible, even in sunlight (Siskind, et al., 1984).

The electron temperature variation for Shuttle ram to wake conditions indicates an enhancement above ionospheric temperatures for "non-wake" measurements (Raitt, et al., 1984). Again, the sensitivity of the Langmuir probe used here was limited in the wake by the low electron density, although the tendency for elevated temperatures into the ram can be clearly seen in the data (Siskind, et al., 1984). This tendency differs from satellite measurements in which an apparent temperature enhancement has been noted in the wake (Troy, et al., 1975). These Shuttle measurements also conflict with theories of electron collisional-cooling rates (Schunk and Nagy, 1978) in the more dense, non-wake plasma, unless there exists an energy source that can heat the electrons. Such a source may also be responsible for the observed plasma density turbulence.

Measurements have been made which indicate that the plasma near Shuttle is in a turbulent state. Raitt and co-workers (Raitt, et al., 1984; Siskind, et al., 1984) have observed this dynamic behavior in the data from a Retarding Potential Analyzer (RPA). This instrument nominally measures the composition and temperature of atmospheric ions, although during STS-3 the ionospheric signatures were often obscured by a plasma turbulence and by the presence of a hot molecular ion contaminant. The former effect was always present, although the level of measured turbulence peaked when the RPA was in the ram. This instrument could not, however, determine whether the turbulence was due to density fluctuations or ion velocity changes. Measurements by Shawhan and colleagues (Murphy, et al., 1983; Shawhan, et al., 1984) suggest that this turbulence is due to the former, while published data from Narcisi, et al., (1983) tend to suggest velocity fluctuations. Murphy, et al., have presented data from an array of broadband electric wave receivers covering the frequency range from 30

Hz to 200 kHz. These emissions are well correlated with plasma density fluctuations as measured with a Langmuir probe on their Plasma Diagnostic Package (PDP). Further, the characteristics of the detected waves are distinct above and below the Lower Hybrid Resonance (f_{LHR} = 7 kHz). Whereas the higher frquency waves appear to be polarized, the lower frequency emissions are apparently not polarized and the signal strength is dependent on attitude, falling below the threshold of detectability when the antenna is in the wake. It would appear from the measurements of Narcisi, et al., (1983) that the oscillations in the RPA data of Raitt, et al., (1984) are due to O^+ velocity fluctuations. Specifically Narcisi et al. claim that the large variations seen in their O^+ current monitor were not associated with any apparent density fluctuations. Interestingly enough, Narcisi, et al., also suggest that under ram conditions their instruments yielded accurate ionospheric data whereas the data of Raitt, et al., in this attitude was often obscured by the intense plasma turbulence.

Turbulent plasma behavior, as a result of spacecraft-environmental interactions, has been noted for several ionospheric satellites. Samir and Willmore (1965) noted VLF oscillations (instrument bandpass 2 kHz to 3.7 kHz) in the wake of Ariel I. More recently, Hanson and Cragin (1981) have reported the measurement of plasma fluctuations on the AE-C and -D satellites when the vehicle velocity is aligned with the local B-field. The above authors, plus others (Lui, 1975), suggest that such turbulent behavior corresponds to locally produced ion plasma waves created as a result of plasma instabilities. The Shuttle measurements differ from these satellite observations in that the most intense turbulence occurs in the ram and is not well correlated to B. It is interesting to note that Papadopoulos (1984) has developed a plasma instability theory to account for the most dramatic ram phenomena--the Shuttle glow (Banks, et al., 1983; Mende, et al., 1984).

The plasma potential and charged particle distributions about a body in LEO has been modeled (Parker, 1977), yet there exist large discrepancies between such results and actual flight data.

4B. References

Note: This general reference contains material relevant to the topic of Shuttle ram/wake effects. References specifically discussing the Orbiter results and predictions are indicated with an asterisk*.

Al'pert, J. L., Space Sci. Rev., $\underline{4}$, 373-415 (1965).

Anderson, D., Bonnedal, M., and Lisak, M., Physics Scripta, $\underline{22}$, 507 (1980).

*Banks, P. M., P. R. Williamson, and W. J. Raitt, "Space Shuttle Glow Observations," Geophys. Res. Lett., $\underline{10}$, 118-121 (1983).

Bourdeau, C. L. and J. L. Donley, Proc. Roy. Soc., $\underline{A281}$, 487-504 (1964).

Fornier, G. and D. Pigache, Phys. Fluids, $\underline{18}$, 1443-1453 (1975).

Gurevitch, A. V., Pariiskaya, L. V., and Pitaevskii, L. P., Sov. Phys. JETP, $\underline{36}$, 274 (1973).

Gurevitch, A. V., Pariiskaya, L. V. and Pitaevskii, L. P., Sov. Phys. JETP, $\underline{27}$, 476 (1968).

Gurevitch, A. V., Pariiskaya, L. V. and Pitaevskii, L. P. Sov. Phys. JETP, $\underline{22}$, 449 (1966).

Gurevitch, A. V., Pitaevskii, L. P. and Smirnova, V. V., Space Sci. Rev., $\underline{9}$, 805 (1969).

Hanson, W. B., and B. L. Cragin, "The Case of the Noisy Derivatives-Evidence for a Spacecraft-Plasma Interaction," J. Geophys. Res., $\underline{86}$, 10022-10028 (1981).

Henderson, C. L. and U. Samir, Planet. Space Sci., 15, 1499-1513 (1967).

Hester, S. D. and A. A. Sonin, Phys. Fluids, 13, 641-648 (1970).

Illiano, J. M. and L. R. O. Storey, "Apparent Enhancement of Electron Temperature Measurements in the Wake of a Spherical Probe in a Flowing Plasma," Planet. Space Sci., 22, 873-878 (1974).

*Katz, I., D. E. Parks, D. L. Cooke, M. J. Mandell, and A. G. Rubin, "Polar Orbit Electrostatic Charging of Objects in the Shuttle Wake," to be published in Spacecraft Environmental Interactions Technology Conference, U.S. Air Force Academy, Colorado Springs, CO, (October 1983).

Kraichnan, Robert H., "Statistical Dynamics of Two-Dimensional Flow," J. Fluid Mech., 67, part 1, 155-175 (1975).

Langhoff, S. R., R. L. Jaffe, J. H. Yee, and A. Dalgarno, "The Surface Glow of the Atmospheric Explorer-C and -E Satellites", Geophys. Res. Lett., 10, 896-899 (1983).

Lui, V. C., "Ionospheric Gas Dynamics of Satellites and Diagnostic Probes," Space Sci. Rev., 9, 423-490 (1969).

Lui, V. C., "A Wave Model of Near Wakes", Geophys. Res. Lett., 2, 485-488 (1975).

Matthews, W. G., Astrophys. J., 165, 147 (1971).

*Mende, S. B., P. M. Banks, and D. A. Klingelsmith III, "Observations of Orbiting Vehicle Induced Luminosities on the STS-8 mission," Geophys. Res. Lett., 11, 527-530 (1984).

*Mende S. B., R. Nobles, P. M. Banks, O. K. Garriott, and J. Hoffman, "Measurements of Vehicle Glow on the Space Shuttle," J. Spacecraft & Rockets, 21, 374-381 (1984).

*Murphy, G. B., S. D. Shawhan, L. A. Frank, N. D'Angelo, D. A. Gurnett, J. M. Grebowsky, D. L. Reasoner, and N. Stone, "Interaction of the Space Shuttle Orbiter with the Ionospheric Plasma," Proceedings of the 17th ESLAB Symposium on Spacecraft/Plasma Interactions and Their Influence on Field and Particle Measurements, Noordwijk, The Netherlands, (13-16 September 1983), ESA SP-198, 73-78 (1983).

*Narcisi, R., E. Trzinski, G. Federico, L. Wlodyka, and D. Delorey, "The Gaseous and Plasma Environment Around Space Shuttle," AIAA Shuttle Environment and Operations Meeting, Washington, D.C., AIAA-83-2659, 183-190 (31 Oct.-2 Nov. 1983).

Oran, W. A., N. H. Stone, and Uri Samir, "The Effects of a Body Geometry on the Structure in the Near Wake Zone of Bodies in a Flowing Plasma," J. Geophys. Res., 80, 207-209 (1975).

*Papadopoulos, K., "On the Shuttle Glow (the Plasma Alternative)," Radio Sci., 19, 571-577 (1984).

*Parish, J. L. and W. J. Raitt, "Turbulent Heating Around Space Shuttle," EOS Trans. Amer. Geophys. Union, 64, 805 (1983).

Parker, Lee W., "Calculation of Sheath and Wake Structure about a Pillbox-Shaped Spacecraft in a Flowing Plasma," in Proceedings of the Spacecraft Charging Technology Conference, (edited by C. P. Pike and R. R. Lovell), AFGL/TR-77-0051 or NASA TMX-73537, 331-366 (1977).

Raadu, M., Plasma Phys., 21, 331 (1979).

Raitt, W. J., Blades, J., Bowling, T. S. and Willmore, A. P., J. Phys. E, 6, 443-452 (1973).

Raitt, W. J., Dorling, E. B., Sheather, P. H. and Blades, J., Planet. Space Sci., 23, 1085-1101 (1975).

*Raitt, W. J., D. E. Siskind, P. M. Banks, and P. R. Williamson, "Measurements of the Thermal Plasma Environment of the Space Shuttle," Planet. Space Sci., 32, 457-467 (1984).

Samir, Uri, "A Possible Explanation of an Order of Magnitude Discrepancy in Electron-Wake Measurements," J. Geophys. Res., 75, 855-858 (1970).

*Samir, Uri, "Bodies in Flowing Plasmas: Spacecraft Measurements," Adv. Space Res., 1, 373-384 (1981).

Samir, Uri and Howard Jew, "Comparison of Theory with Experiment for Electron Density Distribution in the Near Wake of an Ionospheric Satellite," J. Geophys. Res., 77, 6819-6827 (1972).

*Samir, Uri and N. H. Stone, "Shuttle-Era Experiments in the Area of Plasma Flow Interactions with Bodies in Space," Acta. Astronautica, 7, 1091-1141 (1980).

Samir, U. and A. P. Willmore, "The Distribution of Charged Particles Near a Moving Spacecraft," Planet. Space Sci., 13, 285-296 (1965).

Samir, U. and G. L. Wrenn, "The Dependence of Charge and Potential Distribution around a Spacecraft on Ionic Composition," Planet. Space Sci., 17, 693-706 (1969).

Samir, Uri and G. L. Wrenn, "Experimental Evidence of an Electron Temperature Enhancement in the Wake of an Ionospheric Satellite," Planet. Space Sci., 20, 899-904 (1972).

Samir, Uri, L. H. Brace, and H. C. Brinton, "About the Influence of Electron Temperature and Relative Ionic Composition on the Ion Depletion in the Wake of the AE-C Satellite", Geophys. Res. Lett., 6, 101-104 (1980).

Samir, U., M. First, E. J. Mauer, and B. Troy, J. Atmos. Terr. Phys., 37, 577-586 (1975).

Samir, U., R. Gordon, L. Brace, and R. Theis, "The Near-Wake Structure of the Atmospheric Explorer-C (AE-C) Satellite: A Parametric Investigation," J. Geophys. Res., 84, 513-525 (1979).

*Samir, Uri, K. H. Wright, Jr., and N. H. Stone, "The Expansion of a Plasma into a Vacuum: Basic Phenomena and Processes and Applications to Space Plasma Physics," Rev. Geophys. Space Phys., 21, 1631-1646 (1983).

Samir, Uri, P. J. Wildman, F. Rich, H. C. Brinton, and R. C. Sagalyn, "About the Parametric Interplay between Ionic Mach Number, Body Size, and Satellite Potential in Determining the Ion Depletion in the Wake of the S3-2 Satellite," J. Geophys. Res., 86, 11161-11166 (1981).

Schunk, R. W. and A. F. Nagy, "Electron Temperatures in the F Region of the Ionospheres: Theory and Observations," Rev. Geophys. Space Phys., 16, 355-399 (1978).

*Shawhan, S. D. and Murphy, G. B., "Plasma Diagnostic Package Assessment of the STS-3 Orbiter Environment and System for Science," paper presented at AIAA 21st Aerospace Sciences Meeting, Reno, NV (10-13 January 1983).

Siskind, D. E., MS Thesis, Utah State University (1983).

Siskind, D. E., Raitt, W. J., Banks, P. M. and Williamson, P. R., Planet. Space Sci., 32, 881-896 (1984).

Smiddy, M., W. P. Sullivan, D. Girouard, and P. J. Anderson, "Observation of Electric Fields, Electron Densities and Temperature from the Space Shuttle," paper 83-2625 AIAA Shuttle Environment and Operations Meeting, Washington, D.C., (31 October 1 November 1983)

Stone, N. H. and U. Samir, "Bodies in Flowing Plasmas: Laboratory Studies," Adv. Space Res., 1, 361-372 (1981).

*Stone, N. H., Uri Samir, and K. H. Wright, Jr., "Plasma Disturbances Created by Probes in the Ionosphere and Their Potential Impact on Low-Energy Measurements Considered for Spacelab," J. Geophys. Res., 83, 1668-1672 (1978).

Troy, Ballard E., Jr., E. J. Maier, and Uri Samir, "Electron Temperatures in the Wake of an Ionospheric Satellite," J. Geophys. Res., 80, 993-997, (1975).

Whipple, E. C., "Potentials of Surfaces in Space," Reports on Progress in Physics, 44, 1197-1250 (1981).

4C. Key Players

C. Beghin   CNRS
Orleans, France

Dr. William Burke
Space Physics Division
Air Force Geophysics Laboratory
Hanscom AFB, MA 01731

Dr. Henry Berry Garrett
Jet Propulsion Laboratory (144-218)
4800 Oak Grove Dr.
Pasadena, CA 91109
Telephone: (818)-354-2644
            (FTS)-792-2644

Dr. Mike Heinemann
Space Physics Division
Air Force Geophysics Laboratory
Hanscom AFB, MA 01731

Dr. Ira Katz
S-Cubed
Box 1620
LaJolla, CA 92038
(619) 453-0060

Professor Gerry Murphy
Dept. of Physics and Astronomy
University of Iowa
Iowa City, IA

T. Obayashi
ISAS
Tokyo, Japan

Professor W. John Raitt
Center for Atmospheric and Space Sci.
UMC-34
Utah State University
Logan, UT 84322

Dr. Al Rubin
Space Physics Division
Air Force Geophysics Laboratory
Hanscom AFB, MA 01731

Uri Samir
Space Sciences Laboratory
NASA/Marshall Space Flight Center
Huntsville, AL 35812

Dr. Stan Shawhan
Code EE
NASA Headquarters
Washington, D. C. 20546

Appendix A 309

Mr. David E. Siskind
Laboratory for Atmospheric and Space Physics
University of Colorado
Boulder, CO 80309

Dr. Mike Smiddy
Space Physics Division
Air Force Geophysics Laboratory
Hanscom AFB, MA 01731

Dr. P. R. Williamson
Department of Electrical Engineering
Stanford University
Stanford, CA 94305

4D. Existing data

Within the U.S. most of the laboratory measurements and modeling has been done by Samir and Stone at MSFC. However, the plasma ram/wake effect has been extensively studied over the last 20 years by the theoretical group in Russia led by A. V. Gurevitch.

Data on satellite wake measurements in the lower ionosphere have also been analyzed by Samir and co-workers from a number of scientific satellites, the largest of which involved an early rendezvous experiment with an Agena/Gemini combination. More recently measurements on the thermal plasma environment of the Space Shuttle have been made on STS-3 (Banks, Raitt, Williamson; VCAP), (Shawhan, Murphy; PDP), DoD flights (Smiddy), SL-1 (Beghin; PICPAB: Obayashi; SEPAC).

At present most of the data are not in a form that is easily accessible to outside users except some color graphics summary plots of PDP data. The STS-3/VCAP data are all available on a disc pack at Stanford University, although part of the thermal plasma data has been reduced to physical parameters.

The Vehicle Charging And Potential (VCAP) experiment flown as part of the OSS-1 payload on STS-3 contained plasma probes as part of its diagnostics. The main function of these probes was to measure the ambient plasma characteristics prior to electron beam operations. They included a means of providing a high DC bias to back off the electrical charging induced by emitting the electron beam. The probes consisted of a 20-cm gridded spherical retarding potential analyzer biased to measure ionospheric ions up to mass 16, and a spherical Langmuir probe to provide an independent measure of the electron component of the ambient plasma. Both probes used an AC technique to directly measure derivatives of the current voltage characteristics. These types of probes have been flown on unmanned spacecraft, and more details are given elsewhere (Raitt, et.al., 1973, 1975).

The data from the probes flown on STS-3 showed the expected ionospheric data appropriate to the orbiting altitude of STS-3 (250 km) and an expected, marked wake effect. In addition, however, signals before the probe was biased near the drift energy of O^+ ions (5 eV) showed an unexpectedly high signal level indicative of turbulence in the ambient ionospheric plasma at a much higher level than had been seen from smaller unmanned spacecraft. This plasma turbulence was seen all around the orbit except during those periods when the probe was in the wake of the Orbiter structure, at which time the plasma currents were too low to be detected.

A second unexpected effect seen on STS-3 by the plasma probes was the occurrence of high density, hot plasma with indications of a mass number greater than 16 and being "tied" to geomagnetic field lines. This plasma was observed to occur only on the dayside of the orbit and only when there is a component of the atmospheric ram vector into the payload bay. The density of the plasma often becomes sufficiently high that the SRPA is completely saturated, indicating a plasma concentration greater than $10^{13}/m^3$ for molecular ions. The species of the ion is not well defined by the SRPA because the plasma signature is only partially obtained at the end of the probe voltage sweep. However, if they are assumed to be O^{+2} (or N^{+2} or NO^+) and signal levels are matched, an on-scale examples shows a molecular ion concentration of $2 \times 10^{12}/m^3$ with an ion temperature of 2500 K. Coincident with the appearance of first turbulence, then molecular plasma, we see

Appendix A   311

an increasing electron temperature at times reaching values up to 6000° K. These effects have been described more fully by Raitt, et al., 1984; Siskind, et al., 1984; and Siskind, 1983.

In summary, observations indicate that under certain conditions an interaction occurs between the Orbiter and/or its local atmosphere moving at orbital speed through the ionospheric F-region. This interaction results in the generation of heated plasma with electron concentration exceeding $10^{13}/m^3$ and a plasma temperature (Te +Ti) up to 10^4 K. The STS-3 results indicated that sunlight and a ram component into the bay were necessary for the production of this enhanced plasma. A separate plasma effect, seen whether enhanced plasma densities are present or not, is a larger degree of plasma turbulence in the range 0.05 to 1% n/n. The signature of the turbulence can be interpreted as a component at 2.2 kHz, which represents a spatial wavelength of 3.6 m if the turbulent plasma has no drift velocity in the geomagnetic frame. The signature of the enhanced plasma indicates that it too has no drift velocities relative to the geomagnetic field lines.

SYNOPSIS

1. Who has data/models?
   a. Data:
      Prof. W. John Raitt,   Utah State University, STS-3
      Prof. Stan Shawhan[1],   University of Iowa, STS-3
      Dr. Mike Smiddy,   AF Geophysics Laboratory, STS-4
      Dr. R. Narcisi[2],   AF Geophysics Laboratory, STS-4
   b. Models:
      Ira Katz[3],   S-Cubed

   (1) Survey plots from PDP are also available from World Data Center-A Rockets and Satellites (Goddard Space Flight Center)
   (2) R. Narcisi (deceased). Contact Dr. W. Swider, AFGL
   (3) Code under development

2. What format are data/models in?

   a. Data: Published literature (see 1.B, List of References)
   b. Models: Computer model (Contour plots of plasma parameters)

4E. Required data

With regard to ram/wake effects we need data on the following parameters at different locations in and out of the payload bay:

Neutral composition and total pressure
Neutral flow directions
Neutral temperature
Ion composition and total concentration
Ion flow directions
Electron concentration
Plasma temperature
Mapped turbulence distribution in plasma and neutrals.

These data all need to be tagged with time and the Orbiter state vector. Standard software packages should be available to convert the state vector so as to define the velocity and magnetic field vector in the Shuttle coordinate system.

These data should be available through the NASA SCAN network with a user friendly file access system, including a comprehensive HELP file to define the content and format of the files.

The data format should be compact and fast to handle. Character files have the advantage of being printable by common utilities, but in handling large amounts of data the cost of converting back to internal computer format can be very high. Unfortunately, internal floating point formats are not standardized, and again a conversion may be necessary. An option which should be considered is to use 16-bit integers which contain scaled data, so that the numbers stay within the 16-bit range. This is compact and yet gives a wide dynamic range.

The variable scaling can be handled by a separate table contained in the file header. Models should also be available for the parameters listed earlier, probably in the form of parameterized tables rather than programs to solve the basic equations for specific boundary conditions.

4F. Possible experiments

Direct measurements mapped in a region around the Orbiter for specific attitudes and positions of the parameters listed above should be made. These will characterize the ram/wake interaction and will provide necessary data to validate models. The following question needs to be addressed: What are the spatial and temporal dependences of the plasma characteristics about Shuttle?

The output should be: Readily available contour plots of the plasma density, and readily available contour plots of the electric potential.

4G. Mitigation

We should try and reach all groups having thermal plasma measurements and models pertaining to the Space Shuttle surface interactions, and develop a questionnaire to scope what measurements have been made (or are funded for future flights) and see if there are any obvious gaps in "parameter/position/orientation space". Based on the response and distillation of information received (which may involve further paper/telephone iterations), consideration should be given to holding a sponsored Workshop on Surface Interactions, to produce a report for the guidance of NASA/DoD to future activities in this area. This would lead to a much better understanding of these surface effects in good time to have input for Space Station operations.

314　　　　　SPACE SHUTTLE ENVIRONMENT

5. GLOW (by S. Mende, J. Gregory, B. D. Green, and E. Murad)

5A. Description

1. DESCRIPTION OF INTERACTION

Glows in visible and very near infrared wavelengths have been observed on Atmospheric Explorer (AE-E), Dynamic Explorer-B (DE-B), and Space Shuttle spacecraft. A glow associated with a ram effect has been reported on all three spacecraft. The Space Shuttle observations have also reported glows associated with thruster firings.

The physical process leading to the glow phenomenon is poorly understood at present. NASA has many planned scientific investigations involving instrumentation in UV, visible and infrared wavelengths. As we push instruments to higher sensitivity with larger apertures, the instruments become more susceptible to effects from glow. Programs such as Space Telescope, IRT and other optical facilities can plan and optimize around ram glow phenomena, once we understand the physical process. These planning considerations can include operational constraints with respect to the velocity vector at Shuttle orbit altitude, instrument baffle materials and coatings, surface conditioning of the Orbiter (for Shuttle payloads) and related planning. We believe that a great deal more information has to be obtained, and be accessible to our scientific colleagues, so that products from these scientific programs can be optimized.

2. SUMMARY

The AE-E satellite was equipped with a Visual Airglow Experiment (VAE) which observed atomic and molecular features in the Earth's airglow layer. Backgrounds in the photometer filter channels were found to have a variability with ram angle. This was reported by Yee and Abreu (1982, 1983). The data displayed a detectable level of luminosity in the near UV channels of the instrument (3371 Angstroms), with increasing luminosity towards the red wavelengths (7320 Angstroms). The background in all filter channels, when plotted, described a bright ram source, increasing in brightness toward the red wavelengths. The analysis suggested that the glow extended well away from the spacecraft alluding to the probability that the emitter is some metastable species. OH Meinel bands were discussed as being likely candidate species for emission, since the general red character and emission lifetime seemed to fit the evidence. The Yee and Abreu (1982) analysis found a strong correlation between the ram emission intensity and altitude. The emission intensity closely followed the atomic oxygen scale height above 160 km altitude. Atomic oxygen is the probable aeronomical constituent to be a chemical catalyst for whatever process is occurring. Slanger (1983) was among the first to suggest the OH hypothesis.

The AE-B spacecraft was equipped with a high resolution Fabry-Perot interferometer (FPI) (Hays, et al. (1981)). In this instrument, a 7320 Angstrom filter was utilized in series with the Fabry-Perot etalon. Abreu, et al., (1983) reported on the background with ram effect associated with this channel. A ram glow was reported and the inferred etalon spectrum showed similarity with the OH spectrum observed in nightglow from the atmospheric limb. The evidence from these two spacecraft favors the OH hypothesis for the observed glows.

Glow observations have been reported by a number of investigators from Shuttle missions STS-3, -5, -8, and -9. Banks, et al., (1983) reported glow from Orbiter television and still-camera pictures around the aft spacecraft surfaces, and documented glows associated with an electron accelerator experiment on STS-3. Mende, et al., (1983, 1984a, 1984b) have documented ram glows associated with STS-5, -8, and -9 using an intensified camera. On the later missions (STS-8 and -9), objective grating imagery of spacecraft glow from the vertical stabilizer depicted a red structureless glow (Mende, et al. (1984b)). The spectral resolution is on the order of a few hundred Angstroms. The same instrument was used to observe nightglow dispersion using the nightglow layer, as observed from the Orbiter, as the instrument "slit".

The instrument clearly resolves the airglow atomic emissions and OH band emissions as well as a very strong O_2 (0,0) A-band emission at

7620 Angstroms. It is clear, when examining the two spectra with similar resolution, that the OH, if present in the Shuttle glow, is immersed in a spectrum with a lot of other red emitters. Of particular note on STS-8 observations was that glows from surface samples including aluminum, kapton, and Z302 (a polyurethane black paint typically used in low light level detection instrument baffles) were not equally bright. The surface characteristic and/or the material makeup clearly was shown to effect the glow brightness.

High resolution spectral measurements of the ISO spectrometer on Spacelab-1 show the presence of N_2 1PG bands (Torr and Torr, 1984). These observations are indirect in that the spectrometer was pointed in a direction ideally suited to study the airglow and not any Shuttle surface. During these measurements an excess emission was observed. This emission was ascribed to emission from the spectrometer surfaces, in a fashion similar to (but not necessarily the same as) the Shuttle glow. There is also a number of other observed emission features which may be part of the natural aurora/airglow background environment and therefore may not be part of the Shuttle glow. There is no associated imaging data with the ISO measurements; therefore, the precise determination of the source of the emission could be difficult.

Green (1984) recently reviewed the ram glow data and theory for the Shuttle environment. The review described the two classes of mechanisms; one being molecular emission from surface collisions and another due to the plasma critical velocity effect. In this discussion of molecular emission from surface collisions, the Langmuir-Henshelwood process which involves atoms absorbing on a surface, migrating, reacting and escaping in the gas phase was discussed. There is evidence from laboratory studies that O will readily oxidize carbon and come off in the gas phase. If a surface is covered with absorbed atoms, an incident atom can react and escape in the gas phase (a Rideal process). O_2 and OH (both of which can have red emission in excited levels) can be produced through this process (Manella and Hartect, 1961). It has been postulated that N formed by impact will recombine and form N_2 ($a^3\Sigma_u^+$) at high vibrational levels and then

Appendix A   317

radiatively or collisionally go to the $N_2(B^3\Pi g)$ state which emits the 1st positive bands (Green, 1984). Vibrationally excited CO, OH or electronically excited N_2 were postulated as most likely candidates and chemically plausible with the evidence at hand.

A proposed plasma process (Papadopoulos, 1984) involves a two stream instability between the incoming ram material and reflected ions. The ion instability sets up an electrostatic wave which in turn heats the ambient electrons. The "pumped up" electrons can in turn excite *in situ* (and ramming) constituents. Exciting the electrons to tens of eV will allow e + X reactions. The energy is sufficient to excite N_2 to 2PG bands, and ionize to the "first negative" bands (N_2^+). Atomic and molecular ionic UV emissions could arise with this process, whereas the chemical processes postulated will radiate most strongly in the red and infrared region. There are some irreconcilable observations which cannot be explained by Green's hypothesis. The altitude dependence of the glow intensity on AE/BE does not follow the N_2 number density. The properties of the glow associated with thruster firings also raise some questions about a process which involves only the atmospheric N_2 molecules. For plasma processes N_2^+ 1st negative (1,0) at 3914 Angstroms and N_2 2nd positive (0,0) at 3371 Angstroms are expected to be dominant spectral features.

There are also difficulties in reconciling the observations with the suggested plasma mechanism. A review of the plasma hypothesis and the chemical mechanisms so far proposed has been given by Kofsky (1984).

5B. References

Abreu, V. J., W. R. Skinner, P. B. Hays, and J. H. Yee, "Optical Effects of Spacecraft-Environment Interaction: Spectrometric Observations by the DE-B Satellite," AIAA-83-2657-CP at Shuttle Environment and Operations Meeting, *op. cit*.

Banks, P. M., P. R. Williamson, and W. J. Raitt, "Space Shuttle Glow Observations," Geophys. Res. Lett., 10, 118 (1983).

Green, B. D., "Atomic Recombination into Excited Molecular States-A Possible Mechanism for Shuttle Glow," Geophys. Res. Lett., 11, 576 (1984).

Green. B. D., G. E. Caledonia, and T. D. Wilkerson, "The Shuttle Environment," AIAA-AIA-A-84-0546, Reno, NV submitted to J. Spacecraft and Rockets (Jan. 1984).

Hays, P. B., G. Carignan, B. C. Kennedy, G. G. Shephert, and J. C. G. Walker, "The Visible Airglow Experiment on Atmosphere Explorer," Radio Sci., 8, 369 (1973).

Kofsky, I. L., and J. L. Barrett, "Optical Emissions Resulting from Plasma Interactions Near Windward Directed Spacecraft Surfaces," AIAA-83-2661-CP at Shuttle Environment and Operations Meeting, op. cit.

Langhoff, S. R., R. L. Jaffee, J. H. Yee, and A. Dalgarno, "The Surface Glow of the Atmospheric Explorer-C and -E Satellites," Geophys. Res. Lett., 10, 896 (1983).

Manella, G., and P. Hartect, "Surface-Catalyzed Excitations on the Oxygen System," J. Chem. Phys., 34, 2177 (1961).

Mende, S. B., O. K. Garriott, and P. M. Banks, "Observations of Optical Emissions on STS-4," Geophys. Res. Lett., 10, 122 (1983).

Mende, S. B., "Vehicle Glow," AIAA-83-2607-CP, at Shuttle Environment and Operations Meeting, op. cit. (1983).

Mende, S. B., J. A. Hoffman, and P. M. Banks, "Experimental Observations of Orbiting Vehicle Induced Luminosities," presented at Fall American Geophysical Union Meeting, San Francisco, CA (December 1983).

Mende, S. B., "Experimental Measurement of Shuttle Glow," AIAA-84-0550 presented at AIAA 22nd Aerospace Sciences Meeting, Reno, NV (January 1984).

Mende, S. B., G. R. Swenson, and K. S. Clifton, "Preliminary Results of the Atmospheric Emissions Photometric Imaging Experiment (AEPI) on Spacelab-1," submitted to *Science* (1984).

Papadopoulos, K., "On the Shuttle Glow," Radio Science, 19, 571-577 (1984).

Sandie, W. G., S. B. Mende, G. R. Swenson, M. B. Polites, "Atmospheric Emissions Photometric Imaging Experiment on Spacelab-1," Optical Engineering, 22, 756 (1983).

Scialdone, J. J., "Self-contamination and Environment of an Orbiting Spacecraft," NASA TN D-6645 (May 1972).

Slanger, T. G., "Conjectures on the Origin of the Shuttle Glow of Space Vehicles," Geophys. Res. Lett., 10, 130 (1983).

Torr, M. R., P. P. Hays, B. C. Kennedy, and J. C. G. Walker, "Intercalibration of Airglow Observations with the Atmospheric Explorer Satellite," Planet. Space Sci., 25, 173 (1977).

Torr, M. R. and D. G. Torr, "A Preliminary Spectroscopic Assessment of the Spacelab-1/Shuttle Optical Environment," submitted to J. Geophys. Res., (1984).

Witteborn, F. C., K. O'Brien, and L. Caroff, "Measurement of the Night-Time Infrared Luminosity of Spacelab-1 in the H and K Bands," NASA/ Ames, preprint (1985).

Yee, J. H., and V. J. Abreu, "Optical Contamination on the Atmosphere Explorer-E Satellite," *Proceedings of SPIE Technical Symposium*, 338, 120 (1982b).

Yee, J. H., and V. J. Abreu, "Visible Glow Induced by Spacecraft-Environment Interaction," Geophys. Res. Lett., 10, 126 (1983).

Yee, J. H., and A. Dalgarno, "Radiative Lifetime Analysis of the Spacecraft Optical Glow," AIAA-83-2660-CP at Shuttle Environment and Operations Meeting, op. cit.

5C. Key players

Dr. Vincent Abreu
Space Physics Research Lab
2455 Hayward Ann Arbor
MI 48105 USA (313) 764-7220

Dr. Marcel Ackerman
Inst. D'Aeronomie Spatiale
Avenue Circulaire 3
B-1180 Bruxelles BELGIUM 2 374-2728

Dr. Jack Barengoltz
JPL/CALTECH 157/507
4800 Oak Grove Dr.
Pasadena, CA 91109   USA
(818) 354-2516

Dr. William F. Denig
AFGL/PHK Hanscom AFB
Hanscom AFB
Bedford, MA 01731   USA
(617) 861-3989

Dr. B. David Green
Physical Sciences, Inc.
P.O. Box 3100
Andover, MA 01810
USA (617) 475-9030

Appendix A

Dr. John Gregory
Univ. of Alabama
Huntsville, AL 35899
USA (205) 895-6028

Mr. Robert R. Hale
JPL/CALTECH 157/507
4800 Oak Grove Dr.
Pasadena, CA 91109
USA (818) 354-4045

Dr. Karl Knott
Space Science Dept., ESA
Keplerlaan 1 2200 AG
Noordwijk, NETHERLANDS 31-1719

Dr. Steve Mende
Lockheed Palo Alto Dpt 35-12
3251 Hanover St.
Palo Alto, CA 94304
USA (415) 858-4082

Dr. Ed Murad
AFGL/PHK Hanscom AFB
Bedford, MA 01731
USA (617) 861-3046

Dr. John Raitt
Utah State University, CASS UMC-34
Logan, UT 84322
USA (801) 750-2983

Dr. William Swider
AFGL/LID Hanscom AFB
Bedford, MA 01731
USA (617) 861-3891

5D. Existing data

Coarse spectral resolution data and image intensifier photographic data were taken on a number of missions by the Lockheed Palo Alto Research Laboratories. Principal Investigator is Dr. S. B. Mende, Lockheed Palo Alto Research Laboratory, Palo Alto, California. Spectral content of the data has been analyzed and published. Imaging data are being analyzed for the spatial distribution of the glow by Dr. G. Swenson, Lockheed Palo Alto Research Laboratory, Palo Alto, California.

The highest resolution spectrum which is currently available was taken by Utah State University's Imaging Spectrometer Experiment (ISO). Principal Investigator is Dr. Marsha Torr, Utah State University, Logan, Utah. The analysis of the data is in process, and a preliminary publication of results is available (Torr and Torr, 1984).

The Atmospheric Explorer glow was analyzed by Dr. J. H. Yee (Harvard University) and V. J. Abreu (University of Michigan). Glow observations from the Dynamics Explorer (DE) spacecraft were discussed by Abreu of the University of Michigan (Dr. J. H. Yee, Harvard University, V. J. Abreu, University of Michigan, Ann Arbor, Michigan).

The only measurement of Shuttle atmospheric interaction in the infrared was performed by F. C. Witteborn of NASA Ames with a ground-based instrument from Mt Haleakala, Maui, Hawaii.

There is no consensus about the mechanism producing the glow and therefore there are no models which predict spacecraft glow behavior. Two types of modeling are relevant to our current understanding of glow behavior. One type is relevant to generating appropriate molecular band emissions to compare them to observed spectra. Such modeling was carried out for the OH emissions by Langhoff, et al. (1983). (S. R. Langhoff, NASA Ames Research Center, Moffett Field, California).

Another type of modeling related the geometry of the observed light emissions to the spatial characteristics of the emission processes. This type of modeling is carried out by Yee and Dalgarno Harvard University (1983).

Appendix A   323

Theoretical discussion of glow related phenomena has been published by the following individuals:

1. T. Slanger, SRI International, Stanford, California
2. B. D. Green, Physical Sciences, Inc., Andover, Massachusetts
3. D. Papadopolous, SAI, University of Maryland
4. S. Langhoff, NASA Ames Research Center, Moffett Field, California
5. I. L. Kofsky, Photometrics, Inc., 4 Arrow Drive, Woburn, Massachusetts

5E. Required data

1. What information should exist? In order to optimize high sensitivity, spacecraft-borne optical detection instruments, it is desirable to establish the intensity of the spacecraft glow as a function of:

Wavelength
Altitude (or atmospheric density)
Velocity vector
Shuttle surface material
Time variability (interval after launch)
Spatial extent of glow

For a complete determination of the dependence of the intensity of the glow on all these parameters, many measurements have to be made. Instead of making all these measurements, it would be perhaps more reasonable to make a few _key_ measurements directed towards a clear understanding of the glow phenomena. Once an understanding is obtained, the intensity dependence could be obtained by modeling the phenomenological behaviour.

Ideally speaking, there ought to be a set of tables which would provide the intensity of the spacecraft glow as a function of the above variable parameters. The intensity should be given in Rayleighs per Angstrom and ergs per micron for an instrument looking in a direction perpendicular to the Orbiter skin, for a given material consisting of an indefinitely large theoretical surface sample area, as a function of the surface normal and the ram velocity. From these idealized data the glow intensity could be modelled for a realistic spacecraft in a given situation.

Perhaps the most important primary task is the spectral identification of the emitting species. The most important information therefore is the list of identified materials that are emitting in the gaseous phase.

2. What measurements are required?

The glow investigation is still in its infancy. There is a great deal of room for exploratory measurements, especially in the infrared and ultraviolet. To maintain an orderly progress in understanding the phenomena, and to maximize the cost/benefit ratio of the investigation, a program plan is outlined as follows:

a. First Phase of the program
  i. Moderate spectral resolution (30 Angstroms) spectroscopy and imaging in the visible (4000 to 8000 Angstroms)
  ii. Moderate spectral resolution (30 Angstroms) spectroscopy and imaging in the near ultraviolet (2000 to 4000 Angstroms)
  iii. Moderate spectral resolution (50 Angstroms) spectroscopy and simultaneous visible imaging in the infrared (8000 Angstroms to 4 microns)
  iv. High resolution spectroscopy in the visible near infrared (6000 to 8500 Angstroms) to measure vibrational and rotational structure of the known red glow.

Besides the exploratory nature of the investigation under Phase 1 i, ii, and iii, these measurements will give us insight about the glow production processes. It would be highly desirable to give a definite upper limit to the contribution of the O_2 Hertzberg bands in the near UV and perhaps dismiss the direct contribution of O_2 to the glow.

Another important point is to examine whether the N_2 V-K bands exist in the glow. If N_2 1PG bands are a serious contributor to the glow, then copious amounts of V-K bands would be expected in the near UV.

In the visible region it is necessary to make high resolution spectral measurements (with accompanied imaging) to examine the molecular band structure, and simultaneously to image to isolate the source of the light and exclude contamination by other light sources.

Spectral emissions of all the emitting candidates are unique and straightforward to distinguish. O_2 atmospheric, OH Meinel and N_2 1st positive are all red emitters as is CO (n = 6, 7, 8, Wn=5). It is very plausible that more than one of the proposed processes are contributing to the glow.

The process of slow or long surface residence times will characterize the effective rotational distribution of an observed emission. A molecule can take on a wide range of rotational distributions depending on the excitation process involved. If atmospheric N_2 is excited through a fast process, then its rotational distribution would be characteristic of the 1000° K thermospheric environment. If the N_2 was dissociated on impact and the N^2 ($B^3\Pi u$) state was formed through the N_2 ($A^3\Sigma_u^+$), the vibrational population of nitrogen and resultant 1st positive emission would be characterized by the A to B state transition process. A thermalized emission from N_2 ($B^3\Pi u$) on the other hand, would show a rotational distribution of the relatively cold spacecraft surface (300°K). Through the rotation/ vibration distribution, three very different emission characteristics would be expected as described above.

Synthetic band models are a valuable tool to resolve rotational and vibrational information. First positive emission extends out to ~3µm in the infrared. The vibrational distribution is characterized by the emission intensity between 6000 Angstroms and 3 microns.

Besides the spectroscopic studies of the glow in Phase 1, special emphasis should be put on measuring the variation of glow intensity with Shuttle altitudes to establish whether the intensity is a linear function of O or N_2. It would be also important to look for glow emission in neutral gas releases to test the free molecular excitation hypothesis. Further studying of the glow properties of material samples would also be very useful with particular emphasis on glow interaction in the presence of simple molecular gas jets directed on them.

    b. Second Phase of the Experiment

        i. Extend measurement into the far infrared (4 to 20 microns)

        ii. Extend measurement into the vacuum ultraviolet (1100 to 2000 Angstroms)

Better definition for Phase 2 will be obtained after the results of Phase 1 are known.

5F. Possible experiments

The nature of the problem and approach is one of putting together an assembly of instruments and an observational plan for an early mission to learn quickly about the rudimentary facts about spacecraft glow.

First of all it is well established that for the desired spectral coverage it is necessary to mount the instrument outside of the Orbiter cabin. Secondly it is clear that the instrument has to be able to execute some form of pointing to be able to optionally look at glowing spacecraft surfaces or surface samples. From these considerations it is necessary that the instrument field of view can be steered towards glowing surfaces by remote pointing. The instrument has to be configured into a pointing system, or its field of view has to be directed with mirrors.

For Phase 1 we propose the following specific instruments:

INSTRUMENT NO. 1: <u>RED BROADBAND IMAGING SPECTROMETER</u>

Spectrometer capable of operating in imaging as well as conventional spectral mode. Two-dimensional, remote imaging type of detector.

A summary of the optical characteristics includes the following:

Appendix A   327

Field of view (full): 20 degrees
Spectral range (with grating): 4200-8400 Angstroms
Resolution 0.002" slit: 30 Angstroms
Grating characteristics: 300 lines/mm
Filters: 4278 Angstroms (N_2^+ 1st Neg, 1-0 BAND); 7380 Angstroms (N_2 1st Pos, 5-3 BAND); 7774 Angstroms (O^+, 5p-5s)

INSTRUMENT NO. 2: UV BROADBAND

The UV broadband instrument could be identical in electronic, mechanical, and control configuration to instrument 1, except that the UV instrument will be configured with a UV transmissant lens complement to give good imaging characteristics between 2000 Angstroms and 4000 Angstroms, the intended operating range of this instrument. The UV grating will also be optimized for providing dispersion between 2000 and 4000 Angstroms. The front intensifiers will contain an S25 photocathode on quartz for UV response.

The following is a summary of the optical characteristics of this instrument:

Field of view (full): 20 degrees
Spectral range (with grating): 2000 to 4000 Angstroms
Resolution: 60 Angstroms
Grating characteristics: 300 lines/mm
Filters: 3371 (N_2 2nd pos); 2500 Angstroms (2000 to 3000 Å Broadband); 3500 Angstroms (3000 to 4000 Å Broadband)

INSTRUMENT NO. 3: HALF METER EBERT SPECTROMETER - RED

This instrument is well equipped to perform high resolution measurements between 5000 Å and 9000 Å. Wavelength information will be horizontal, while spatial information will be in the vertical scan of the imager.

A priority should be given to performing good resolution on the red visible first, where imagers given the spatial extent of the glow. A brief summary of desirable spectrograph characteristics is

328   SPACE SHUTTLE ENVIRONMENT

given as follows:

    Field of view: 2 x 0.1°
    Spectral range: 5000 to 9000 Å/200 Angstroms per grating
                     increment
    Resolution: 1 Angstrom
    Grating character: 1200 lines/mm

INSTRUMENT NO. 4: <u>CIRCULAR VARIABLE FILTER - INFRARED</u>

This broadband IR instrument will be used to gather course spectra between 1.2 and 4+ microns and will concentrate on relative intensities in this region with respect to the measured intensities at lower wavelengths.

The detector could be a thermo-electrically cooled detector whose line of sight is oriented with the other instruments. The variable filter would rotate and scan to high and then lower wavelengths continuously. Detector current and filter position information would be instrument outputs. A summary of instrument characteristics is as follows:

    Field of view: 1°
    Spectral range: 1.2 to 3 microns
    Resolution: 0.07 micron
    Filter: 4", < 2 percent BW
    NEP: TBD
    Aperture: 50 mm

5G. Mitigation (TBD)

6. <u>SURFACE EROSION</u> (by L. Leger)

6A. Description

    Data obtained on recent low Earth orbit flight experiments indicate that environment/spacecraft surface interactions are important considerations in optical system performance. In fact, with the development of large scale, highly sensitive instrument and flight opportunities presented by Space Shuttle, a new, more refined understanding of the environment and its interaction with spacecraft surfaces is rapidly evolving. These developments are central to several Space Station design issues such as solar arrays, scientific observing platforms and stability of exterior coatings.

One of the important surface interactions is the recession or removal of mass from surfaces during flight. This environmental exposure effect was observed during early Shuttle flights (Ref. 1-9) and has been investigated on two dedicated experiments conducted on STS-5 and STS-8. Surface changes are seen only on the ram or forward facing surfaces apparently due to impingement of atmospheric constituents on these surfaces at high velocity (7.5 km/sec). Since the major component of the low Earth environment is atomic oxygen, a strong oxidizing agent, and since the impingement flux is approximately 10^{15} atoms/cm^2 sec, organic materials undergo significant surface changes.

The amount of surface change is proportional to the total integrated atomic oxygen impingement flux or fluence which is, in turn, dependent upon parameters such as surface flight attitudes, altitude, exposure time and solar activity because of its relationship to ambient atmospheric density. Models for determining the fluence for a given set of spacecraft orbital parameters are being developed by several organizations.

6B. References

Banks, B., M. Mirtich, S. Rutledge, and D. Swel, "Sputtered Coatings for Protection of Spacecraft Polymers," NASA Technical Memorandum 83706 (April 1984).

Fraundorf, P., D. Lindstron, N. Pailer, S. Sandford, P. Swan, R. Walker, and E. Zinner, "Erosion of Mylar and Protection by Thin Metal Films," AIAA Shuttle Environments and Operations Meeting (October 1983).

Leger, L. J., "Oxygen Atom Reaction with Shuttle Materials at Orbital Altitudes," NASA TM 58246 (May 1982).

Leger, L. J., "Oxygen Atom Reaction with Shuttle Materials at Orbital Altitudes--Data and Experiment Status," AIAA 21st Aerospace Sciences Meeting (January 1983).

Leger, L. J., I. K. Spiker, J. F. Kuminecz, T. J. Ballentine, and J. T. Visentine, "STS Flight Leo Effects Experiments--Background Description and Thin Film Results," AIAA Shuttle Environment and Operations Meeting (October 1983).

Liang, R. and A. Gupta, AIAA Shuttle Environment and Operations Meeting (October 1983).

McCargo, M., R. Dammann, J. Robinson, and R. Milligan, Proceedings of International Symposium on Environmental and Thermal Systems for Space Vehicles, Toulouse, France (October 4-7, 1983).

Park, J. J., T. R. Gull, H. Herzig, and A. R. Toft, "Effects of Atomic Oxygen on Paint and Optical Coatings," AIAA Shuttle Environment and Operations Meeting (October 1983).

Peters, P. N., R. C. Linton, and E. R. Miller, "Results of Apparent Atomic Oxygen Reaction on Ag, C and Os Exposed During the Shuttle STS-4 Orbits," J. Geophys. Res., $\underline{10}$, 569-571 (July 1983)

Whitaker, A., "LEO Atomic Oxygen Effects on Spacecraft Materials," AIAA Shuttle Environment and Operations Meeting (October 1983).

Appendix A   331

6C. Key players

Lubert Leger
NASA Johnson Space Center

Ann Whitaker
NASA Marshall Space Flight Center

John Park
NASA Goddard Space Flight Center

Bruce Banks
NASA Lewis Research Center

Wayne Slemp
NASA Langley Research Center

Ranty Liang
NASA Jet Propulsion Laboratory

Wayne Stuckey
Aerospace Corp., El Segundo, Calif.

C. Radley
British Aerospace Corp., United Kingdom

Lyle Bariess
Martin Marietta, Denver, Colorado

John Gregory
University of Alabama, Huntsville, Alabama

Ernst Zinner
Washington University, St. Louis, Missouri

6D. Existing data

1. Flight Experiment Data

Most of the data relating to the behavior of materials in the atomic oxygen flight environment has been obtained on two experiments conducted on STS-5 and STS-8. Both flights provided an exposure of 41 hours to atmospheric ram conditions. However, due to altitude, attitude and solar activity differences the total fluence was different: 1.0×10^{20} atoms/cm^2 on STS-5 and 3.5×10^{20} atoms/cm^2 on STS-8. Additionally, the attitude of the vehicle was such that the velocity vector rotated about the exposed surfaces once every orbit for the STS-5 case and was held normal to the surface on STS-8.

Samples exposed under the conditions described above were studied post flight for changes. Since the exposures resulted in significant loss of material (as much as 12 microns of surface thickness), mass change measurements of the flight samples provide an excellent assessment of reactivity in the environment. Most of the data obtained are reported in terms of reactivity defined as thickness of material loss normalized to total oxygen fluence (obtained from atmospheric models, spacecraft velocity and exposure history). If the particular material in questions contained fillers which were stable to atomic oxygen, the matrix material is shadowed or protected from impingement by the fillers and, consequently, mass loss rate changes with time. For these samples, mass loss per unit area is typically reported.

In addition to mass loss measurements, the surface morphology of exposed samples was extensively examined using, primarily, scanning electron microscopy (SEM). Surface compositional changes were investigated using techniques such as ESCA, EDAX, and Raman spectroscopy.

A summary of data obtained on experiments conducted to date is shown in Table I. The reaction rate range includes data for both sweeping and fixed velocity vector conditions. The reaction rates can be used for general assessment of effects on spacecraft surfaces. For more detailed assessment of spacecraft exposure problems, the source of the data as noted in Table I should be contacted for detailed discussion. Sufficient data exist at this time to permit generalizations which can be used in gross assessment of surface effects. These generalizations are listed below:

a. Unfilled organic materials containing C, H, O, N, and S, only, react with approximately the same reaction efficiency (2-4x10^{-24} cm^3/atom).

b. Prefluorinated carbon based polymers and silicones have lower reaction efficiencies by a factor of ten or more than organics.

c. Filled or composite materials have reaction efficiencies that are strongly dependent upon the characteristics of the fillers.

d. Metals, except for silver and osmium, do not show macroscopic changes. Microscopic changes have been observed and should be investigated for systems very sensitive to surface properties. Silver and osmium react rapidly and are generally considered unacceptable for use in uncoated applications.

e. Magnesium fluoride and oxides in various forms show good stability.

As mentioned previously, the specific effects on a given surface depend upon the exposure parameters, solar activity and the specific reaction efficiency in question. These parameters have been combined in nomographic form in Ref. 4 and are reproduced in Fig. 1. This nomograph relates surface loss rate to altitude, attitude, and solar activity for two reaction efficiencies. The specific attitude as noted is for an inertially fixed surface which is unprotected, resulting in exposure to both sides during a given orbital pass. Surfaces which are continuously pointed to ram conditions will have a factor of 2 greater loss than shown in the nomograph.

Environment Simulation

Several organizations have developed apparatus over the last two or three years in an attempt to simulate the low Earth orbit neutral environment to study surface interactions. The apparatus basically produces neutral atomic oxygen at thermal (300° K) or suprathermal energies, or atomic oxygen ion beams at various energies. Thermal (0.7 eV) atomic oxygen is produced by microwave excitation of molecular oxygen typically at a pressure of 1 torr. About 10 percent atomization occurs from the excitation process. Materials are exposed, therefore, to a 90 percent molecular and 10 percent atomic oxygen mixture. Reaction rates of organic materials to atomic oxygen as determined from this approach are a factor of 100 lower than rates determined from flight experiments. Qualitative effects of the atomic oxygen produced by this approach have been reported by M. McCargo. Measurement of atomic oxygen concentration, which allows a quantitative measurement of reaction rates, has been made by the author at Johnson Space Center.

Suprathermal atomic oxygen beams have been produced from arc discharge methods by Arnold (Ref. 11). This apparatus which produces about 1 eV atomic oxygen at approximately the orbital flux has been used to study reaction with carbon and Kapton. Reaction rates of these two materials reported by Arnold (Ref. 12) agree with flight data from STS-5.

Atomic oxygen ion beam surface interaction studies have been conducted at Lewis Research Center (Ferguson, Ref. 13), Ames Research Center, Martin Marietta/Denver, and General Electric Company. Etching of organic surfaces can easily be produced using this technique; however, reaction rates are approximately 100 times higher than flight rates. Energies typical of the atomic oxygen ions produced by this technique range from tens to thousands of electron volts.

Other technques for generating a 5 eV atomic oxygen beam are being pursued (most notably Los Alamos National Labs/Jon Cross) for mechanistic studies.

Appendix A

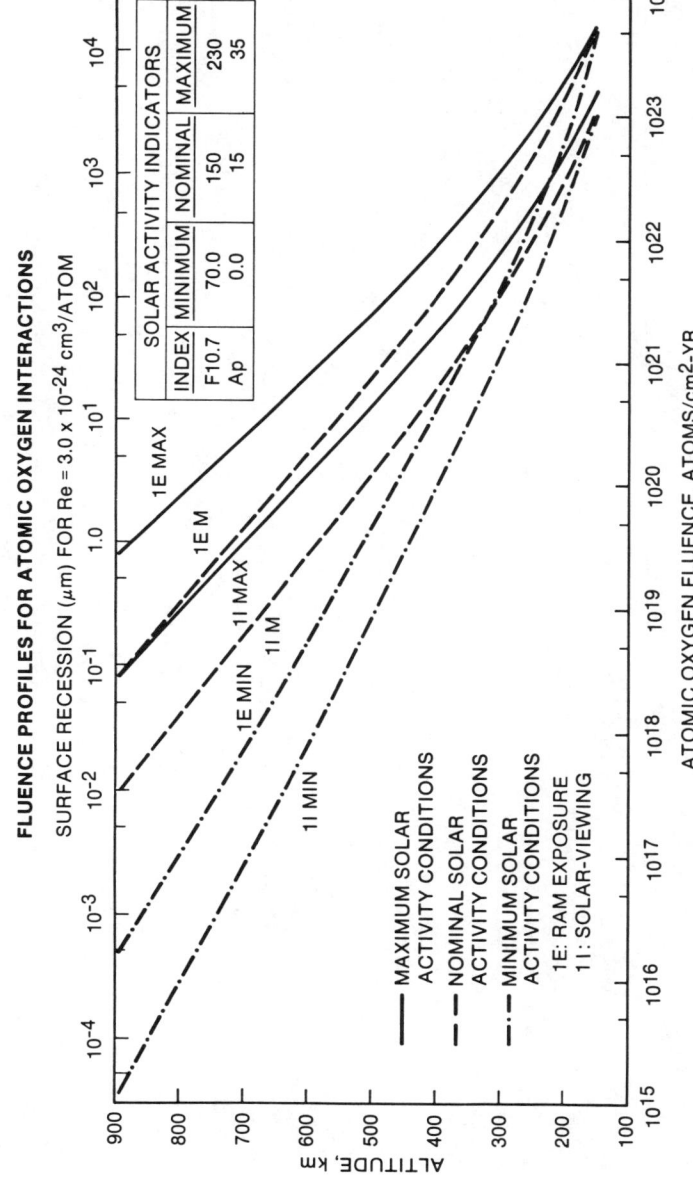

Fig. 1. Fluence for Various Solar Activity Conditions and for Two Flight Attitudes, Ram and Solar Oriented

Table I
EFFECTS OF ATOMIC OXYGEN ON MATERIALS

	REACTION EFFICIENCY SWEEPING VELOCITY VECTOR (STS-364)	DIRECT VELOCITY VECTOR (STS-8)	AVG.	RANGE OF REACTION EFFICIENCY	NO. OF* SAMPLES (EST.)	OPTICAL PROPERTY CHANGES $\Delta\alpha_s$	$\Delta\epsilon$	ΔR_s	COMMENTS	DATA SOURCE
	($\times 10^{-24}$ CM^3/ATOM)									
TEFLON										
FEP & TFE	<0.2	<0.03	<0.1	0-<0.2	>22					JSC
25% POLYSILOXANE POLYIMIDE		0.3	0.3							LaRC
TEDLAR, WHITE	0.4	0.6	0.5	0.4-0.6	>11					JSC/MSFC
7% POLYSILOXANE POLYIMIDE		0.6	0.6							LaRC
CARBON		1.2	1.2	0.9-1.7	≥ 4					LaRC/MSFC/UA
POLYBENZIMIDAZOLE (PBI)		1.5	1.5							LaRC
EPOXY		1.7	1.7							JPL
POLYSTYRENE		1.7	1.7							JPL
KAPTON, BLACK	1.4	2.2	1.8	1.4-2.2	>13			-5.1		JSC/MSFC
TEDLAR, CLEAR	1.3	3.2	2.3	1.3-3.2	16					JSC
POLYSULFONE		2.4	2.4							JPL
KAPTON	2.5	3.0	2.6	2.2-3.1	>81			-2.4		JSC
MYLAR	2.3	3.9	2.8	1.8-4.5	>46					JSC
PMMA		3.1	3.1							JPL
POLYIMIDE		3.3	3.3	2.3-4.7	10					LaRC
POLYETHYLENE		3.7	3.7	3.3-4.1	≥ 2					JSC/JPL
Reaction Efficiencies Not Determined										
POLYESTER			Heavily Attacked							WU
POLYESTER WITH ANTIOXIDANT			Heavily Attacked							WU
MYLAR WITH ANTIOXIDANT			Heavily Attacked							WU

* For most materials without an estimate of the number of samples tested, the efficiency OR mass loss data is based on only one (1) sample.

Appendix A

Table I (continued)

EFFECTS OF ATOMIC OXYGEN ON MATERIALS

COMPOSITES	AVG. REACTION EFFICIENCY (X10^{-24} CM^3/ATOM)	RANGE OF REACTION EFFICIENCY	NO. OF SAMPLES (EST.)	OPTICAL PROPERTY CHANGES $\Delta \alpha_s$	$\Delta \epsilon$	ΔR_s	COMMENTS	DATA SOURCE
GR/AL	-0-							LaRC
GR/MG	-0-							LaRC
1034C EPOXY	2.1							LaRC
5208/T300 EPOXY	2.6							LaRC
Reaction Efficiency Not Determined								
POLYIMIDE PMR-15 W/CELION 6000							No discernible change (limited exposure)	JPL
POLYSULFONE P1700 W/CELION 6000							No discernible change (limited exposure)	JPL
KEVLAR FIBER	—						Fibers tested showed 30 percent strength reduction	A

337

Table I (continued)

EFFECTS OF ATOMIC OXYGEN ON MATERIALS

METALS (NOT USED AS COATINGS)	AVG. MASS LOSS mg/cm²	MASS LOSS RANGE	NO. OF SAMPLES (EST.)	OPTICAL PROPERTY CHANGES $\Delta\alpha_s$ $\Delta\epsilon$ ΔR_s	COMMENTS	DATA SOURCE
PLATINUM	-0-					MSFC
LEAD	-0-					MSFC
MAGNESIUM	-0-					MSFC
MOLYBDENUM	-0-					MSFC
HOS-875 BARE AND PREOXIDIZED	-0-					MSFC
TOPHET 30 BARE AND PREOXIDIZED	-0-					MSFC
TUNGSTEN	-0-				Changes in visible transmission	UA
NICKEL	-0-				Changes in visible transmission	UA
COPPER	0.05				Changes in visible transmission	UA/MSFC
Mass Loss Not Determined						
GOLD	Appears resistant			-0-		GSFC
PLATINUM	Appears resistant			-0-		GSFC
TANTALUM	Appears resistant					GSFC
NIOBIUM	--				Changes in visible transmission	UA
IRIDIUM	--				Changes in visible transmission	UA
ALUMINUM	--				Changes in visible transmission	UA
OSMIUM	Heavily attacked					GSFC
SILVER	*					MSFC

*Gains weight from oxidation--loses weight from spalling (black oxide film formed)

Appendix A 339

Table I (continued)

EFFECTS OF ATOMIC OXYGEN ON MATERIALS

COATINGS	AVG. MASS LOSS mg/cm²	MASS LOSS RANGE	NO. OF SAMPLES (EST.)	OPTICAL PROPERTY CHANGES			COMMENTS	DATA SOURCE
				$\Delta \alpha_s$	$\Delta \varepsilon$	ΔR_s		
YB-71 (ZOT)	-0-			-0-	-0-	-0-		LaRC
S13-GLO WHITE	-0-			-0-	-0-	-0-		LaRC
RTV-670	-0-			-.004		+0.1		MSFC
RTV-615 BLACK CONDUCTIVE SILICONE	-0-			-0-	-.005		71% increase in resistance/unit area	GSFC
28-20-1	-0-							MSFC
100Å NiCr	-0-							MSFC
GSFC GREEN	-0-			-.002				MSFC
150Å ALUMINUM	-0-			-0-	-0-	-0-	100°A alum sample had slight mass loss	MSFC
650 Å S_iO_2	<0.002			-0-	-0-	-0-	UV/IR transmission unaffected	LeRC
650 Å of 96% SiO_2 +4% PTFE	<0.003			-0-		-0-		LeRC
M_gF_2	0.006						UV/IR transmission unaffected	LeRC
1000Å T_iO_2	<0.013							LeRC
DC1-2577 SILICONE	<0.02		12					JSC
RTV-650 + T_iO_2	0.02			+.001	-.01			GSFC
DC6-1104 SILICONE	0.02			-.011	-.01			GSFC
SILICONE T-650	0.02	0-0.03	>17	-0-		+0.2	A very thin coating showed much greater mass loss	JSC/MSFC
1000Å Mo	0.023							LeRC
1000Å Cu	0.023							LeRC

Table I (continued)
EFFECTS OF ATOMIC OXYGEN ON MATERIALS

COATINGS	AVG. MASS LOSS (mg/cm^2)	MASS LOSS RANGE	NO. OF SAMPLES (EST.)	OPTICAL PROPERTY CHANGES $\Delta\alpha_s$	$\Delta\varepsilon$	ΔR_s	COMMENTS	DATA SOURCE
ELECTRODAG 402	0.04							JSC
RTV-3145	0.05							MSFC
INDIUM TIN OXIDE	0.15	0-0.38	≥ 4					BAE/MSFC/LeRC
BLACK CONDUCTIVE URETHANE	0.30			+.042	+.55		196% increase in resistance/unit area	GS
401-C10 FLAT BLACK	0.30			+.005				MSFC
Z-853 YELLOW	0.32			-.034				MSFC
CHEMGLAZE Z306 FLAT BLACK	0.35	0.3-0.4	≥ 3	+.031	-.02			MSFC/LaRC/MM
CHEMGLAZE A276 WHITE	0.38	0.35-0.4	≥ 2	-.005	+.03	-3.9		MSFC/LARC
PV100	0.45							BAE
RTV-S695	0.67							BAE
ELECTRODAG 106	0.75							JSC
AQUADAG E	>0.96							JSC
CARBON/KAPTON 100XAC37	1.08							BAE
CHEMGLAZE H322	1.46							BAE
CHEMGLAZE Z302 GLOSSY BLACK	2.03			+.043		-4.2		MSFC

Appendix A 341

Table I (continued)

EFFECTS OF ATOMIC OXYGEN ON MATERIALS

COATINGS	AVG. MASS LOSS	MASS LOSS RANGE mg/cm^2	NO. OF SAMPLES (EST.)	OPTICAL PROPERTY CHANGES $\Delta\alpha_s$	$\Delta\epsilon$	ΔR_s	COMMENTS	DATA SOURCE
Mass Loss Not Determined								
ANODIZED ALUM (CHROMIC ACID PROCESS)	—			-0-		-0-		LaRC
AlMgF$_2$	—			-0-		-0-		MSFC
513-C10 FLAT WHITE	—						No change in optical properties	MSFC
S1023 SILICONE	—			-.022	-.02			MM
CHEMGLAZE A276 (WITH MODIFIERS)	—			-.006 to +.016	+.02			GSFC
V-200 WHITE URETHANE	—			+.020	+.02			GSFC
V-200 (WITH MODIFIERS)	—			+.024 to +.097	+.02			GSFC
Z-301 BLACK	—						α/ϵ Unaffected	MSFC
125Å CHROMIUM	Partially eroded			-0-		-0-		LeRC
700Å Al$_2$O$_3$	Partially eroded			-0-		-0-		LeRC
ZnS	Partially oxidized						No change in IR transmission	A
ThF$_4$	Partially oxidized						No change in IR transmission	A
2.5 uM PARYLENE	Eroded away							WU
2 uM APIEZON GREASE	Eroded away							A
60 uM SILICONE GREASE	Intact, but oxidized							A
D-111 BLACK	Flaked off substrate							LaRC

6E. Required data

Data obtained to date represent a sizable sampling of materials typically used in spacecraft systems. Relative reaction efficiencies as obtained from a given mission can be used with no limitation. One should be careful in cross comparing data for the generation of reaction rates because of various exposure conditions. For example, the reaction efficiency for Kapton is a factor of two higher on STS-8 than on STS-5. These differences may be due to atmospheric density variations not accounted for, or to reaction rate dependence on impingement angle.

As mentioned earlier, total fluence for all the Shuttle experiments was determined from atmospheric density data generated from the GSFC/MSIS model. Any errors in predicting ambient density are introduced into the rate data. Because of the relatively good agreement (factor of 2) between data generated over four missions for some materials, it appears that errors introduced by this source are not large. Since the reaction efficiency data will be used for future hardware design, however, verification of these data with special flight measurements is mandatory.

6F. Possible experiments (TBD)

6G. Mitigation

1. Raise orbit above 400 km.
2. Proper selection of materials.

APPENDIX B
DISCUSSIONS IN MEETINGS OF THE SUBPANEL
ON ELECTROMAGNETIC INTERFERENCE

The following EMI Tasks have been identified as guidelines for operation of this subpanel. Completion of these tasks will provide the natural and induced environment panel with a useable data base which can be programmed on the computerized data base.

- Evaluate Available Cargo Bay EMI Data
- Determine How to Use the Available Data
- Identify Future Data to be Obtained
- Determine Data Deficiencies
- Identify Means for Providing Data to Eliminate Deficiencies
- Prepare Available Data for Inclusion in Data Base
- Provide Guidelines for Use of Data Base

Following evaluation of the existing data and what is planned for the future, data deficiencies will be identified. If possible, a means to eliminate these deficiencies will be recommended. Guidelines for use of the Electromagnetic Environmental Data Base will be provided for the first time user.

ORBITER CARGO BAY EMI ENVIRONMENTAL DATA

- ORBITER EMI ENVIRONMENT NOT MEASURED IN FLIGHT

- SAIL EMI DATA (ORBITER)

  * UNINTENTIONAL RADIATED EMISSIONS
    - Broadband
    - Narrowband
    - AC Magnetic Fields

  * CONDUCTED EMISSIONS
    - DC Power Bus Ripple
      · Narrow Band
      · Wideband
    - AC Power Bus Ripple

  * COMMON MODE
    - 576 Bulkhead to 1307 Bulkhead
    - Hydraulic Pump Effects

  * POWER SOURCE IMPEDANCE

- LIGHTNING TEST DATA

  * MAGNETIC FIELDS
  * DIFFERENTIAL MODE (WIRE PAIR)
  * COMMON MODE (WIRE TO ORBITER SKIN)

The overall electromagnetic environment has never been measured in flight. NASA/JSC has spent considerable efforts in the Shuttle Avionics Integration Laboratory (SAIL) to identify the orbiter unintentional electromagnetic environment. These data included classic engineering definition of radiated emissions, conducted and common mode noise, and circuit impedance definition.

Lightning tests were conducted by Rockwell at their Palmdale facility. A Marx generator was used to develop the high voltage which was imposed on an insulated OV-101 vehicle through a spark gap. Magnetic fields and coupling modes were measured.

Appendix B

ORBITER CARGO BAY EMI ENVIRONMENTAL DATA

- INTENTIONAL RADIATED EMISSIONS DATA
  * S-BAND
    - High Power Antenna Patterns
    - Low Power Antenna Patterns
    - Hemi Antenna Patterns
    - Payload Interrogator Antenna Patterns
  * UHF
    - EVA Operations
  * Ku BAND
    - Antenna Pattern Ground Tests

- PLASMA DIAGNOSTICS PACKAGE DATA (STS-3)
  * BROADBAND RADIATED EMISSIONS
  * MAGNETIC FIELD INTENSITY
  * S-BAND EMISSIONS

- EED TESTING DATA (FRANKLIN INSTITUTE)
  * NSI DEVICES

Intentional radiated emissions have been measured by NASA/JSC at a number of facilities. The cargo bay has been well characterized for normal operations for all radiators. Fields outside the cargo bay have been characterized to some degree. Ku band radiation patterns are shown on a later chart; however S-Band patterns outside the cargo bay have not been identified by JSC.

On STS-3, The DoD sponsored a Plasma Diagnostic Package (PDP) experiment which partially characterized the orbiter S-Band radiation. Additional PDP experiments confirmed some radiated and magnetic fields in the cargo bay.

Electroexplosive device testing has long been the responsibility of the Franklin Research Center Division of Franklin Institute in Philadelphia, PA. They have completed several extensive studies on the NASA Standard Initiator (NSI) which is the recommended device for Shuttle payloads.

SPACE SHUTTLE ENVIRONMENT

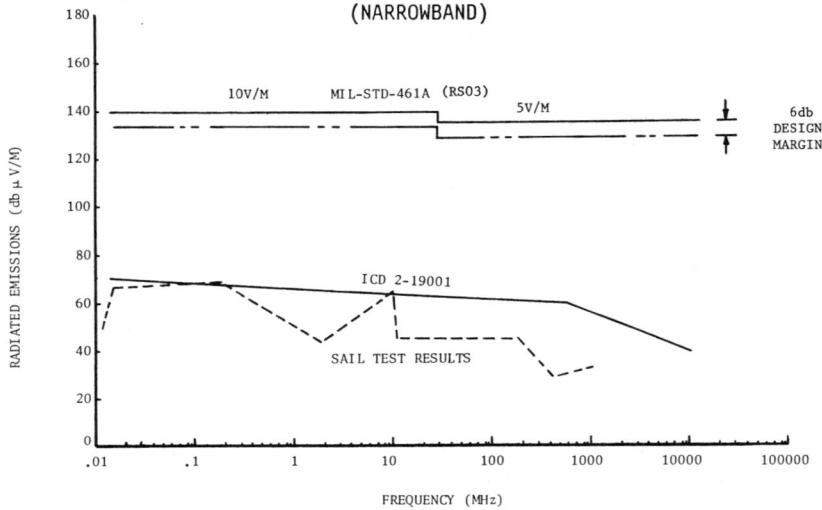

1. The orbiter narrowband unintentional radiated emissions from ICD 2-19001 are shown on the chart. The levels measured during SAIL tests correlate well with the ICD requirement.

2. The DoD requirement for testing to MIL-STD-461A is shown as 10 volts per meter up to 30 MHz. From 30 MHz to 10 GHz the requirement is 5 volts per meter.

3. A 6 db design margin is shown in accordance with MIL-STD-1541. The overall margin appears to be in the order of 70 db.

Appendix B

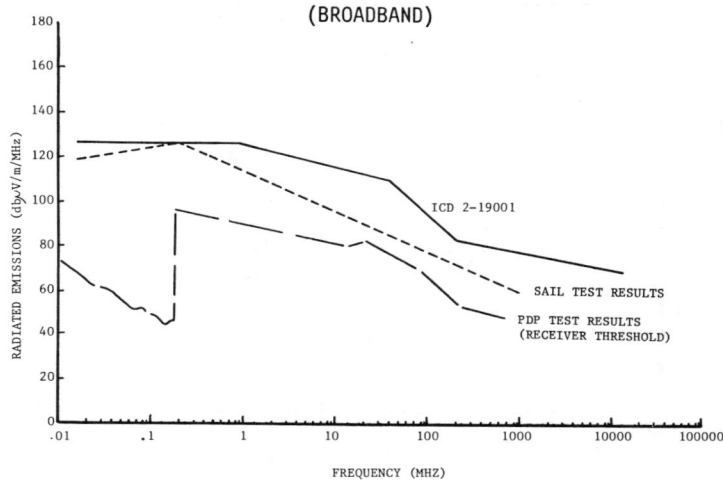

ORBITER CARGO BAY UNINTENTIONAL RADIATED EMISSIONS
(BROADBAND)

1. The orbiter narrowband unintentional radiated emissions from ICD 2-19001 are shown on the chart. The levels measured during SAIL tests correlate well with the ICD requirement.

2. The levels measured during the PDP operations on STS-3 are shown somewhat below the ICD requirement. The level appears strange; however, the level results from a composite of the lower threshold levels of the PDP receivers. Therefore, we do not know the actual levels, but they are somewhere below the curve presented which is significantly below the ICD level.

3. There is no MIL-STD-461A requirement for measurement of broadband unintentional radiation.

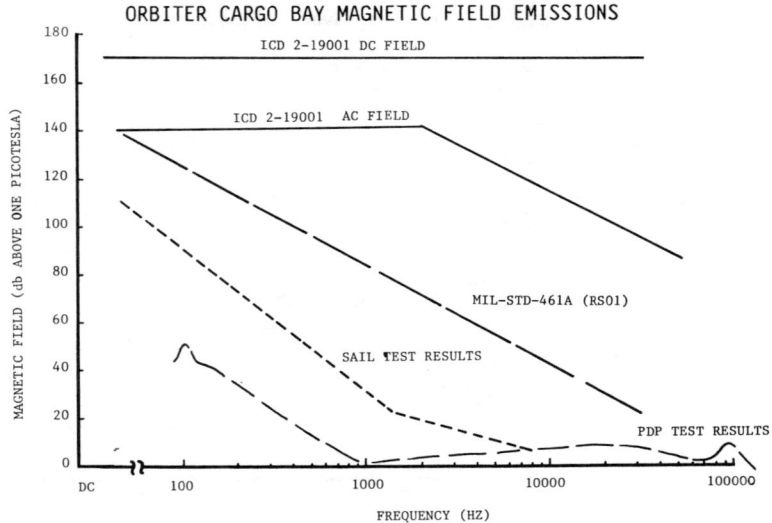

1. The orbiter magnetic field emissions requirements from ICD 2-19001 are shown on the chart for both AC and DC fields. These levels are specified near the orbiter power buses ($y_0 = \pm 79$, $z_0 = 349$, x_0 = any position in cargo bay) and are quite high. An equation is presented in the ICD for determination of levels at other locations in the cargo bay, e.g.,

$$db \text{ (Reduction)} = 20 \log_{10} (57R^2)$$

where R (meters) is the radial separation in the y-z plane from the nearest port or starboard bus.

2. The DoD requirement for testing per MIL-STD-461A (RS01) is shown on the chart.

3. SAIL and PDP flight test results are also shown and are significantly below the ICD requirements.

4. As a result of these data it is evident that each spacecraft has to determine its sensitivity to electromagnetic fields at the location within the orbiter cargo bay. If they are sensitive, they should tailor the MIL-STD-461A requirements to be compatible with those identified in ICD 2-19001.

Appendix B

ORBITER CARGO BAY Ku BAND ENVIRONMENT

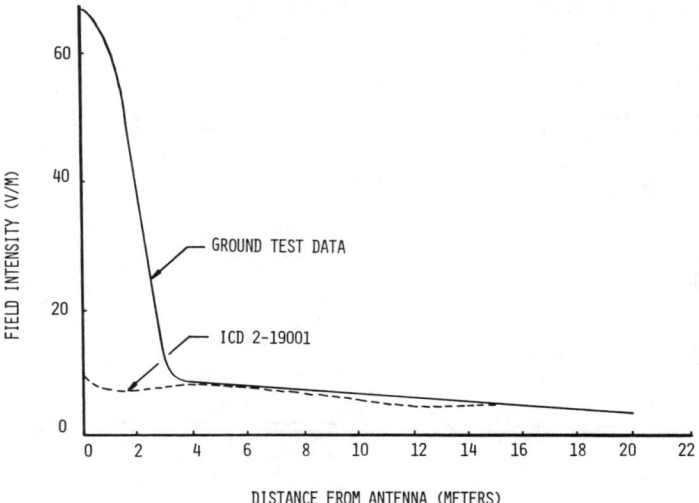

1. The cargo bay Ku band environment is shown in the chart. Ground testing at Rockwell has indicated levels that are considerably higher than ICD 2-19001 at a close proximity to the antenna. However, levels drop rapidly due to imposition of a computer "obscuration" program which limits the radiaton into the bay.

SPACE SHUTTLE ENVIRONMENT

ORBITER EXTERNAL Ku BAND ENVIRONMENT

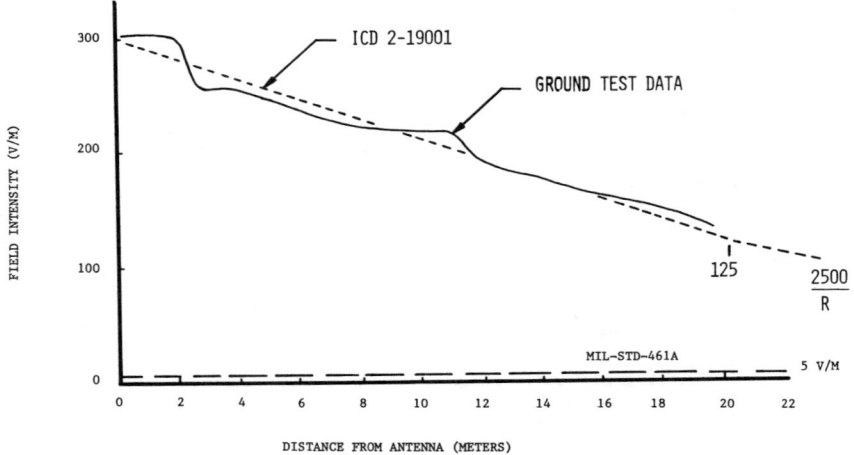

1. The Ku band radiation external to the cargo bay is another matter. ICD 2-19001 shows a level of 254 volts per meter as a maximum level outside of the cargo bay. This level decreases with distance as shown. The other curve on the chart was developed by Rockwell from ground tests and is applicable on the Ku band radar dish centerline out to approximately 22 meters.

2. The DoD position is that a spacecraft must endure these levels when outside the cargo bay. In addition, contractually the spacecraft must test to provide proof of environmental operation; however, RF generators are not available for test of a complete spacecraft at a 6 db margin.

3. Most contractors to date have been requesting that the Ku band system be off during their deployment sequence. Thus far JSC appears to be receptive to this concept as long as the total turnoff time is not extensive.

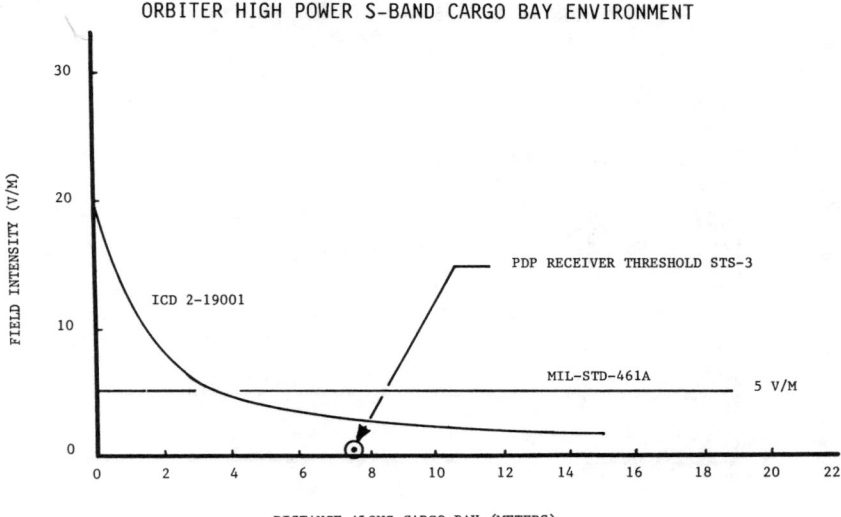

1. The cargo bay S-band intentional radiation environment is shown on the chart. The ICD 2-19001 indicates a requirement which is nearly 20 volts per meter near the foreward bulkhead but drops quickly to below 5 volts per meter.

2. The specification MIL-STD-461A indicates a RS01 level of 5 volts per meter.

3. During the STS-3 mission, the PDP attempted to measure the S-band radiation at the PDP location in the cargo bay. The level was somewhere below the S-band receiver threshold of 0.32 volts per meter.

ORBITER HIGH POWER S-BAND EXTERNAL ENVIRONMENT

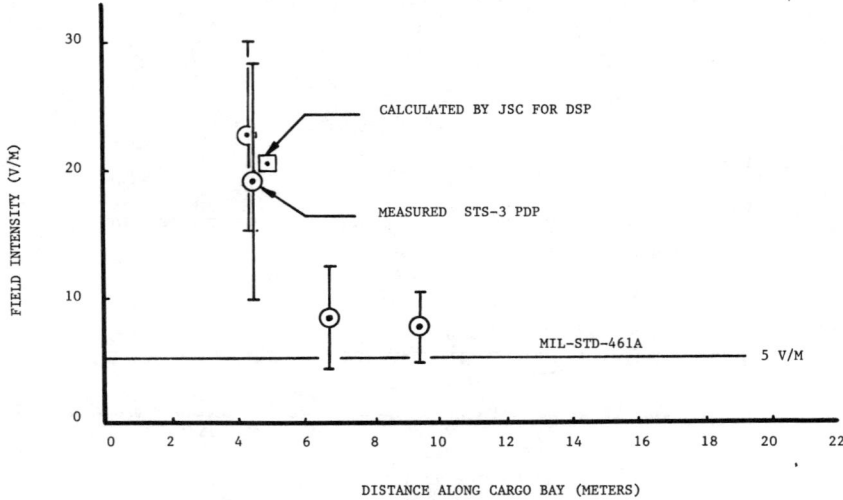

1. The S-band radiation outside the cargo bay is somewhat higher. The four circled data points were observed by the PDP on STS-3. The error bars on these measurements include instrument calibration errors, and effects of other antennas and S-band links. These are larger than previously anticipated and were identified post-flight. Future measurements of the PDP S-band should be better characterized.

2. Again, the MIL-STD-461A level is noted as 5 volts per meter.

3. One calculation, made by JSC for the DSP program, is shown in the square. It appears that the PDP test data may not contain significant errors in measurement.

Appendix B

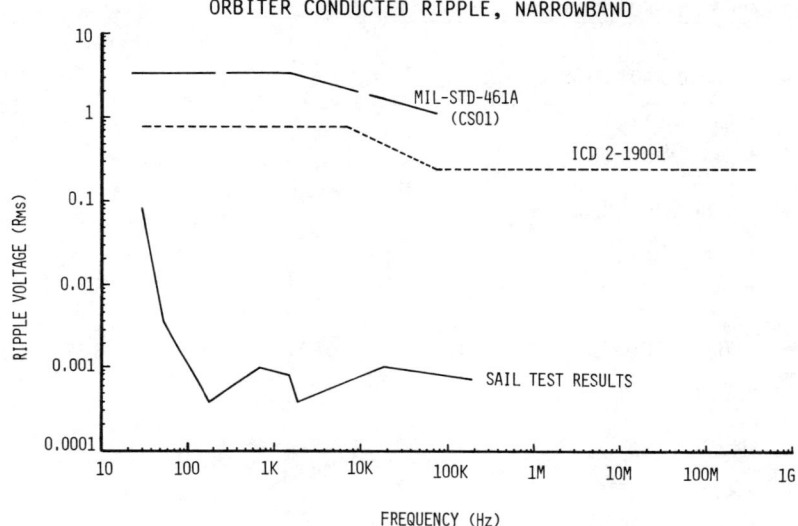

1. Orbiter power system conducted ripple requirement is shown as identified in ICD 2-19001. The DoD requires testing to MIL-STD-461A which shows a margin in the area of 5 db.

2. The SAIL test results show that expected levels are significantly lower than the ICD.

FUTURE CARGO BAY EMI ENVIRONMENTAL DATA

- Ku BAND RADIATION LEVELS
    * CARGO BAY
    * EXTERNAL
- S-BAND HIGH POWER LEVELS
    * CARGO BAY
    * EXTERNAL
- PAYLOAD INTERROGATOR LEVELS
    * EXTERNAL
- SPACELAB II
    * STS-24 (51F)
    * APRIL 1985

The DoD has contracted with the University of Iowa to make additional measurements of the S-band and Ku Band RF environment around the cargo bay on the Plasma Diagnostics Package which will fly on SPACELAB II. As this PDP is a free flyer as well as operating on the end of the RMS, a considerable amount of near and far field RF measurements will be made.

SHUTTLE-PRODUCED PAYLOAD BAY RADIATED NARROWBAND EMISSIONS

The next nineteen charts reflect the electromagnetic environmental requirements of the Shuttle orbiter/cargo standard interfaces as shown in JSC 07700 Volume XIV, attachment 1 (ICD 2-19001) revision H, change 46, dated 31 May 1984. These charts were presented by Mr. Roger Henkle of Martin Marietta Corporation.

Appendix B

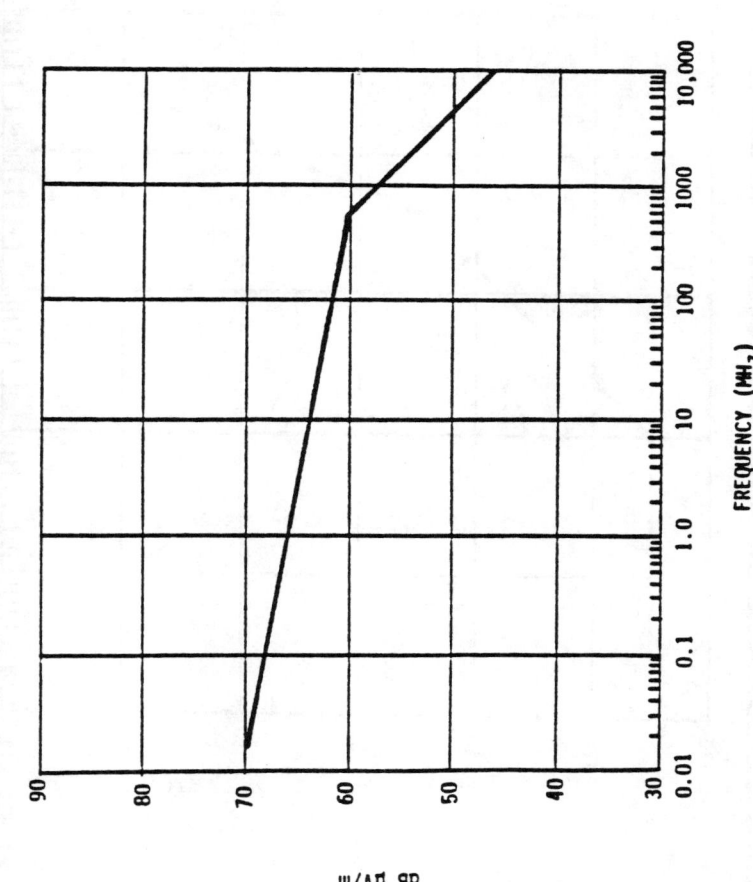

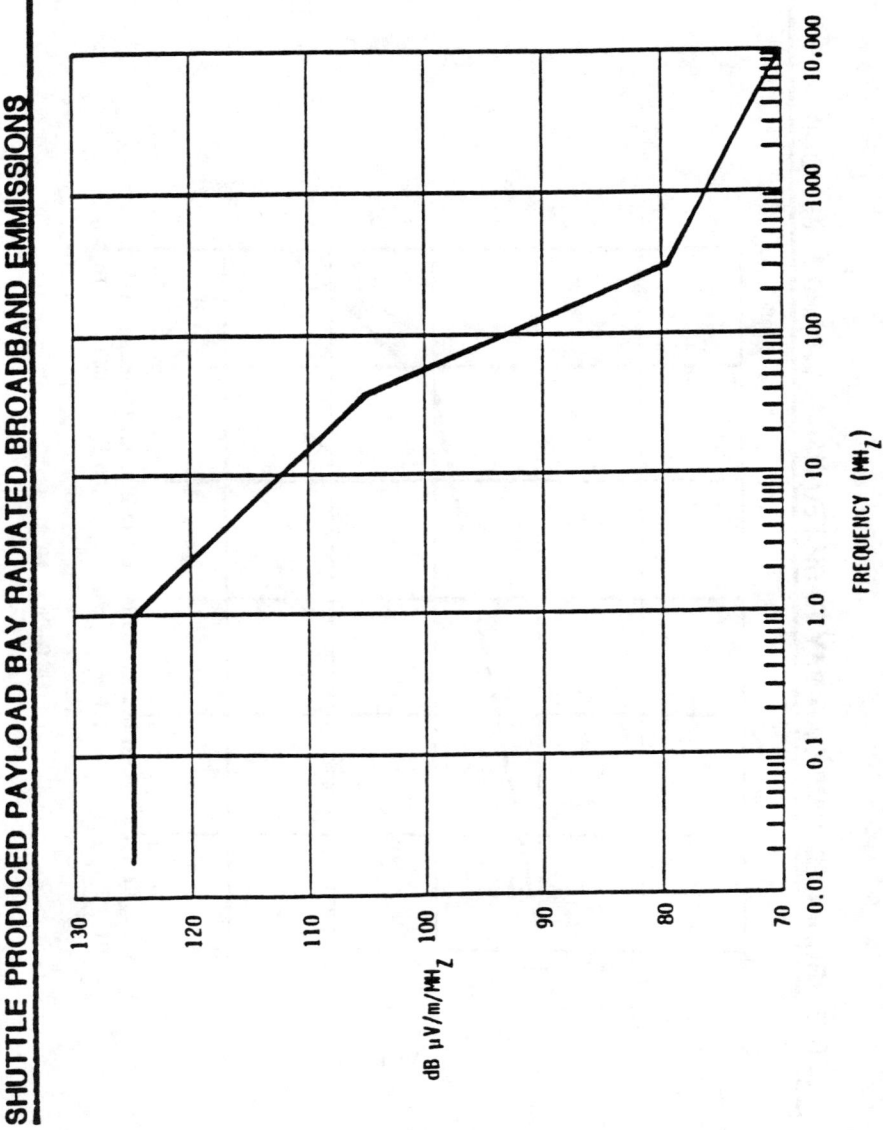

Appendix B

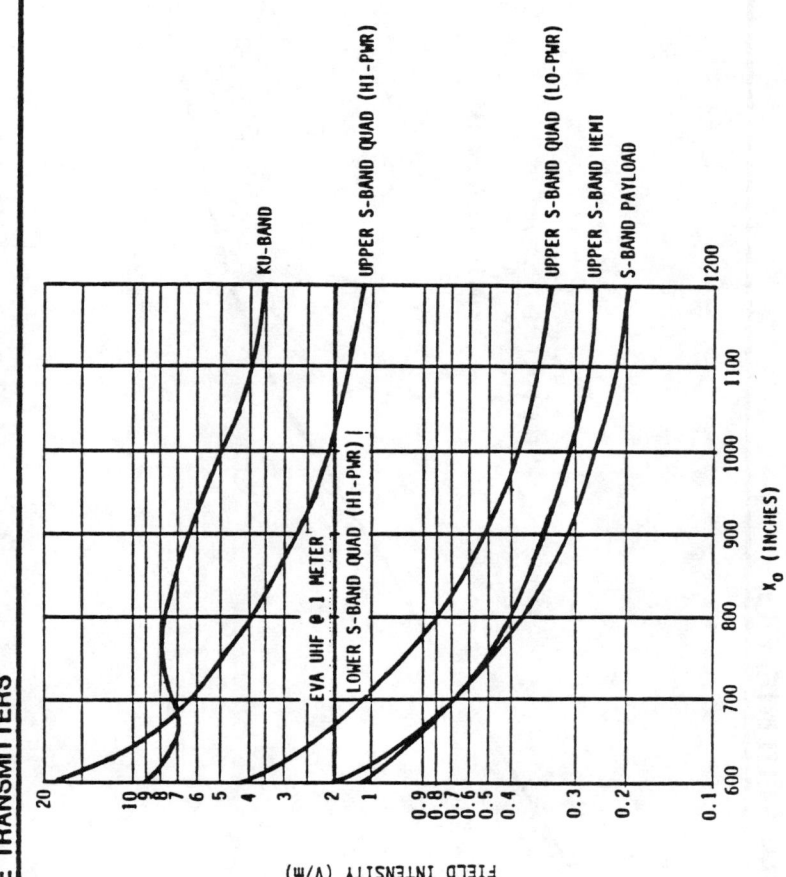

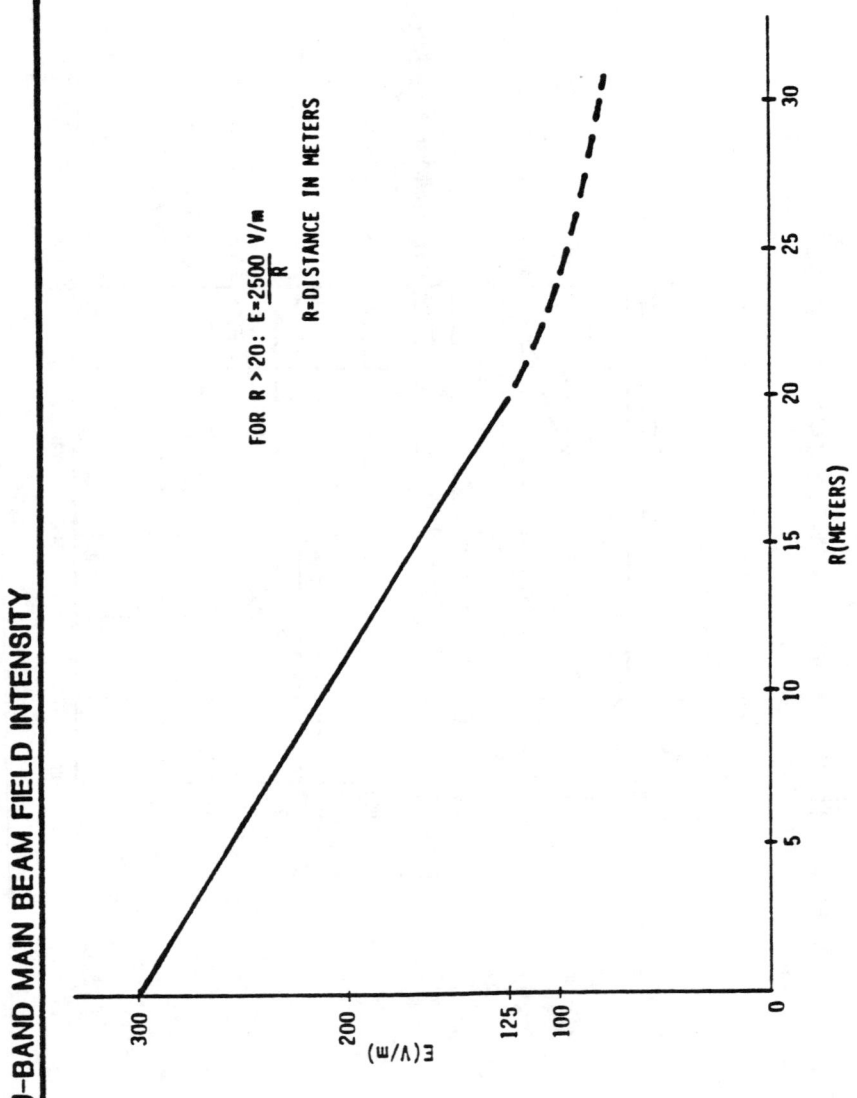

Appendix B

CARGO ALLOWABLE RADIATED NARROWBAND EMISSIONS

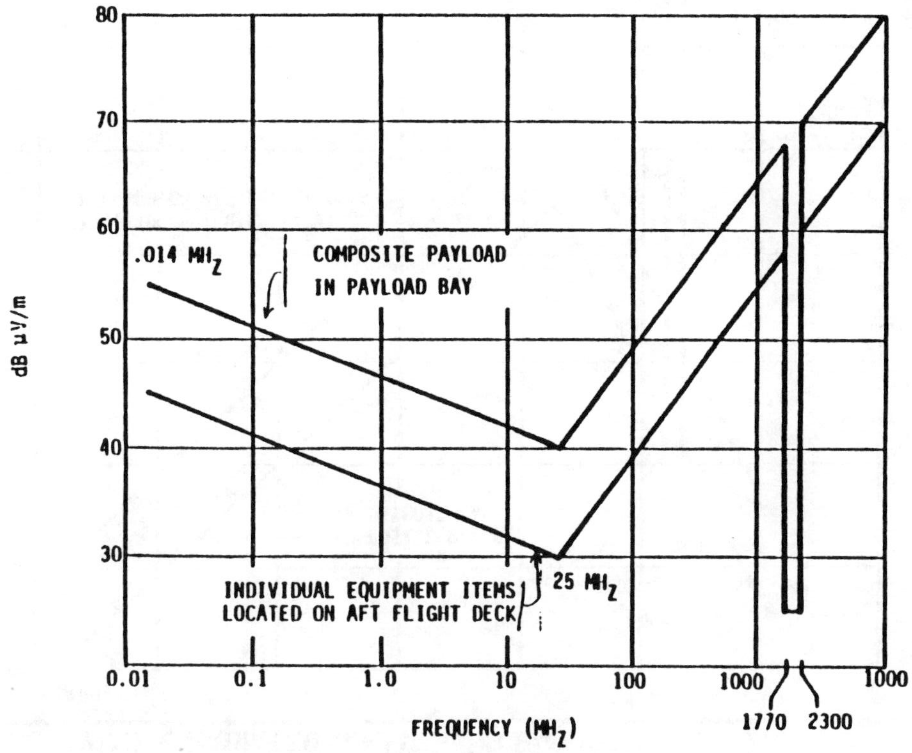

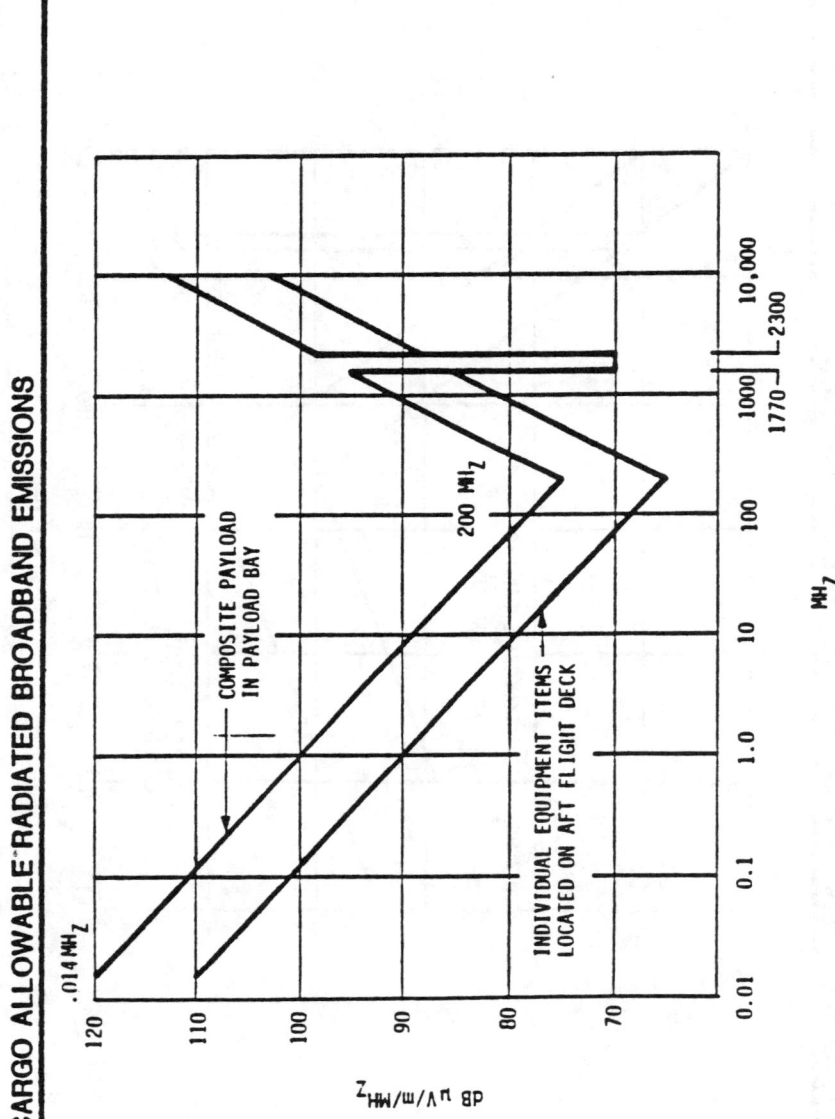

Appendix B 361

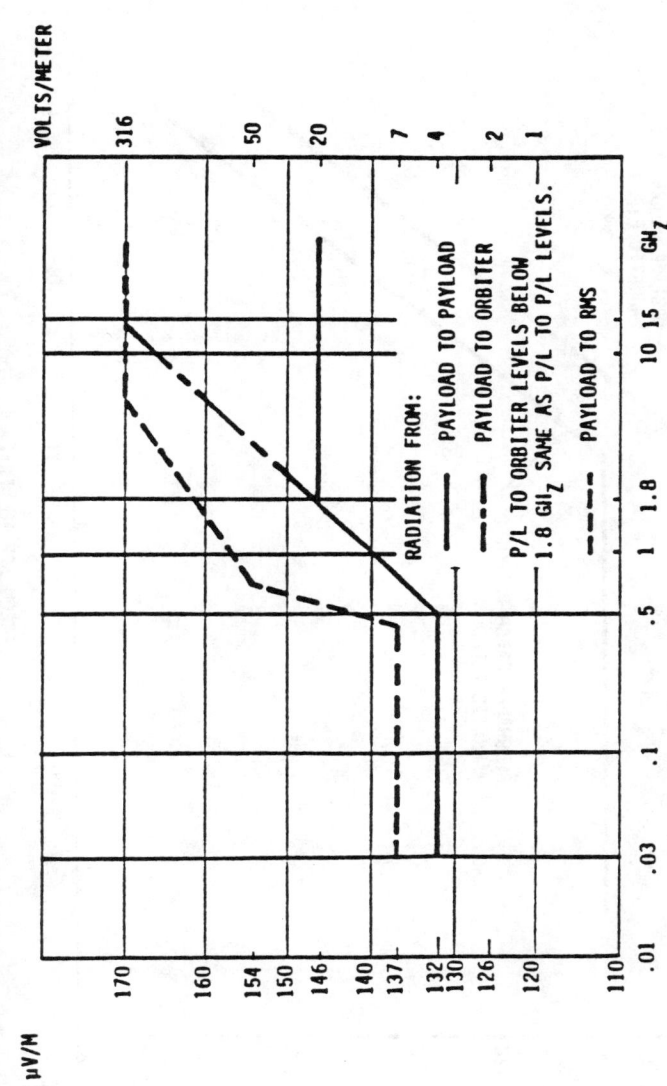

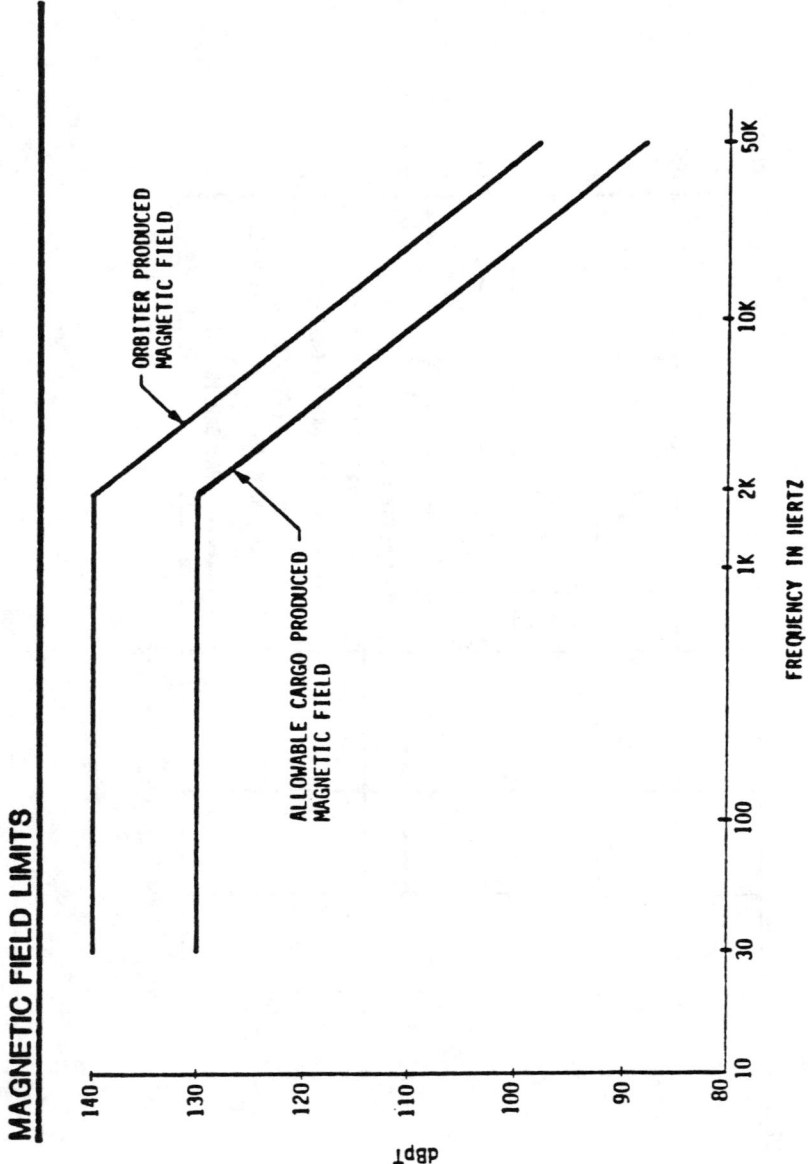

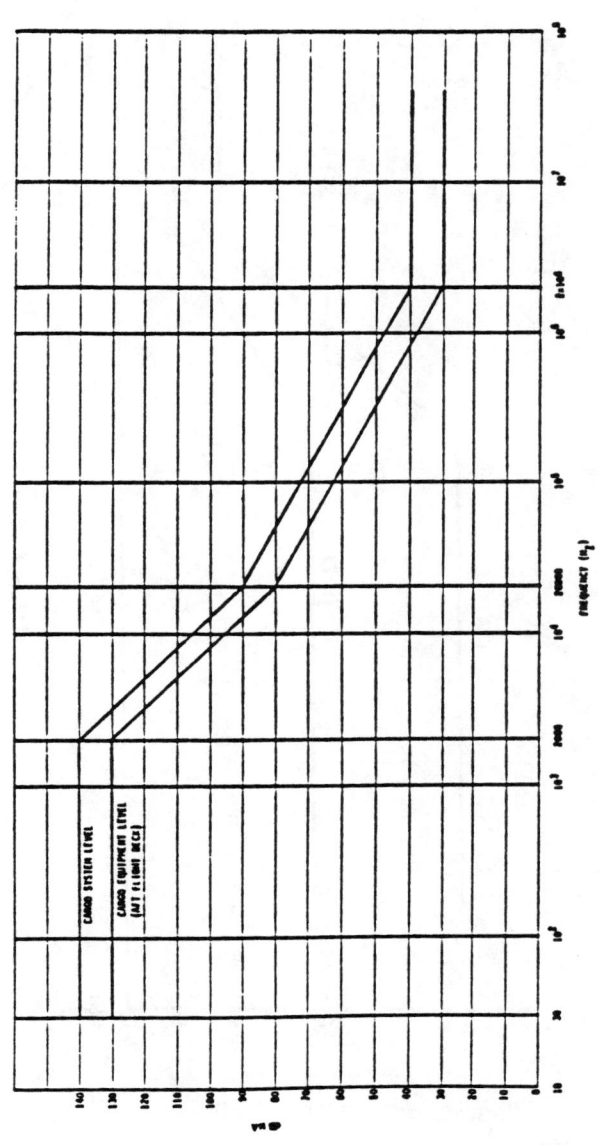

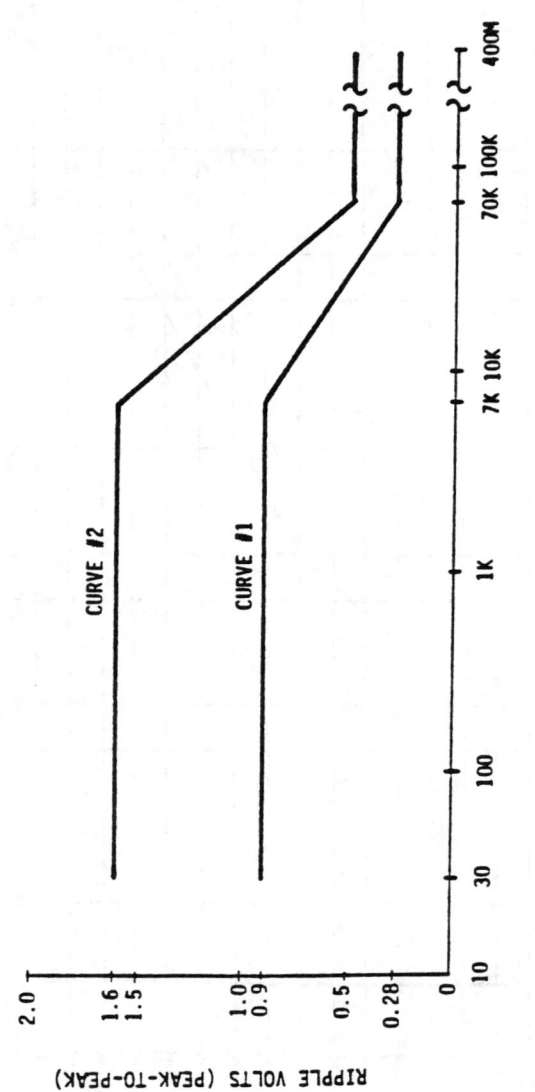

SHUTTLE PRODUCED RIPPLE ON IN-FLIGHT DC POWER BUS

Appendix B 365

RIPPLE ON GROUND DC POWER (VIA ORBITER EPDS)

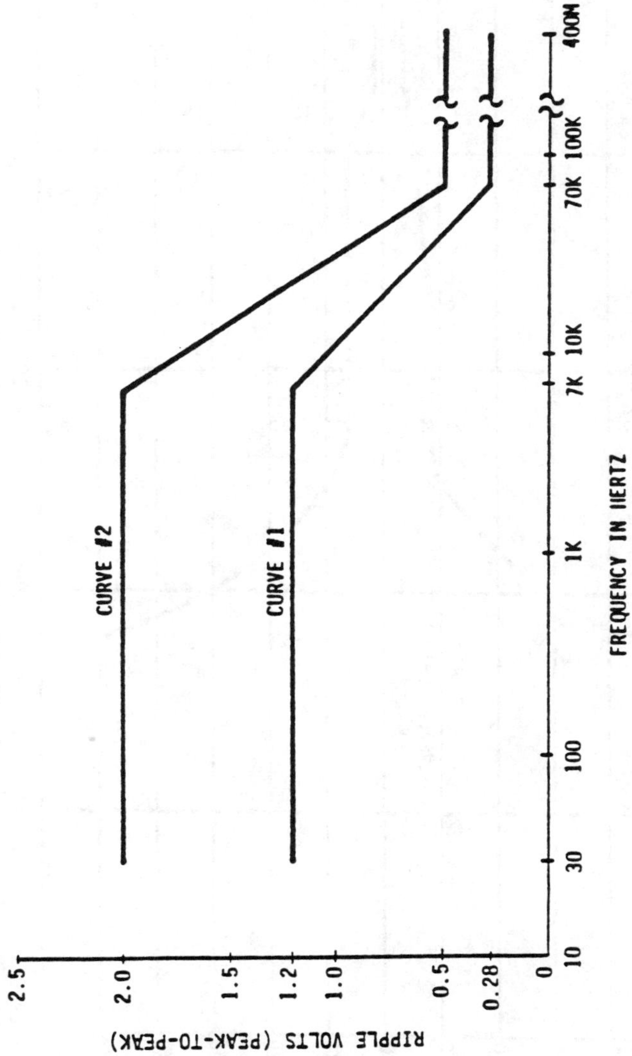

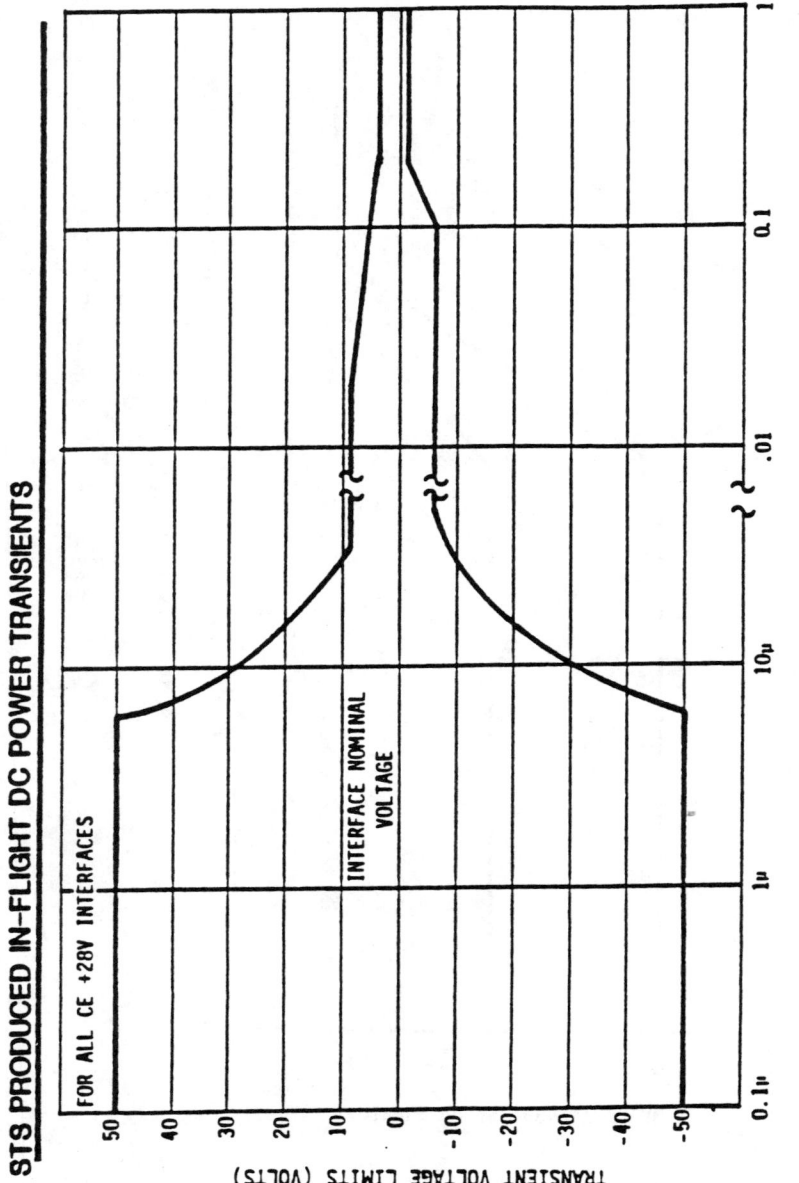

Appendix B

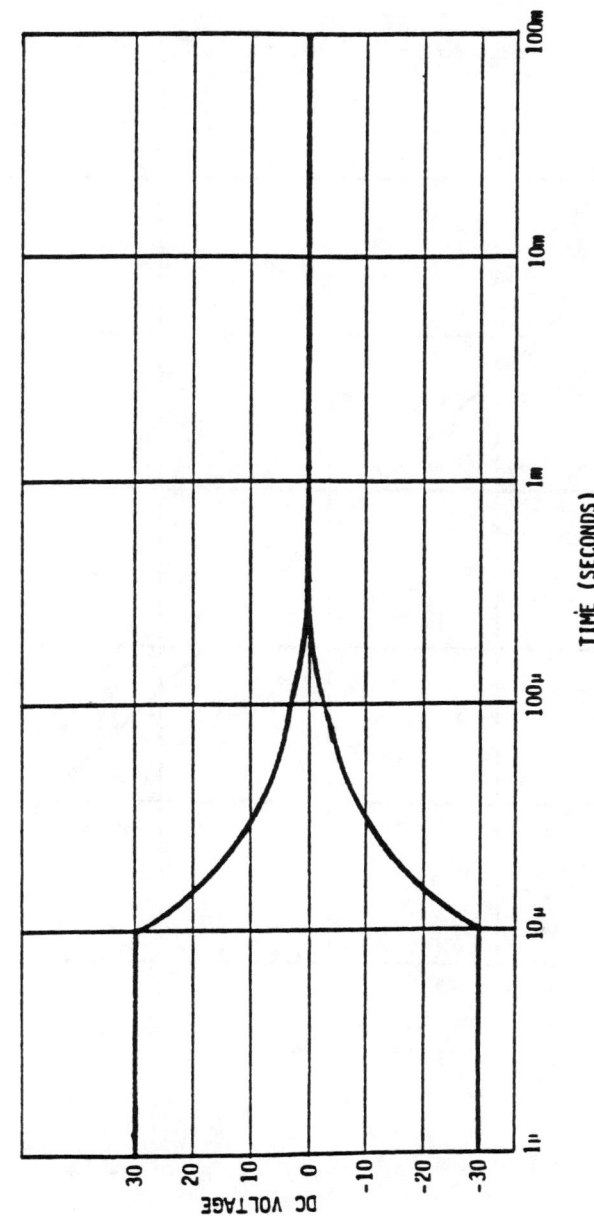

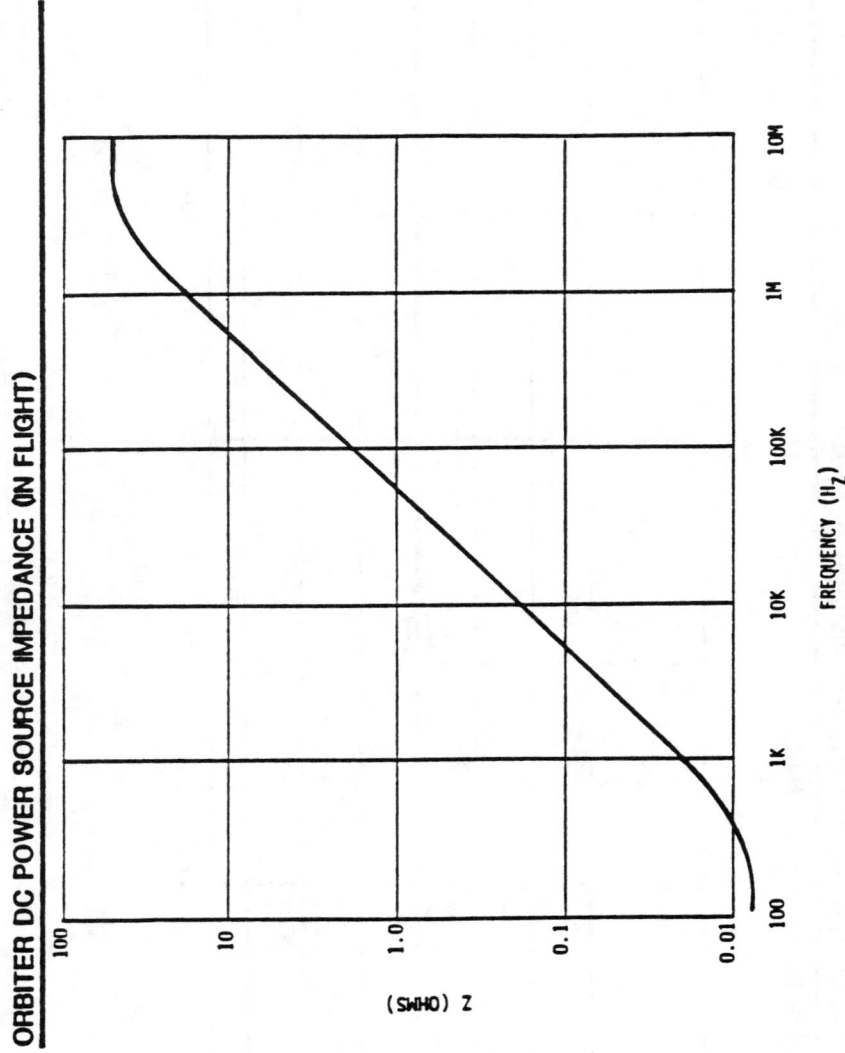

Appendix B

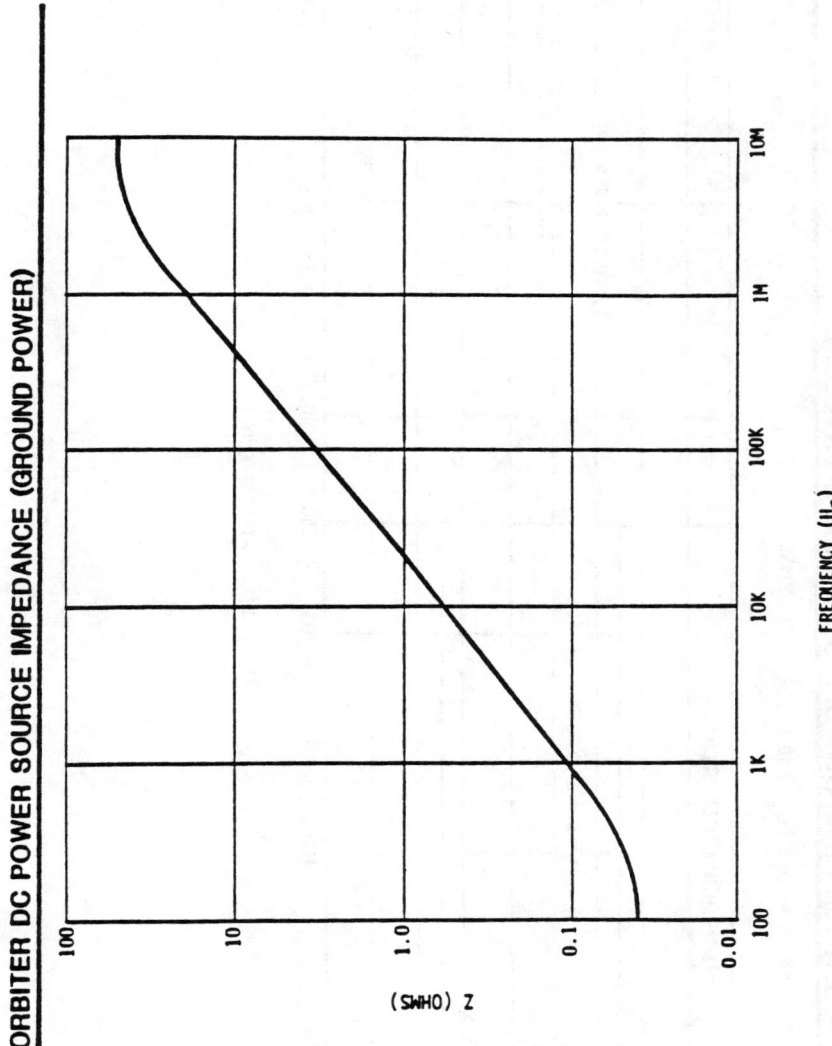

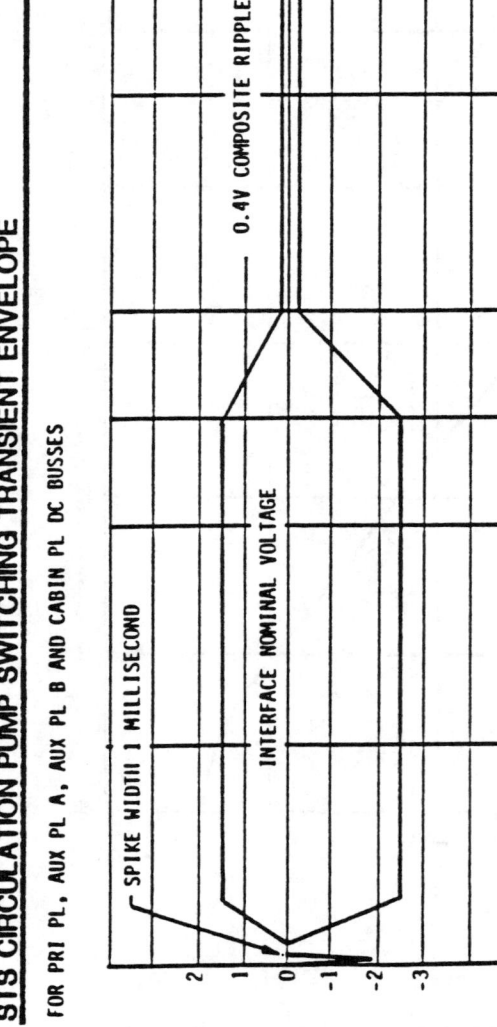

Appendix B

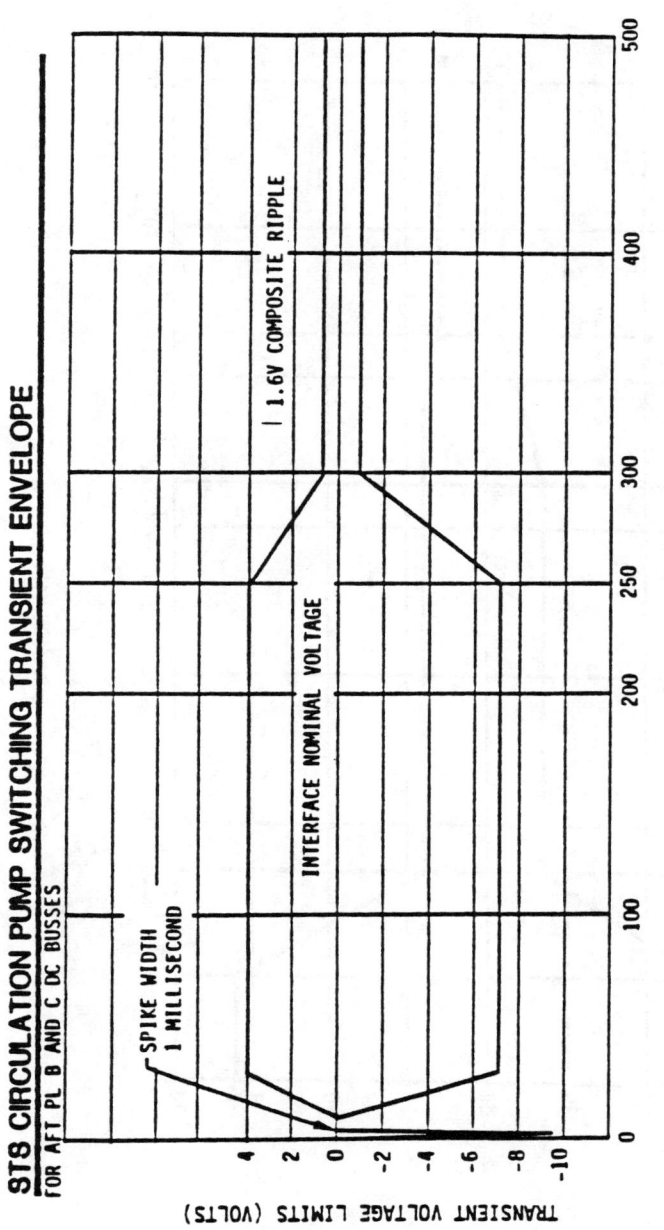

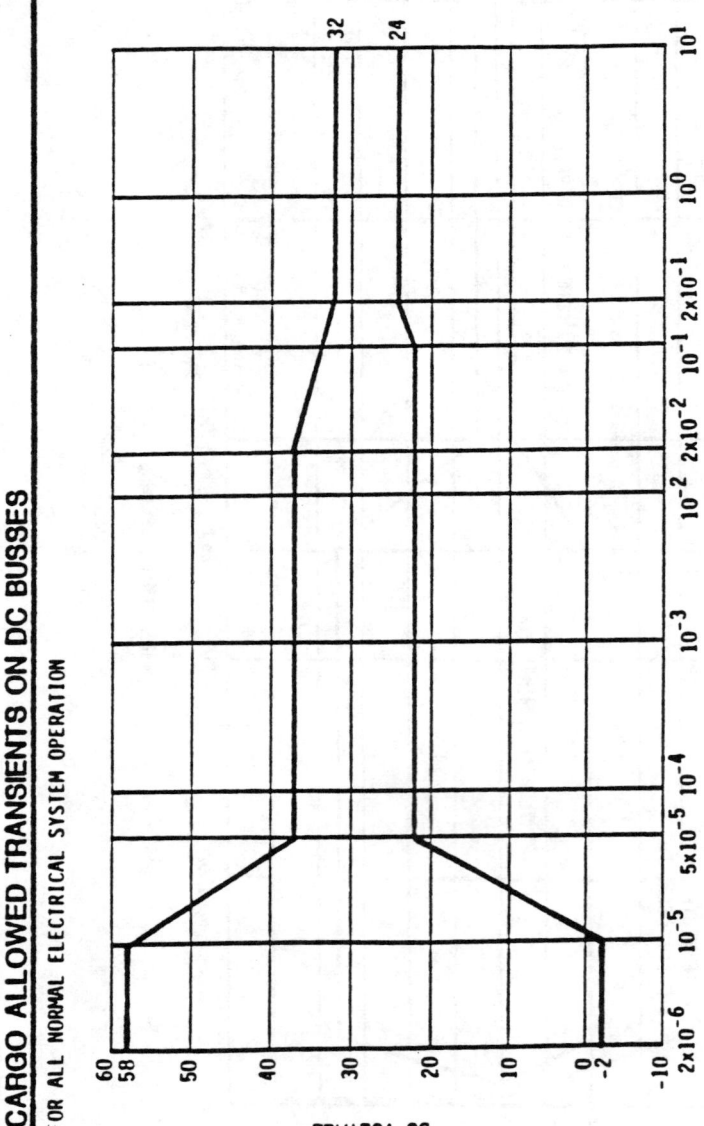

Appendix B 373

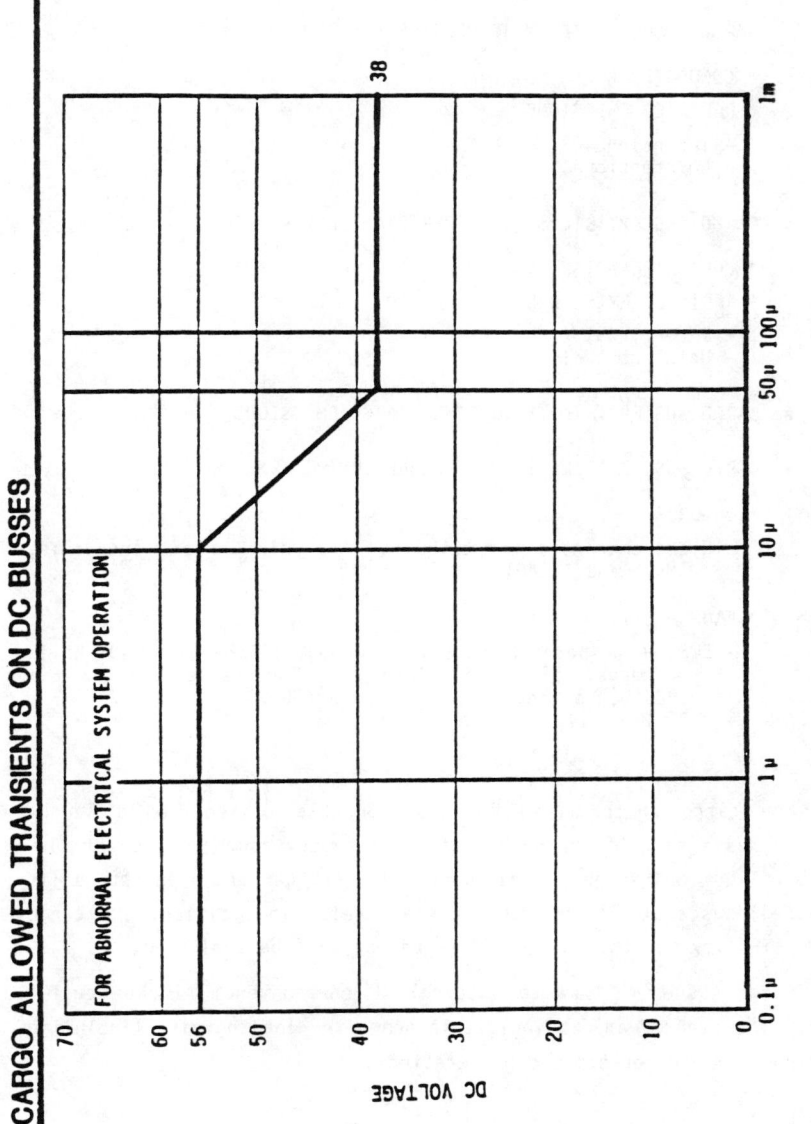

CARGO EMI ENVIRONMENTAL REQUIREMENTS

- CARGO SUSCEPTIBILITY TO ORBITER ENVIRONMENT
  * CONDUCTED NOISE
  * RADIATED EMISSIONS
    - Intentional
    - Unintentional

- CARGO SUSCEPTIBILITY TO UPPER STAGE ENVIRONMENT
  * CONDUCTED NOISE
  * RADIATED EMISSIONS
    - Intentional
    - Unintentional

- CARGO SUSCEPTIBILITY TO OTHER CARGO EMISSIONS

- CARGO SUSCEPTIBILITY TO EXTERNAL ENVIRONMENT
  * ON-ORBIT
    - Electromagnetic Compatibility Analysis Center, Annapolis, Maryland 21401
  * LAUNCH SITE
    - ELS; Aerospace Corporation TOR-0084(4338-42)-1 "RF Environment, STS, ELS" Aerospace Corporation, P. O. Box 92960, Los Angeles, California, 90009
    - WLS, TBD

Any cargo which uses the Space Shuttle orbiter as a launch vehicle has a significant number of separate environments imposed during preflight and flight operations. Both orbiter and upper stage (if required) susceptibilities must be evaluated. In addition, the cargo susceptibility to other cargo (if required) must be evaluated.

Cargo susceptibility to external RF environments must be evaluated. These environments include launch site, landing site (including contingency), and on-orbit considerations.

Appendix B

CARGO EMI ENVIRONMENTAL REQUIREMENTS (Cont.)

- CARGO SUSCEPTIBILITY TO EXTERNAL ENVIRONMENT
    * LANDING SITE
        - TBD
    * CONTINGENCY LANDING SITE
        - TBD
- ORBITER LIMITATIONS TO CARGO INITIATED ENVIRONMENT
    * CONDUCTED NOISE
        - Narrowband
    * POWER BUS TRANSIENTS
    * RADIATED EMISSIONS
        - Intentional
        - Unintentional
            • Broadband
            • Narrowband

In addition to cargo susceptibility, the orbiter must evaluate the cargo-produced electromagnetic environment on their systems. If the cargo meets all of the requirements in ICD 2-19001, no problem will exist. If the cargo environment exceeds the ICD environment, an evaluation must be conducted by JSC.

EMI ENVIRONMENTAL DATA BASE CONCLUSIONS

- EXISTING DATA SUPPORT ICD 2-19001 REQUIREMENTS
- USE ICD 2-19001 REQUIREMENTS IN ENVIRONMENTAL DATA BASE
- NO OTHER DATA REQUIRED
    * Assumes Future Measurements Support ICD 2-19001 Requirements
- OTHER CONCLUSIONS
    * Open

Review of the limited electromagnetic environment data available has been shown to support the ICD 2-19001 requirements. For an engineering definition of this environment, this is all that is required. This however, assumes that the planned future measurements continue to support the ICD requirement.

Other conclusions from this subgroup remain open. Participants have identified a need for definition of the lowest level of electromagnetic environment available in the cargo bay to support definition of their scientific requirements. No previous requirements for this "Scientific Electromagnetic Environment" have been identified, and no plans exist for definition of this environment.

Appendix B

HOW TO USE EMI ENVIRONMENTAL DATA BASE

- EMC SYSTEM ANALYSES

  * RAND CORPORATION REPORT R-3046-RF "TECHNIQUES FOR ANALYSIS OF SPECTRAL AND ORBITAL CONGESTION IN SPACE SYSTEMS"
    - RF Cull and Coordination Models
    - Space Systems Site Analysis
    - Intrasystem EMC Analysis
    - Intersystem EMC Analysis
    - Electromagnetic Vulnerability Analysis
    - Multipurpose EMC Analysis
    - Available From:
      Defense Technical Information Center
      Building 5, Cameron Station
      Alexandria, Virginia 22314

The next seven pages give guidelines on how to use the data provided to complete an electromagnetic compatibility (EMC) analysis. References are included which provide details on system analyses, EED analyses, and orbiter interface analysis. Recommendations are included for EMC testing to prove compliance with the electromagnetic requirements.

HOW TO USE EMI ENVIRONMENT DATA BASE (Cont.)

- EED ANALYSIS

  * FRANKLIN RESEARCH CENTER DOCUMENT, "MONOGRAPH OF COMPUTATION OF RF HAZARDS",
    - Available From:
      Franklin Research Center
      Divison of the Franklin Institute
      Benjamin Franklin Parkway
      Philadelphia, Pa   19103

  * MARTIN MARIETTA DOCUMENT, "USERS MANUAL FOR ELECTROEXPLOSIVE DEVICE ANALYSIS TOOL (EEDAT)"
    - HP-85 Implementation of Franklin Institute Monograph
    - Available From:
      Martin Marietta Corporation
      P.O. Box 179
      Denver, Colorado   80201

- INTERFACE ANALYSIS

  * TECHNIQUES FOR INTERFACE ANALYSIS PROVIDED AS AN ATTACHMENT
    - System Sensitivity
    - Receiver Coupling

- EMC TESTING REQUIREMENTS

  * DETERMINE CARGO SUSCEPTIBILITIES

  * IDENTIFY TEST LEVELS FROM DATA BASE

  * TEST PER STANDARD METHODOLOGY
    - MIL-STD-462

  * DEMONSTRATE SAFETY MARGIN OVER EMI ENVIRONMENT IDENTIFIED IN THE DATA BASE

Appendix B

EMC SYSTEM INTERFACE ANALYSIS

- REQUIREMENT

    * Provide confidence that equipment will meet Orbiter ICD requirements
        - Assess Each Interface Against Requirement

    * Early identification and prevention of EMC problems

- DEFINITION OF EMC ENVIRONMENT

    * Internal
        - Payload to Payload

    * External
        - Orbiter ICD
        - Upper Stage/Orbiter ICD
        - Launch Area
        - Worldwide

- DEFINITION OF SYSTEM CONFIGURATION

    * Internal
        - Spacecraft Design
            · Interface sensitivities
            · Coupling loops
            · Receiver sensitivities
            · Other internal sensitivities
        - Spacecraft Operations
            · Ground
                ○ Checkout requirements
                ○ Timelines
                ○ Locations
            · Flight
                ○ Trajectory
                ○ Ground track
                ○ Altitude
                ○ Inclination
                ○ Contingency

    * External
        - Ground Operations Equipment
        - C^3 Equipments

EMC SYSTEM INTERFACE ANALYSIS (Cont.)

- EVALUATION OF EFFECTS OF EMI ENVIRONMENT

  * Internal
    - Circut Sensitivities
    - Intra-system Coupling
    - Receiver Sensitivities
    - EED

  * External
    - Conducted Emissions Susceptibility
    - Power line/AC/DC/ground
      · Ripple
      · Transients
      · Common mode
    - Radiated Emissions Susceptibility
      · AC Magnetic Fields
      · Electric Fields
    - Lightning
      · Magnetic Fields
    - EED Susceptibility

- EVALUATION OF EMI ENVIRONMENT PRODUCED BY SPACECRAFT

  * Conducted Emissions Generation
    - Power Bus (DC/AC)
      · Ripple emissions
      · Transients
      · Impedance
    - Interface Circuts

  * Radiated Emissions Generation
    - Unintentional
      · Electric field emissions
      · AC magnetic field emissions
    - Intentional
      · Transmitters

Appendix B

EMC SYSTEM INTERFACE ANALYSIS (Cont.)

* Signal and Control Line Noise
* Electrostatic Discharge
  - Static Electricity Buildup Requirements
- MINIMUM SAFETY MARGINS

  * 6db Critical Circuts
  * 20db EED

- EMC TESTING/DEMONSTRATION

  * Identification of Susceptible System Elements
  * Test by Standard EMC Methodology
    - MIL-STD-462
  * Demonstrate Safety Margin Over EMI Environment

- REFERENCES

  1. JSC 07700, Volume XIV, Attachment 1 (ICD 2-19001), "Shuttle, Orbiter/Cargo Standard Interfaces," Revision H., Change 44, dtd 13 January 1984.

  2. MIL-STD-462, "Measurements of Electromagneic Interface Characteristics," dtd. 31 July 1967.

  3. Library of Congress Catalog Card No. 76-39643, "EMI Control Methodology and Procedures," by D. R. J. White.

  4. MDCE 1929, "Integrated Circut Electromagnetic Susceptibility Handbook," dtd August 1978.

  5. TOR-0084(4338-42)-1, "Radio Frequency Environment STS Eastern Launch Site," dtd 1 December 1983.

  6. AFSC Design Handbook 1-4, "Electromagnetic Compatibility," Fourth Edition, dtd 2 March 1984 (available only to U. S. personnel _per_ ITAR/EAR Regulations, Export Administration Act of 1979)

PROPOSED NASA PROGRAM TO SEARCH OUT AND IDENTIFY THOSE PAYLOADS THAT COULD MEASURE SHUTTLE EMI

DWIGHT L. FORTNA, NASA/GSFC, CODE 302 (AUGUST 1984).

Comprehensive electromagnetic emission measurements have not been performed in the payload bay of the Space Shuttle. This includes electric field as well as magnetic field emissions and is true for both ground and inflight measurements. Limited EMI measurements of the Shuttle have been made. These include a preliminary measurement program conducted in 1976 on Orbital Vehicle OV101 at Palmdale. However the data are probably not applicable since that Orbiter was not equipped with final Orbiter electrical systems. EMI measurements in orbit were performed by the University of Iowa's Plasma Diagnostics Package (PDP) on the STS-3 flight. The PDP will again make EMI measurements on the Spacelab II flight next year. These measurements are limited in scope and do not provide sufficient data for a complete data base. Proposals have been made to conduct comprehensive EMI measurements of the Shuttle but they have not been funded. It is my understanding at this time that no measurement programs are planned for the future except for the Spacelab II flight mentioned previously.

A data base which defines the EMI levels in and around the Shuttle is needed by the payload and scientific communities. This data base should define the frequencies, the locations, and the operational modes of the Shuttle where the emission levels are both maximum and minimum. Scientists planning sensitive experiments could then use this information to optimize the location and operational parameters of their experiments.

A number of payloads and instruments, some of which have already flown in the Shuttle, have the capability for measuring electric and magnetic fields as part of their normal functional capability. Some payloads include magnetometers in payload pointing systems and spacecraft stabilization and control systems. It is expected that some scientific instruments will also have EMI measurement capabilities. Some examples of these capabilities are the following:

Appendix B

1) The ERBS spacecraft provides telemetered outputs from three magnetometers during its "in-bay" checkout. The data are available for the asking.

2) The U. S. Navy is funding scientists in the Geophysics Branch at GSFC to study a SPARTAN Magnetic Field Measurement Mission. The Navy would like to fly a magnetometer on a SPARTAN carrier twice a year for 10 years to gather data for their world magnetic field charting program. This program could provide much data on Shuttle produced magnetic fields.

It is proposed that NASA set up a program to search out and identify those payloads and instruments that could provide Shuttle EMI data. The program should provide planning and coordination with the projects, scientific investigators, etc. in order to maximize the data return. Some level of funding should be available to provide for calibrations and limited additional wiring and circuitry to make data available and for data processing. This program is not intended as a substitute for a well engineered EMI measurement program, but the bits and pieces of data gleaned over the years would add up to a significant data base of Shuttle EMI information.

TYPICAL RECEIVER COUPLING CALCULATIONS

- Identify Incident Field Intensity--E(v/m)
    * Source Data
        - Nominal operations
        - Maximum (damage)

- Calculate Average Incident Power Density--P(w/m^2)
    * Assume isotropic radiator
    * $P = E^2/377$

- Identify Peak Antenna Gain-G(dbi)
    * Determined at frequency of operation
    * In direction of receiver

- Determine Maximum Effective Aperture at Frequency of Interest--A(m^2)
    * $A = G \lambda^2/4\pi$
    * Where $\lambda = 300/f$(MHz) for frequency of interest

- Determine Maximum Signal Power Intercepted by the Antenna of Interest--P_{max}(watts)
    * $P_{max} = P \times A$

- Calculate Receiver Power Budget
    * P_{max} intercepted by Antenna   _____dbm
    * Plus receiver/filter/diplexer losses   _____dbm
    * Equals maximum pwr available at receiver input _____dbm
    * Less receiver sensitivity   _____dbm
    * Equals safety margin   _____dbm

Repeat calculation for maximum power available to identify damage margin.

APPENDIX C
THERMAL AND HUMIDITY ENVIRONMENT

P. Tulkoff
Goddard Space Flight Center
National Aeronautics and Space Administration

The Thermal and Humidity Subpanel Consisted of:
    James F. Clawson (Chairman) (Rockwell-Houston)
    Donald Bartelson (Lockheed-Kennedy Space Center)
    James Fu (Jet Propulsion Laboratory)
    David J. Russell (Rockwell Houston)
    Philip Tulkoff (Goddard Space Flight Center)
    Frederick Wenkstern (McDonnell Douglas Aircraft Corporation-
                      Kennedy Space Flight Center)

The panel met several times with various Shuttle users and experimenters over the period of August 5th-10th, 1984. The primary goals of the meetings were as follows:

* Determine the information needs of Shuttle users and experimenters for thermal and humidity data
* Construct a preliminary outline for the data base
* Locate existing information and information sources
* Evaluate existing information
* Formulate future plans for accumulating and integrating information for the data base

SPACE SHUTTLE ENVIRONMENT

The following is an outline of presentations delivered at the subpanel <u>adhoc</u> meetings.

Presentation	Presenter
Thermal and Humidity Outline	Clawson
KSC Environments	Bartelson and Wenkstern
Orbiter Purge System	Clawson
Payload Core ICD (Thermal)	Russell
Solar/Earth Environments	Clawson
Orbiter Interim Thermal Constraints	Clawson
On-Orbit Payload Bay Temperatures	Russell
STS Payload Bay Thermal Environments Summary	Fu
Integrated Payload Bay Models	Russell
Orbiter Thermal Control Correlation Findings	Clawson
Transparent Material Coatings	Clawson
The Need for Additional Thermal Environment Flight Data	Fu
Simplified Thermal Design Process	Fu
JPL Thermal Environment Design Approach	Fu

Appendix C   387

1. THERMAL AND HUMIDITY OUTLINE (J.F. Clawson)

Summary: A proposed detailed outline for the thermal and humidity data base

Launch Site Accomodations
    OPF
    VAB
    Other Checkout Facilities
    VAFB Facilities

Prelaunch Pad Conditions
    Purge System Capabilities
        Temperature
        Humidity
        Timeline
        Purge/Payload Interaction
    Orbiter Prelaunch External Environments
        Temperature
        Humidity
        Winds
    Orbiter Response to Prelaunch Environments/Purge

Ascent Phase Conditions
    Orbiter/Payload Bay Temperatures
    Ascent Pressure Decay

On-Orbit Conditions
    External Environments
        Solar Flux
        Albedo
        Planetary Infrared Radiation
    Orbiter Temperatures
    Orbiter Thermal Constraints
    Payload Bay Optical Properties
    Empty Payload Bay Temperatures
    Integrated Payload Effects

        Solar Trapping
        IR Interaction
        Blockage Effects
    Payload Bay Pressures
    Spacelab Environments
        Pressurized Module
        Pallets
    Crew Cabin Environments
        Payload Interaction
        Cabin ARS Capability
    ATCS/Payload Carriers
    Payload Carriers
        Typical Payload Thermal Constraints
        RCS Thruster Plume Heating Environments

Entry Conditions
    Orbiter Thermal Response
    Payload Bay Thermal Response
    Pressure Profile
    Humidity Profile

Post Landing conditions
    Environments
        Thermal
        Humidity
        Timeline
    Purge System Capabilities
        Temperature
        Humidity
        Timeline
    Orbiter Thermal Response to Post Landing

Ferry Flight Conditions
    Altitude/Pressure Profile
    Temperature
    Humidity

Appendix C

Orbiter Instrumentation

Mathematical Models
    390 Node Model
    SOTS Model
    Cabin Model

Payload Model Requirements

2. **KSC ENVIRONMENTS** (D. Bartelson and D. Wenkstern)

Summary: Monthly and annual average ambient environments are based on a 14-year data base. Measurements include mean daily maximum and minimum temperature, mean relative humidity, and mean precipitation. Also included in the presentation were the internal environmental data for canister #1 (canisters are the containers used to transport payloads to and from various payload facilities and the orbiter) that were collected over a three day period during July 1984.

OPERATIONAL CONSIDERATIONS
WEATHER

KSC AMBIENT ENVIRONMENT

MONTH	J	F	M	A	M	J	J	A	S	O	N	D	ANNUAL
TEMPERATURE (°F) MEAN DAILY MAXIMUM	69.8	69.8	73.4	77.0	82.4	86.0	87.0	87.8	86.0	80.6	75.2	69.8	78.8
TEMPERATURE (°F) MEAN DAILY MINIMUM	51.8	53.6	57.2	62.6	66.2	71.6	73.4	73.4	73.4	69.8	60.8	53.6	64.4
(%) MEAN RELATIVE HUMIDITY	80	80	78	75	77	81	83	84	83	79	79	79	80
(INCHES) MEAN PRECIPITATION	2.95	3.40	4.13	2.01	1.80	4.23	5.70	5.97	8.85	5.10	3.45	1.58	49.17 CUM.

SOURCE: KSC-FINAL ENVIRONMENTAL IMPACT STATEMENT-1979; BASED ON A 14 YEAR DATA BASE.

390 SPACE SHUTTLE ENVIRONMENT

KSC Ambient Environment
Canister #1
Environmental Data, July 1984

	Expected	Actual
Temperature	71 ±6°F	69-72°F
Humidity	30-50%	38-44%

3. **ORBITER PURGE SYSTEM CAPABILITIES** (J.F. Clawson)

Summary: A pictorial and numerical representation of all cargo bay purge vent locations. Also included were low and high purge rate tables for various orbiter processing facilities. Data included temperature, humidity, and max/min flow rates.

ORBITER VENT LOCATIONS

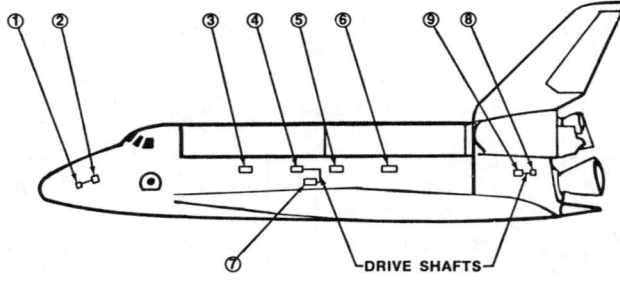

ALL VENTS SHOWN ARE LH AND RH
- VENTS 1 AND 2 ARE COUPLED
- VENTS 4 AND 7 ARE COUPLED
- VENTS 8 AND 9 ARE COUPLED
- 18 TOTAL VENTS ARE REQUIRED

VENT NO.	COMPARTMENT VENTED	C/L VENT LOCATIONS		
		X_o	Y_o	Z_o
1	FWD RCS	383.05	75.27	371.01
2	FWD FUSELAGE PLENUM	399.13	79.24	374.58
3		765.12	±105	385.43
4	MID FUSELAGE	904.70	±105	385.43
5	(CARGO BAY AND LOWER MID-FUSELAGE)	995.50	±105	385.43
6		1127.84	±105	385.43
7	WING (DEDICATED)	934.12	±105	356.19
8	OMS POD (DEDICATED)	1429.29	116.49	355.50
9	AFT FUSELAGE	1389.63	112.70	357.82

Appendix C   391

4. **CORE INTERFACE CONTROL DOCUMENT (ICD) SECTION 6.0** (D.J. Russell)
Summary: Chapter 6.0 of the core ICD is the environmental control interface section which contains all aspects of the thermal interfaces and environments that may be encountered while in the shuttle. Topics include the orbiter hold durations, space flux environments, structural attachment thermal interfaces, grapple fixture, etc.

6.1.1.3   Space environment, the numerical values of the parameters defining the space environment shall be as follows:

a.	Solar Radiation (hot case)	444 BTUH/Ft^2
b.	Earth Albedo	30% of solar radiation
c.	Earth Radiation	77 BTUH/Ft^2
d.	Space Sink Temperature	0° Rankine

6.1.4.1-1 Cargo Bay Wall Temperatures

Condition	Temperature	
	Minimum	Maximum
1. Prelaunch	+40°F	+120°F
2. Launch	+40°F	+150°F
3. On-Orbit (doors opened)	-250°F	+200°F
4. Entry Post-landing	-50°F	+220°F

5. **SOLAR/EARTH ENVIRONMENTS** (J.F. Clawson)
Summary: A discussion of values for the solar constant's variation over the year, global average albedo assumptions and their maximum and minimum values, and a derivation of a value for earth infrared emitted radiation.

Solar Constant

429 BTUH/Ft^2 + 3.43% (perihelion)=443.6 BTUH/Ft^2
429 BTUH/Ft^2 - 3.26% (aphelion)=415 BTUH/Ft^2

Albedo

0.30 is currently the best assumption globally
Ranges from 0.19 near the equator (in the summer) to 0.06 at the South Pole

Planetary Infrared

77 BTUH/Ft^2 is the most common current value--it was used for all Shuttle thermal design and math model correlation activities

6. ORBITER INTERIM THERMAL CONSTRAINTS (J.F. Clawson)
Summary: Durations for orbiter attitudes are based on various assumptions and specifications. Parameters include beta angle, orbiter attitude, and orbiter vehicle. (See Thermal ICD for more concise table of restrictions)

7. ON-ORBIT PAYLOAD BAY TEMPERATURES (D.J. Russell)
Summary: A comprehensive study of payload bay temperatures at various locations for six shuttle attitudes based on a correlated math model. The study compares empty bay with an integrated cargo bay.

Payload Effects
* Bay temperatures generally warmer with payload
   -Solar entrapment
   -Decreased view to space

* Local hot spots greater than 300°F possible
  -Payload liner
  -Payload to bulkhead
  -Payload to payload

* Cradles, wire trays, closeouts reduce effects

* Deployable mission timeline response depends on payload
  -30 minute top sun
  -90 minute bay to space

8. **STS PAYLOAD BAY THERMAL ENVIRONMENTS SUMMARY REPORT** (J. Fu)
Summary: A report to be published in FY 85 by JPL to include STS 1 through STS 5 flight data. An attempt will be made to verify existing thermal math models using existing and future flight data. The report will summarize thermal measurement locations, techniques, and will evaluate data validity. Timelines, trajectories, mission objectives, and thermally significant events will be listed for the first five shuttle flights. The final report will be user oriented and will contain a complete payload bay thermal summary.

9. **INTEGRATED PAYLOAD BAY THERMAL MODELS** (D.J. Russell)
Summary: An overview of the 390 node orbiter thermal math model and the SOTS model. The report contains a pictorial description of all nodes and a general model description. Also included is a list of modelling options and user considerations.

<u>390 Node Thermal Math Model</u>

User Considerations
* Payload Bay Renodalization Required
  -Liner, bulkheads, wingbox covers
  -Especially for sun in bay attitudes
  -Minimum four node circumferentially(six with wire trays)
  -Eight nodes (longitudinal) under large cylindrical payload

* Model options
  -Wire trays
  -Aft fuselage
  -Retention fittings
  -Stowed deployed radiators, flowing or not flowing

* Atmospheric model version
  -Prelaunch, ascent, on-orbit, entry, post landing
  -External structure driven during ascent and entry
  -Purge/spigot capability

* Atmospheric model inputs
  -Mission timeline(purge on/off, liftoff, entry interface, vent doors
    open, radiator flow off/on, touchdown, orbiter payload facility
    entry)
  -Ascent/entry drivers (structural, vent air flow)
  -Purge flow rates, inlet temperatures
  -Internal air pressure and volume
  -External environment (ground, sky, air, solar, planetary)

SOTS Description

* Payload bay
  -Bay divided into two areas of uniform thermal properties (**thermal**
    zones)
  -Allows user to define nodal breakdown to suit the payload

* External surfaces-two options
  -Boundary driver option-uses boundary nodes, driven to simulate
    effect of external environments
  -External surface option-external surface modelled, with temper-
    atures calculated based on external environment
  -Both options built with similar boundary locations and geometry
    for easy interchangeability

Appendix C

10. ORBITER THERMAL CONTROL CORRELATION FINDINGS (J.F. Clawson)
Summary: A listing of various modeling assumptions that were changed or added in order to correlate the math models with flight data.

Major Orbiter Thermal Control Systems Correlation Findings
* Thermal math model when originally built tended to include only 50-60% of total mass

* Thermal protection systems have different thermal conductance values when in a 10^{-6} vacuum

* Internal heat dissipation from relatively small sources can be significant
* Small gaps in insulation blankets can double heat losses

* Achieving good Multi Layer Insulation blanket performance is difficult
* Low conductivity structural materials require additional modeling attention

11. TRANSPARENT MATERIAL COATINGS (J.F. Clawson)
Summary: Beta cloth is a material used as the outer layer of many Multi Layer Insulation (MLI) blankets and as the liner for the Shuttle bay. Due to its transparent nature, care must be taken in choosing a backing material since the effective solar absorptance will be dependent upon the properties of both the beta cloth and the backing material. This presentation gives measurement techniques and theoretical calculations for effective absorbency and temperatures.

Summary of Findings

* Beta cloth outer surface temperatures insensitive to second surface temperature, only to solar optics of second (and sometimes third) layer(s) (for net heat transfer inwards, i.e. adiabatic conditions).

* Basic solar absorptance of beta cloth is quite low (approx. 0.09) but the absortance of the cloth/backing system is in the range of previously assumed properties for all payload analyses performed to date (i.e. <= 0.32). Note that previous measurements of absorption were indeed system values, not the true beta cloth values.

* Therefore, payload environmental temperatures are about as predicted. The recorded 260° F recorded during STS-3 top sun exposure was due to the greenhouse trapping effect under the beta cloth (Sensor was under a transparent beta cloth/tedlar patch).

* When simulating interactions between payloads, the system absorption is the proper value to use when the transparent materials are modeled as opaque surfaces.

* Even though payload bay surface temperatures are insensitive to the second surface material, the second surface temperature can be quite high (in the case of a metallic film), thus increasing the heat load to the orbiter or experiment (if similar MLI system used). This heat load increase can be quite dramatic.

12. **THE NEED FOR ADDITIONAL THERMAL ENVIRONMENT FLIGHT DATA** (J. Fu)
Summary: Defines future data desired from future flights to verify existing models for high beta angle launches. Desired measurements include temperatures, fluxes, and thermal optical property degradation.

The Need for Additional Thermal STS Payload Bay Environment Flight Data

* Limited flight data do not provide sound statistical basis for future flight environment predictions.
* Data needed at high beta angles for on-orbit, descent, and post-landing phases.
* Data needed on heat fluxes, including heat flows across MLI.
* Data needed on surface degradation vs. number of flights.

Appendix C

13. SIMPLIFIED THERMAL DESIGN PROCESS (J. Fu)

Summary: Outlines additional data needed to simplify modeling and design of payloads as well as various existing options to simplify the analysis and design process.

Simplified Thermal Design Process

* Simplified thermal design process will provide significant cost savings for future STS users.

* The thermal design of the payloads can be accomplished without the use of large mathematical models.

* The JPL thermal summary report should provide an effective way of using existing flight data in support of a simplified design approach.

* Continuing analysis and data monitoring, summarization, and characterization of future data will enhance the confidence level of the simplified analytical design approach.

14. JPL THERMAL ENVIRONMENT DESIGN APPROACH (J. Fu)

Summary: Outlines JPL testing levels for various types of acceptance.

JPL Thermal Environmental Design Approach

* Basis: allowable flight temperatures (aft)
* Margins for flight acceptance: ±5°C above aft
* Margins for flight qualification: ±25°C above aft
* Margins for overall design: ±35°C above aft
* The above applies for operating and non-operating conditions.
* Number of cycles in T/V or T/A tests: 1
* Durations of T/V T/A tests: 68 hrs minimum for acceptance tests

Summary

Good data are available from KSC for prelaunch environments and payload facilities (temperature, humudity, purges, etc.). On-orbit payload bay temperatures are available from flight data and correlated models for low beta angles. Lift-off, re-entry, and post landing conditions have also been calculated from correlated models. Data from the spacelab missions should eventually become available through Marshall via flight data or model predictions.

There currently exist two well-documented correlated Shuttle models (390 node model SOTS). They are both capable of predicting on-orbit, launch and re-entry conditions. Unfortunately both models would be difficult for small users to implement. There exists a need to develop a simplified "equivalent sink" program that would utilize on-orbit Shuttle bay temperatures and experiment locations for inputs. This would require some type of additional funding which at this time is not available. There was also a request for cabin temperatures and requirements for various cabin and mid-deck locations. We believe that the requirements are documented and that there is probably an existing model for predicting cabin temperatures.

Issues

There was a concern that there might be a misinterpretation of key data by "novice" users and that data may be misapplied to analyses. A suggestion was made to include warnings about directly applying numbers without first examining all relevant assumptions that were made when the data were either predicted or measured. Another concern was that small users could have to design for all possible in-flight thermal extremes, since they would not have a flight assignment far enough in advance to design for specific cases. There should be an attempt to bound the design process, in order to reduce thermal analysis costs.

APPENDIX D

ORBITER MOTION*

INTRODUCTION

This is the background information relevant to the subpanel on Orbiter Motions (see Chapter 2). The Space Shuttle employs both primary and vernier reaction jets for attitude control and has two orbital maneuvering engines. The autopilot is nonlinear and digital. The control of the rigid body of the Space Shuttle includes automatic rotation modes, commanded translation modes and earth-, inertial-, and orbital-pointing modes. Control has small bandwidth (< 0.1 Hz) and moderate (0.5 deg) accuracy. The location in the Space Shuttle of the principal control elements is illustrated in Figure 1.

Attitude control employs an inertial measurement unit (IMU) as its primary reference. The position of the IMU is updated at 8-12 hour intervals. The modes of the attitude control which are not inertial require navigation information. Bandwidth of the attitude control is constrained to minimize bending interactions: bandwidth of the Primary Reaction Control System (PRCS) is limited to 0.1 Hz; that of the Vernier Reaction Control System (VRCS) to 0.04 Hz. The reaction control systems have a small duty cycle (< 1%); the average pulsing period varies from 15 sec for the VCRS to more than 100 sec for the PRCS. Sources of pointing error may include navigation errors, alignment instabilities, and errors of control dynamics. The locations in the frequency spectrum of major equipment resonances affecting attitude controls are indicated in Figure 2.

---

*This suplementary information to the Orbiter Motion subpanel paper in Chapter 2 was prepared by Dr. Alan Grobecker, visiting scholar at the University of California at Los Angeles.

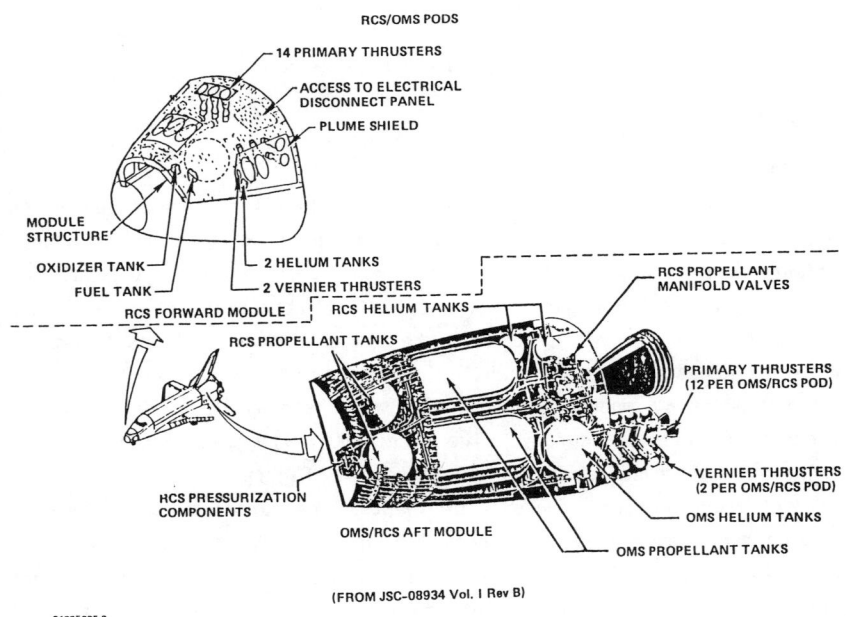

Fig. 1. Location of Attitude Controls of Space Shuttle

Appendix D 401

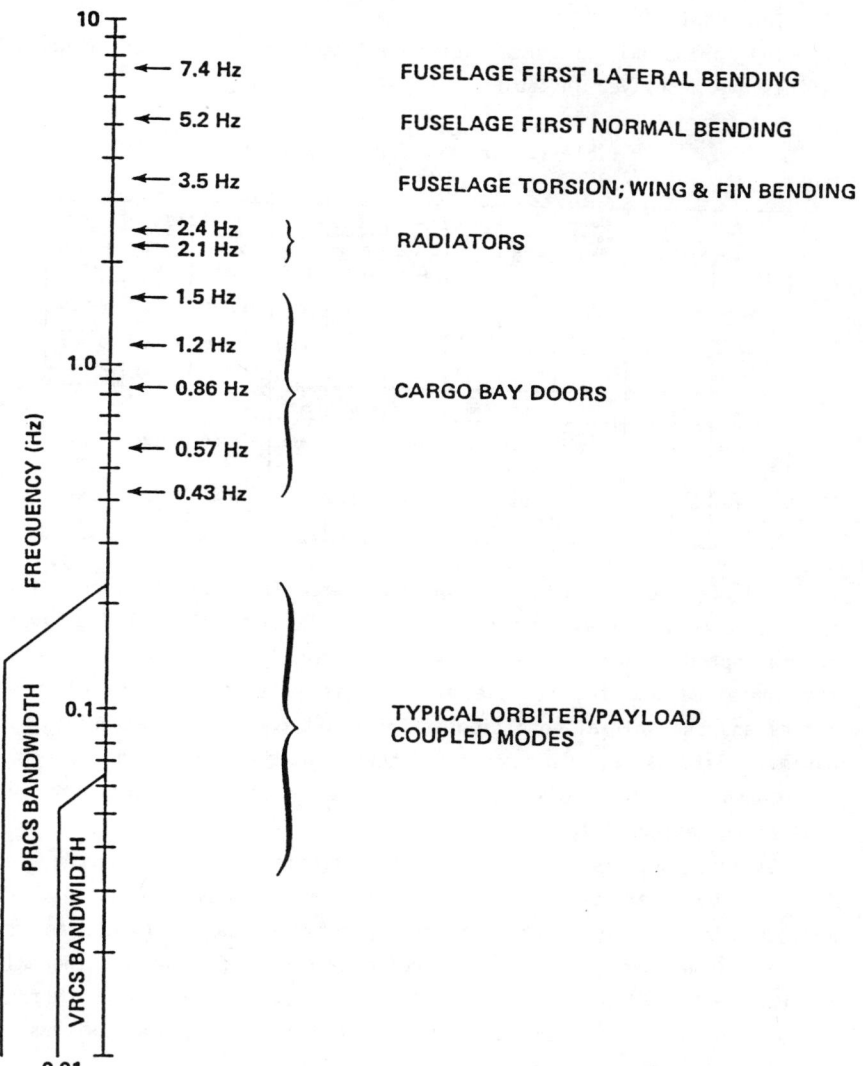

Fig. 2. Resonance Frequencies of Space Shuttle

TRACKING CONFIGURATIONS

The principal characteristics of the Space Shuttle control systems are displayed in Table 1.

Table 1. Orbiter Control Systems

	System	Actuators No.	Red?	Controlled Axes	Accel	Minimum Pulse Width	Minimum Rate Change	Disturbance Angular Accel	Disturbance Linear Accel	Control Performance Rate Typ/Min	Control Performance Angle Typ/Min	Maneuver Rate Typ/Max
					deg/s^2	ms	deg/s	deg/s^2	ft/s^2	deg/s	deg	deg/s
Rotation	PRCS	38	Y	3	1.0	80	0.08	0.07	0.1	0.50/0.20	2.0/1.0	0.8/2.0
Rotation	VRCS	6	N	3	0.02	80	0.002	0.01	<0.001	0.02/0.01	1.0/0.03	0.1/0.2
					ft/s^2	ms	ft/s	deg/s^2	ft/s^2			ft/s
Translation	PRCS	38	Y	3	0.50 (0.25)	80	0.04 (0.02)	0.3	0.04	N/A	N/A	0.1/20
Translation	OMS	2	Y	1(X)	2.0	80	3.0	0.7	0.2	N/A	N/A	3/500

The modes of the reaction control system (RCS) attitude controls can be classified as three: drift mode, universal pointing with several options, and a manual pulse mode. In the drift mode, the jets are inhibited and the vehicle attitude is affected only by disturbances and the initial conditions. The drift mode is characterized by minimum disturbance, minimum contamination by jet gases, the absence of closed loop controls, and seems most useful for sensing the environment external to the vehicle.

Several options for the universal pointing mode include an inertial maneuver or hold (reference frame is non rotating), a maneuver or hold in a rotating reference frame (called LVLH), and a rotation about the X-axis of the orbiter employed for passive thermal control (PTC) options. A constant angular rate of rotation of the Space Shuttle may be realized by employing one of the universal pointing mode options of inertial, LVLH or PTC.

The manual pulse mode is open-loop, in which the pulse is initiated by means of a hand controller and has a duration keyed in by the Space Shuttle crew. Maneuvers involve discrete attitude changes and the tracking of landmarks.

Appendix D

FREE DRIFT CHARACTERISTICS

The characteristics of the vehicle motion when in a condition of free drift are an indication of natural frequencies of response to disturbances. Figure 3 displays the roll error due to gravity gradient and drag effects. The period of oscillation is about two hours (frequency of $1.4 \, (10^{-4})$ Hz). Figure 4 displays the roll rate response of the shuttle vehicle to a one second pulse of roll command: the oscillation represents the superposition of frequencies of 1/7 and 1/3 Hz.

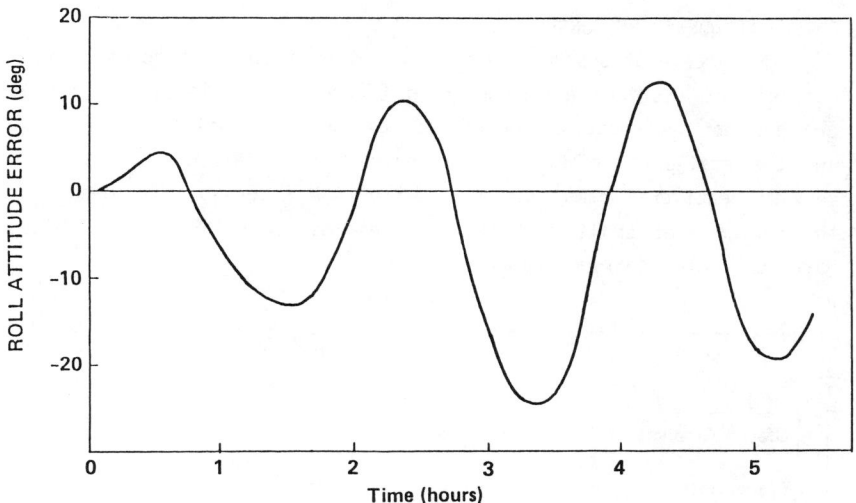

Fig. 3. Free Drift in Gravity Gradient

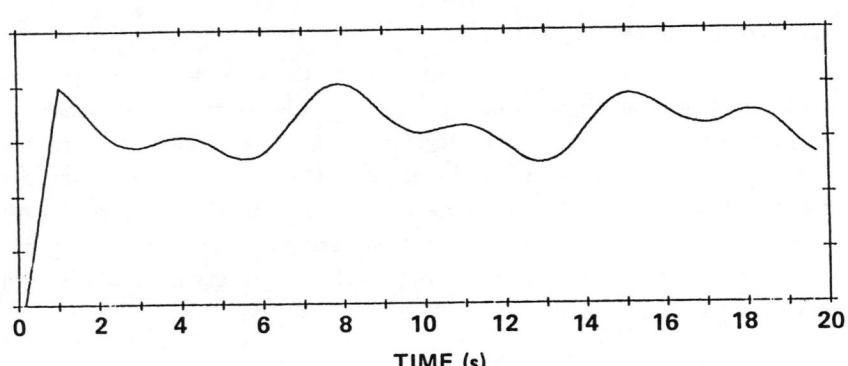

Fig. 4. Orbiter Roll Rate Response to 1s Vernier Roll Pulse with SPAS-01 in Starboard Overhead Position

FORCED FREQUENCY RESPONSES

The accelerations of the vehicle in response to forcing disturbances are described graphically in Figure 5. The largest responses are due to the reaction control systems and activities of the crew, and the smallest are due to the environment. The operation of the vernier reaction control system (VCRS), with a zero-pitch command, in the presence of small disturbances, results in a variation of pitch rate for a 0.01 deg/sec deadband (Figure 6).

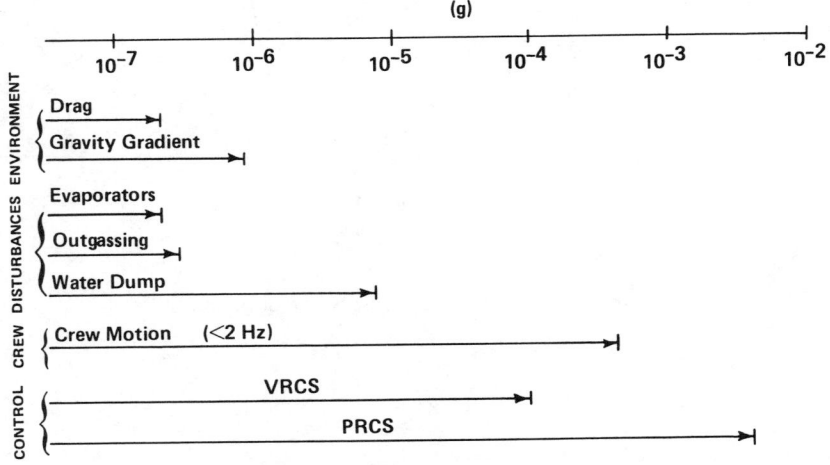

Fig. 5. Disturbance Accelerations

Appendix D

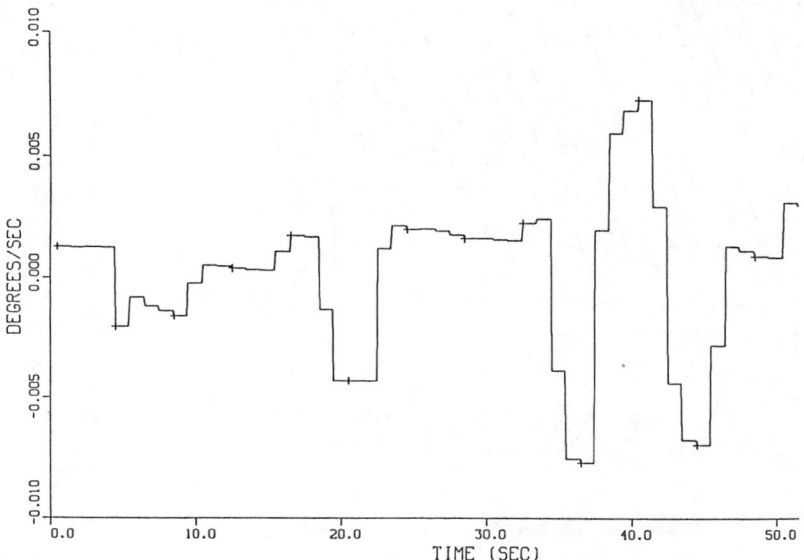

Fig. 6. Pitch Rate Trajectory--VRCS Narrow Deadband Test

Manual control by the keyboard available to the crew is best done by combinations of maneuver rate and attitude deadband (inside of which the vehicle can wander) which are shown in Figure 7. The limits of keyboard-controlled attitude deadband are a minimum of 0.01 and a maximum of 40 deg. The limits of keyboard-controlled maneuver rate-of-change are a minimum of 0.002 deg per sec and a maximum which varies linearally between 0.015 deg per sec at the lower keyboard attitude deadband limit of 0.01 deg and 1.0 deg per sec at the upper keyboard attitude deadband limit. It is possible to pulse the vernier reaction control system (VCRS) to produce maneuver rates even larger than the preferred limits indicated.

Four types of acceleration are experienced by the Space Shuttle. The steady low-level accelerations, of magnitude 10^{-5} g or smaller, result from aerodynamic drag or the gradient of the gravity field along the orbit trajectory. Compensated transient accelerations usually have the form of a positive acceleration pulse followed by a negative pulse of equal magnitude: many experiments may be insensitive

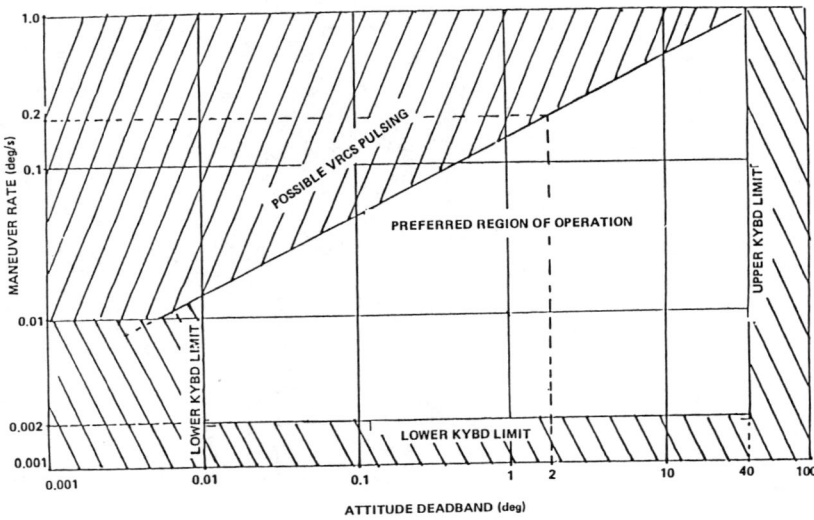

Fig. 7. Preferred Combinations of VCRS Manuever Rate vs Deadband

to such compensated transients. Vehicle rotation, necessary for temperature control and instrument pointing, causes accelerations to be experienced at locations away from the axis of rotation. Uncompensated transient accelerations are caused by firing of thrusters, dumping of water, deployment of satellites and other functions or activities of vehicle and crew. Accelerations encountered or expected in Spacelab, of both residual and perturbative origin, are summarized in terms of units of gravity at the earth surface (g = 32 ft s^{-2} = 981 cm s^{-2}) in Table 2. Some sources of vibration which have been observed in a manned orbiting spacecraft are listed in Table 3. The relative magnitudes of accelerations contributed by several sources are shown simply as functions of frequency in the line plot of Figure 8. These do not include the effects of crew activities and of thruster reactions. The sources contributing frequencies smaller than 1 Hz yield accelerations less than 10^{-5}g; the accelerations encountered between 1 and 100 Hz generally are smaller than values indicated by the straight line between 10^{-5}g at 1 Hz and 10^{-3}g at 100 Hz. and those encountered at frequencies greater than 100 Hz are smaller than 10^{-3}g.

Appendix D

Table 2
Summary of Residual and Perturbative Accelerations in Spacelab

Source	Acceleration (in units of g)	Remarks
Gravity gradient (at 400 km altitude)	Residual acceleration: 10^{-6} max.	"tidal" field
Solar radiation pressure and micro-meteoriod fluxes	$4 \cdot 10^{-9}$ max.	slow time variation, diurnal
Atmospheric drag (neutral component) at 400 km altitude	$2 \cdot 10^{-8} - 10^{-7}$	dependent on attitude of orbiter relative to velocity vector
Internal self-gravitational field of Shuttle/SL-System	$<10^{-9}$ ($10^{-8} - 10^{-7}$ max., in very extreme situations, which are hardly of practical importance)	stationary field
Venting activities	$\sim 10^{-7}$	at irregular intervals in Shuttle and SL.
Vernier attitude control thruster activity	$3 \cdot 10^{-4}$ max.	for translation motion as well as for tangential component of rotation (at a distance of 5 m from axis of rotation)
SL Life support system and auxiliary equipment	Measured results: $3.6 \cdot 10^{-3}$ $4 \cdot 10^{-4}$ $4 \cdot 10^{-5}$	peak value (time domain) spectral value < 100 Hz spectral value < 10 Hz
Astronaut motion and other crew activities	Measured results: $2 \cdot 10^{-2}$ $6 \cdot 10^{-3}$ $1 \cdot 10^{-3}$	peak value (time domain) spectral value < 100 Hz spectral value < 10 Hz
Experiments and their auxiliary equipment	quantitatively still to be determined for D-1 experiments	local effects

Table 3
On-Orbit Vibration Sources of a Manned Spacecraft

	Spacelab Subsystem	Component	Vibration mechanism
1	ECLS	Cabin Fan Package	aerodynamic, vibro-acoustic mechanical vibration
2		CO_2 Control Assembly	aerodynamic, vibro-acoustic
3		Humidity and Temperature Control Assembly (Heat Exchanger)	aerodynamic, vibro-acoustic
4		By-Pass of Humidity/Temperature Control Assembly	aerodynamic, vibro-acoustic
5		Cabin Air Ducts, Bends	aerodynamic
6		Outlet Diffuser	aerodynamic, vibro-acoustic
7		Water Separator	aerodynamic, vibro-acoustic
8		By-Pass Valve Actuator Motor	mechanical, vibration intermittent
9		Avionics Fan Package	aerodynamic, vibro-acoustic mechanical vibrations
10		Avionics Heat Exchanger	aerodynamic, vibro-acoustic
11		Avionics Air Ducts/Bends	aerodynamic, vibro-acoustic
12		Rack Inlet Diffuser	aerodynamic, vibro-acoustic
13		Rack Return Ducts	aerodynamic, vibro-acoustic
14		Upper Feedthrough Valves	shocks, exceptional
15		Smoke Detectors	mechanical vibrations
16		$O_2 N_2$ Panel Valves	shocks, exceptional
17		Monitoring & Control Panel Switches	shocks, exceptional
18	TCS	Water Pump Package	mechanical vibrations
19		Freon Pump Package	mechanical vibrations
20		Tubing	mechanical vibrations
21		Fluid Heat Exchanger	mechanical vibrations
22		Cold Plates	mechanical vibrations
23	EPDS	400 Hz Inverter	mechanical vibrations
24		Luminary	mechanical vibrations
25		Monitoring & Control Panel Switches	shocks, exceptional
26	CDMS	Mass Memory Unit	mechanical vibrations, intermittent
27		Data Display Unit/Keyboard	mechanical vibrations, intermittent, small
28		High Data Rate Recorder	mechanical vibrations, shocks due to start and stop
29		Intercom & Remote Loudspeaker	vibro-acoustic, intermittent
30	SAM	RAAB Converter	mechanical vibrations
31		Caution & Warning Loudspeaker	vibro-acoustic, exceptional
32	Airlock	Pressure & Depressure Valve	mechanical vibrations, shocks
33	Spacelab Transfer Tunnel (STT)	Fan/Airducts	aerodynamic, vibro-acoustic, mechanical vibrations
34	Shuttle	via Trunnions	mechanical vibrations
35	Payload	Experiments	mechanical vibrations

Appendix D

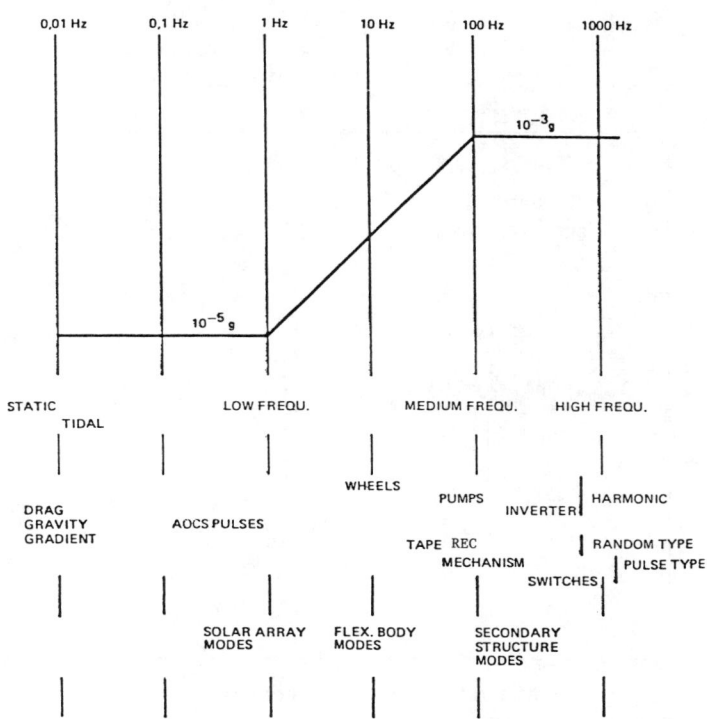

Fig. 8. Relative Contribution to the Microgravity Environment

SPACE SHUTTLE IN-FLIGHT DATA

Some time sequences of accelerations typical of space shuttle in-flight data are given with remarks to illustrate the microgravity environment of STS-7, STS-9 and STS-11. In these remarks and figures, the Z-direction is down, the X-direction is forward, and the Y-direction is right laterally (to starboard). In Figure 9 is shown a 500 sec interval of microgravity measurements made on the X-axis of STS-7. In three experiment sequences made by the Low-G Accelerometer System (LGAS) on that flight, the valid acceleration data varied within the range of 100 100 micro-g, with noise levels of ± 15 micro-g and accuracy of ±20 micro-g. The data showed no correlation with thruster firings, since the recorded 1-second averages of the accelerometer output mask the 0.08 VRCS firings.

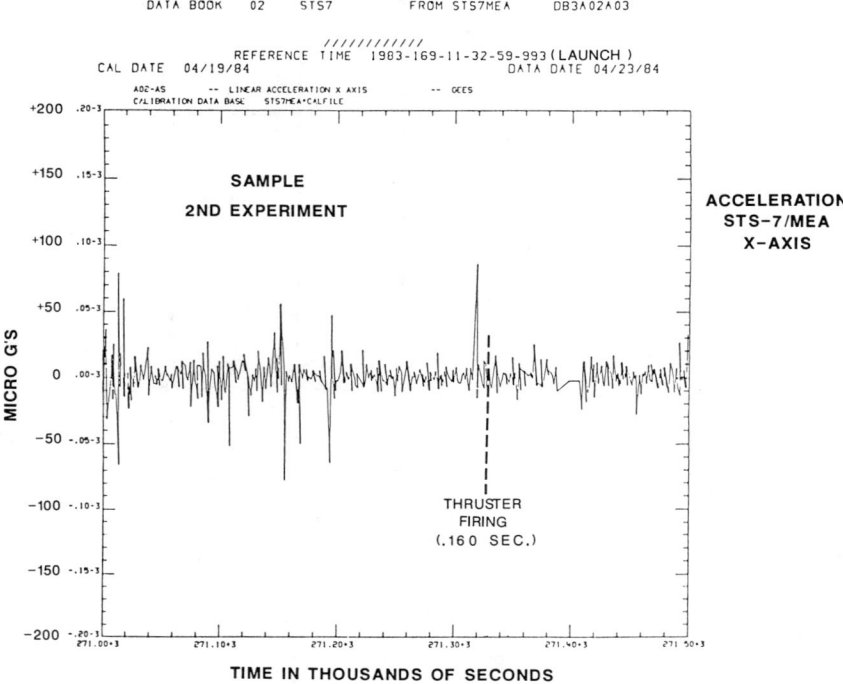

Fig. 9. Accelerations--STS-7/MEA x-axis

The activity of the crew as a source of small accelerations of the spacecraft is graphically displayed by a time history of observations along the Z-axis of STS-7, given in Figure 10. The sequence shows a sleeping interval of 0.3 time units, followed by a first working period of 0.9 time units, another sleeping interval followed by a second working period, and a third sleeping period followed by a working period. Although the magnitude of the accelerations is not apparent from the labelling, the relatively quiet sleeping periods are easily distinguished from the active working periods.

Appendix D   411

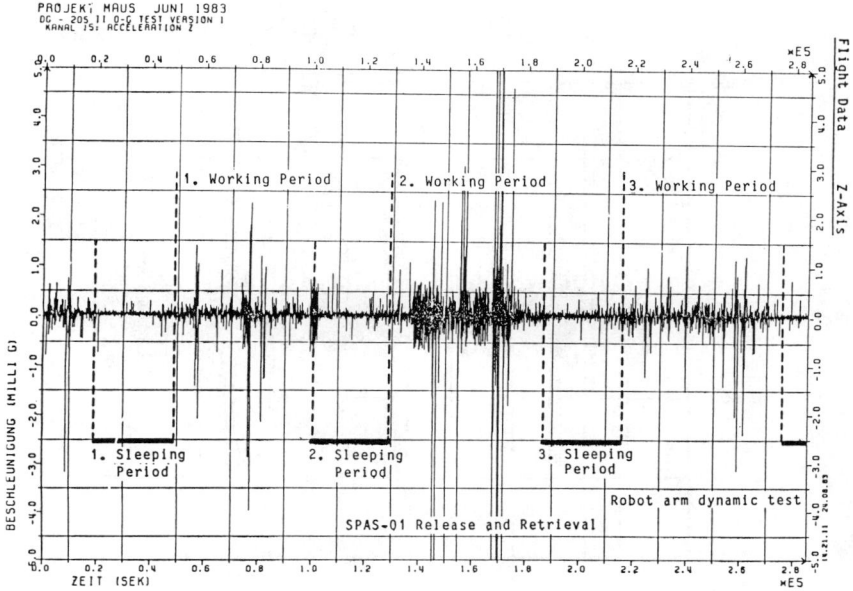

Fig. 10. Accelerations--STS-7/MAUS z-axis

Shown in Figure 11 is the time history of microgravity observations made in the forward (X) direction over a 10 minute interval in Space Shuttle STS-9. The acceleration activity is particularly intense during the firing of the orbiter thrusters; other large peaks result from crew motion, heating of the payload bay and other unidentified activities.

Shown in Figure 12 is a 15.5 sec record of observations along, each of two axes of the vehicle STS-11. The ordinate is labelled in 10^{-3}g, and the abscissa in seconds. Although the recording noise level was 100 micro-g, a number of accelerations larger than that appear in the record.

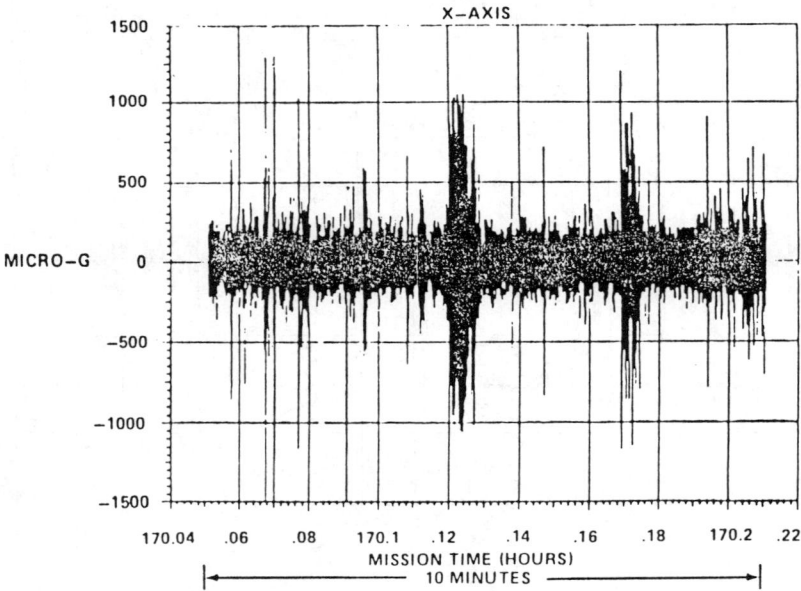

Fig. 11. Acceleration, STS-9 x-axis

Appendix D 413

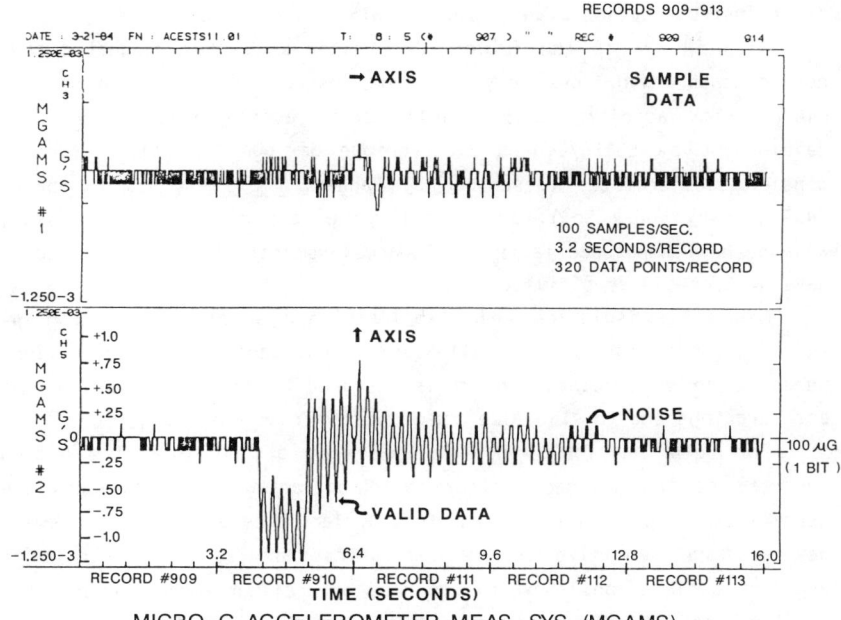

Fig. 12. Acceleration--STS-11/MGAMS x- and z-axis

IN-FLIGHT INSTRUMENTATION

Even after conclusion of the Orbital Flight Test (OFT) program, utilization of in-flight instrumentation continues as a valuable part of the Space Shuttle test activity. Instrumentation has often provided information in areas other than those planned for investigation. For example, the High Resolution Acceleration Package (HIRAP), developed for identification of aerodynamic coefficients during reentry, was also used to dynamically characterize the forward control system (FCS) while the vehicle was in orbit. The strain gauges for the reaction measuring system (RMS), originally intended for loads analysis, were subsequently used also for identification of the RMS structural model. Some sensors originally developed as instrument payloads, such as the SPAS-01 accelerometers and a redundant inertial measurement unit (IMU), became useful in supplementing other instrumentation planned for the STS.

With an eye to the future, Dr. Walter Knabe has suggested some developments that one day may be useful for observation of microgravity accelerations: a solid state accelerometer (now under development by ESTEC/Centre Electronique Horloger (CEH)); a surface tension accelerometer (described by Padday, Fluid Physics in Space, published by Kodak Co.); an optical range rate measurement by laser, and the Bell Aerospace Textron MESA accelerometer which is expected to have a threshold sensitivity of $10^{-8}g$.

A payload isolation and stabilization system has been conceptually designed for the Marshall Space Flight Center. Called the Suspended Experiment Mount (SEM) it is designed to provide isolation from accelerations and to stabilize the viewing direction of a payload.

For both the low-g and viewing types of payloads, the motion imparted to the payload platform is important. Low-g payloads are sensitive to both linear and to angular accelerations. Viewing payloads are sensitive to both the first and second integrals of angular accleration. Several sources of disturbance prevent the orbiter vehicle itself from achieving an ideal, disturbance-free environment for such payloads. Crew motion and the operation of the attitude control system cause the most significant of the low-g accelerations. Reducing the integrals of all the low-g accelerations, which contribute to instability of pointing direction and pointing jitter, is limited by the drift rate and deadband of the attitude control system.

The characteristics of pointing accomodations currently available for use in Space Shuttle are listed below.

DYNAMIC CHARACTERISTICS:

	Hardmounted	Isolator Mounted
SEM stability	±2 arc min	±1 arc sec
SEM stability rate (jitter)	±1 arc min/s	±1 arc sec/s
Orbiter relative motion	none	±3 arc min
Orbiter relative translation	none	±2 cm

CMG MOMENTUM:

	Gravity Gradient Only	+20% Aero
Orbiter X-axis perpendicular to orbit plane	2350 nms/orbit	2820 nms/orbit
Orbiter X-axis in orbit plane	7650 nms/orbit	9180 nms/orbit

OPERATIONAL TIME UNTIL CMG DESATURATION:

Using orbiter VCS thrusters	2 to 6 orbits
Using gravity gradient dump with offset pointing systems mechanisms	indefinite

They either are of low performance (the orbiter itself), are elaborate for low-g payloads and expensive (the instrument pointing system (IPS)), or impose great demands of design and testing effort on the part of the experimenter providing the payload.

The concept of SEM consists of a flexible suspension system and payload-mounted control moment gyros (CMG). The suspension system, which is rigidly locked for ascent and descent, isolates the payload from high frequency disturbances in its unlocked mode; the control moment gyros stabilize the payload orientation.

A comparison of the characteristics of pointing control dynamics when the payload is mounted on the SEM and when it is hard mounted to the orbiter vehicle platforms without benefit of an SEM is listed below. Also listed is a comparison of the momentum accumulation of the control moment gyros under four conditions of relatively steady low-g forces, and the operational time for removal of momentum saturation from the control moment gyro system.

Orbiter
- o Accuracy - 1 degree
- o Stability - arc minutes (determined by deadband setting)
- o Hard-mounted payload
    - Full orbiter resources available
    - Relatively simple integration
- o Accuracy can be improved by use of payload attitude sensor
- o Target change requires orbiter maneuver

IPS
- o Accuracy - 2 arc seconds
- o Stability - 1-5 arc seconds
- o Pointing cone - 60° half angle
- o Limited electrical power across gimbal
- o Limited signals across gimbal
- o No thermal control fluids across gimbal
- o Extensive integration effort

Payload Provided
- o Optimized for payload
- o Increases payload complexity and cost

Saturation is a concern for any momentum-based attitude control system. The SEM is capable of operating for several orbits before desaturation is required. Desaturation is accomplished by applying an external torque to cancel the accumulated angular momentum absorbed earlier from the orbiter. One possible source of external torque is the use of the orbiter thrusters; another possibility is the use of gravity gradient torques either by flying with an attitude that produces the negligible secular torques or by periodically maneuvering to an attitude at which gravity gradient and aerodynamic torques cause an accumulation of angular momentum in a direction opposite to that of the accumulated saturation.

For low-g payloads viewing only in a single direction (not switched between several directions systematically), the CMG system may be used both to control the payload and to control the orbiter through the suspension system. Such orbiter control, by minimizing orbiter firings, reduces the disturbances to the payload and the contamination due to gases emitted by excessive firings.

In short, the applications of the SEM are twofold. It may be used as a passive isolation system which isolates low-g payloads from disturbances of the Space Station or orbiter platform, attenuating high frequency disturbances due to crew motions, thrusters and docking and berthing maneuvers. It may accommodate some viewing payloads having moderate requirements for accuracy and short frame times. A second application is as an isolation system employing control moment gyros (CMG). This application provides a stable viewing platform suitable for payloads having only a single viewing direction, but would involve an interaction with the primary Space Station control system and require distributed controls not yet completely investigated.

The SEM provides pointing for pallet-mounted experiments and control of the Shuttle orbiter without employment of the reaction control system, thereby reducing contamination by jet exhausts and improved viewing capability of the experiments. During launch and reentry phases of Shuttle flight, the loads are carried by a solid structure; during in-orbit operations, the experiment base is stabilized with 1 arc sec stability (compared to the 2 arc min stability of a hard mounted pallet) and with reduction of g-disturbances of at least an order of magnitude. Interfaces with the Shuttle vehicle and for integration of experiments with stringent pointing requirements are more easily satisfied with respect to independence of experiment mounting, simplicity of electrical and fluid feed lines to the pallet, and easing of constraints on the envelope of the center of gravity motion of the experiment.

R. Vandervoort discussed techniques available for the engineering analysis involved in coupling an experimenter's instrument to the Shuttle environment by means of an isolation mechanism. He used as examples the Pointing/Occulting Facility (P/OF) and the K_u-band antenna for the Shuttle.

NEEDS OF MATERIAL SCIENTISTS

Y. Malmejac (France) introduced considerations of micro-gravity levels which were important to materials science investigations in orbit. His categorization of the effects of several types of low-level accelerations experienced in the Space Shuttle is given in Table 4. Compensated transients are relatively unimportant for the success of contained solidification and of fluid experiments. For containerless experiments, only the three other types of acceleration (low level steady, uncompensated transients and rotation-induced flows) were unimportant in hazarding success.

Table 4
Hazards of Low-g Accelerations on Materials Science Experiments

	Contained Solidification	Quasi-Containerless Solidification	Containerless Experiments	Fluid Experiments
Low-Level Steady Accelerations	Possibly Serious	Possibly Serious	Unimportant	Possibly Serious
Compensated Transient Accelerations	Relatively Unimportant	Possibly Serious	Possibly Serious	Relatively Unimportant
Uncompensated Transient Accelerations	Possibly Serious	Possibly Serious	Relatively Unimportant	Possibly Serious
Rotation-Induced Flows	Should be Avoided	Should be Avoided	Unimportant	Should be Avoided

STATUS

The present status of orbiter motion controls was summarized by Kevin Daly as follows:

1. All basic orbiter control capabilities have been validated by flight experience.
2. Payloads themselves can significantly influence control performance and stability as shown by reaction measurement system (RMS) activities, by the use of flexible mounting of payloads, and by operations of experiments in too close proximity.
3. For evaluation of controls, a limited amount of instrumentation is currently available.
4. Full maturity of control systems for orbital operations is expected in 1987-1988.

SUMMARY

In general terms, the spokesmen for the Orbiter Motion Panel summarized their conclusions as follows:

1. The dynamic environment in the Space Shuttle orbiter is significantly different from that of other satellites in our experience.
2. Operations of payload experiments can significantly influence the orbit dynamics.
3. Desirable performance represents a balanced choice among many different (sometimes opposed) considerations of operational flexibility, accuracy, utilization of fuel, contamination and necessary orbiter activities.
4. The micro-g environment to be provided to experiments depends upon an assessment of future requirements, analyses of compatibility of experiments to be carried by the same vehicle, verification on the ground of equipment and instrumentation characteristics, in-orbit implementation and measurement of actual micro-g characteristics.
5. Reduction and alleviation of disturbances already encountered in orbit should include the study of feasibility (with experimental verification) of promising concepts for suspension of experimental equipment.

6. The present operational Spacelab Module is endowed with a micro-gravity environment of good quality. The observed g-jitter contributions from crew activities and from operations of experiments and auxiliary equipments must be kept as low as possible in order to retain the low acceleration levels of the basic system-generated background.

7. Although some micro-g levels desirable for the conduct of important materials science experiments are lower than can be realized on board a manned vehicle, there are still many important investigations that can be accomplished in the Space Shuttle. They should be planned for operation with suitable acceleration-reduction isolation mechanisms.

APPENDIX E
ABBREVIATIONS AND ACRONYMS

AE	Atmosphere Explorer (satellite, NASA)
AFB	Air Force Base
AFCRL	Air Force Cambridge Research Laboratories (now US Air Force Geophysics Laboratory)
AFGL	Air Force Geophysics Laboratory
AGCY	agency
AIAA	American Institute of Aeronautics and Astronautics
ALT	altitude
AM	Amplitude Modulation
a.m.	ante meridian
AMU	Astronaut Maneuvering Unit
APL	Applied Physics Laboratory of Johns Hopkins University
ARC	Ames Research Center (NASA)
ARS	Advanced Solid-State Recording System
atm	atmosphere
ATS	Applications Technology Satellite (NASA)
avg	average
BE	Beacon
Btu	British thermal unit
BTUH	British thermal unit-hour
C	Celsius; coulomb
CCE	Charge Composition Explorer (satellite, AMPTE program)
CG	center of gravity
CRRES	Combined Release and Radiation Effects Satellite (joint NASA/USAF mission)

DATE	Dynamic, Acoustic, and Thermal Environments
DFI	Development Flight Instrumentation
DMSP	Defense Meteorological Satellite Program (DoD)
DoD	Department of Defense
EED	ElectroExplosive Device
ELDO	European Launcher Development Organization
EMC	Electromagnetic Compatibility Study
EOS	Earth Observation Satellite (NASA)
ERBS	Earth Radiation Budget Satellite (NASA)
ESA	European Space Agency; electrostatic analyzer
ESRO	European Space Research Organization (now ESA)
ESTEC	European Space Technology Center (ESA)
EURECA	European Retrievable Carrier (spacecraft)
EXOSAT	European X-ray Observation Satellite (ESA)
FCS	Forward Control System
FSC	Fleetsat Com
FWC	Filament Wound Case
GAS	Get Away Special
GIRL	German Infrared Laboratory
GOES	Geosynchronous Operational Environmental Satellite (NASA-NOAA)
GSFC	Goddard Space Flight Center (NASA)
HCS	Helium Control System
HIRAP	High Resolution Acceleration Package
ICD	Interface Control Document
IECM	Induced Environment Contamination Monitor
IML	International Microgravity Laboratory

IMU	Inertial Measurement Unit	
IOCM	Interim Operational Contamination Monitor	
IPS	instrument pointing system	
IRS	Information Retrieval System	
ISAS	Institute of Space and Aeronautical Science	
ISO	Imaging Spectral Observatory	
ISPM	International Solar Polar Mission (ESA)	
IUS	intermediate upper stage, Inertial Unit Special	
JHU	Johns Hopkins University	
JPL	Jet Propulsion Laboratory (Cal Tech & NASA)	
JSC	Johnson Space Center (NASA)	
KSC	Kennedy Space Center (NASA)	
LAMAR	large area modular array of reflectors	
LaRC	Langley Research Center (NASA)	
LDEF	Long-Duration Exposure Facility	
LeRC	Lewis Research Center (NASA)	
LFC	large format camera	
LVLH	Low Velocity Low Height (a mode of reaction control)	
MAPS	Measurement of Air Pollution from Satellite	
MPL	Materials Processing Laboratory	
MSC	Manned Spacecraft Center (now Johnson Space Center)	
MSFC	Marshall Space Flight Center (NASA)	
MSL	Materials Science Laboratory	
NASA	National Aeronautics and Space Administration (Washington, D.C., Headquarters)	
NASCOM	NASA Communications Network	

NOAA	National Oceanic and Atmospheric Administration (formerly ESSA)
NRC	National Research Council
OEM	Optical Effects Module
OFT	Orbital Flight Test
OMS	Orbital Measurement System
OMV	Orbital Maneuvering Vehicle
OSF	Office of Space Flight
OSSA	Office of Space Science and Applications (NASA)
OSTA	Office of Space and Terrestrial Applications
OTV	Orbital Transfer Vehicle
PAM	Payload Assist Module
PDP	Plasma Diagnostic Package; Passive Dosimeter Packet
PIP	Program Interaction Plans
POCC	Payloads Operations Control Center
PRCS	Primary Reaction Control System
PTC	Pointing and Tracking Controls
RAHF	research animal holding facility
RCS	Reaction Control System
RMS	remote manipulator system, Reaction Measurement System
RTG	Radioactive Thermal Generator
SAIL	Shuttle Avionics Integration Laboratory
SAR	synthetic aperture radar; search and rescue
SBS	Satellite Business Systems
SCEE	Shuttle Spacelab Contamination Environment and Effects
SHEAL	Shuttle High Energy Astrophysice Laboratory
SMACS	Spacecraft Maximum Allowable Concentrations
SMM	Solar Maximum Mission (satellite, NASA)

Appendix E   425

SPAS 01	an accelerometer instrument
SOT	Solar Optical Telescope (satellite)
SPF	Specific Pathogen Free
SPL	Space Plasma Laboratory
SRB	Solid Rocket Boosters
SRL	Shuttle Radar Laboratory
SS	Space Shuttle
SSIP	Shuttle Student Involvement Program
SSME	Space Shuttle Main Engine
STEP	Space Technology Experiment Program
STS	Space Transportation Program
TDRSS	Tracking and Data Relay Satellite System
TOS	Transfer Orbit Stage
TSS	Tethered Satellite System
UIT	Ultraviolet Imaging Telescope
UV	ultraviolet
VAPEPS	Vibroacoustic Payloads Environment Prediction System
VFI VRM	Venus Radar Mapper
VRCS	Vernier Reaction Control System
WUPPE	Wisconsin Ultraviolet Photometer Polarimeter Experiment

APPENDIX F
CONVERSION UNITS

To assist readers in converting between the systems of units used in data brought to the Workshop, Dr. Alan Grobecker has suggested the following conversion tables, organized according to the presentation in which the units were employed.

SURFACE INTERACTIONS, APPENDIX A (p. A-1)

Unit of	Brit	cgs	Mult. by	To get mksa (SI)
Length				
velocity		cm s^{-1}	10^{-2}	m s^{-1}
		micron	(10^{-6})	m
		km	(10^{+3})	m
	mil		$2.5(10^{-5})$	m
	naut mi.		$1.8(10^{-3})$	m
wave length Å			10^{-10}	m
Frequency				
rot. rate cps			1	Hz
cps per µamp per cm^2			(10^{-10})	Hz A^{-1} m^{-2}
Current				
	µ amp		(10^{-6})	A
Area				
per area		cm^{-2}	10^{-4}	m^{-2}
Volume				
per volume		cm^{-3}	10^{-6}	m^{-3}
Energy				
	eV		$1.6(10^{-19})$	joule
	kilocal mole^{-1}		$4.19(10^{+3})$	joules mole^{-1}
Temperature				
	°C		add 273	Kelvin (K)

ELECTROMAGNETIC INTERENCE (p. 161)

unit of	Brit	cgs	Mult. by	To get mksa(SI)
Resistance		emu	10^{-9}	ohm
Potential	μV		(10^{-6})	Volt (V)
		emu ab volt	(10^{-8})	V
Current	db(mA)		$(10^{-3+db/20})$	ampere (A)
	db(μA)		$(10^{-6+db/20})$	A
		emu ab amp	10	A
Elec. Field		emu	(10^{-6})	V m^{-1}
	μV m^{-1}		(10^{-6})	V m^{-1}
	db(μV m^{-1})		$(10^{-6+db/20})$	V m^{-1}
spectral, db(μV/m) per MHz			$(10^{-12+db/20})$	V m^{-1} Hz^{-1}
Magn.Field				
magn. intens. (B)		Gauss	(10^{-4})	Tesla (T)
magn. fld str. (H)		Oersted	$(4\pi)^{-1} \times 10^3$	ampere turn m^{-1}(A m^{-1})
inductance		emu	(10^{-9})	Henry (H)
		pT	(10^{-12})	T
		db(pT)	$(10^{-12+db/20})$	T
Frequency				
rep. rate kpps			(10^{+3})	s^{-1}

Appendix F

INTERFACE CONTROL DOCUMENT (JSC) ICD 2-19001

Resistance

	milliohm	(10^{-3})	ohm (Ω)
	megohm	(10^{+6})	ohm
res. per length	ohm cm^{-1}	(10^{+2})	ohm m^{-1}
resistivity	ohm cm	(10^{-2})	ohm m

Potential

	mV		(10^{-3})	Volts (V)

Current

	kA	(10^{+3})	Amperes (A)
curr. density	A mm^{-2}	(10^{+6})	A m^{-2}
db(mA)		$10^{-3+db/20}$	A
db(μA)		$10^{-6+db/20}$	A

Elec. Field

	emu	(10^{-6})	V m^{-1}
μV m^{-1}		(10^{-6})	V m^{-1}
db(μV m^{-1})		$(10^{-6+db/20})$	V m^{-1}
spectral, db(μV/m) per MHz		$(10^{-12+db/20})$	V m^{-1} Hz^{-1}

Magn. Field

magn. intense. (B)	Gauss	(10^{-4})	Tesla (T)
magn. fld. str. (H)	Oersted	$(4\pi)^{-1} \times 10^3$	,ampere turn m^{-1} (A m^{-1})
inductance	emu	(10^{-9})	Henry (H)
	pT	(10^{-12})	T
	db(pT)	$(10^{-12+db/20})$	T

Frequency

| rep. rate kpps | | (10^{+3}) | s^{-1} |

APPENDIX G
THE SHUTTLE ENVIRONMENT DATA BASE (ENVIRONET)

ENVIRONET is an on-line information system that allows users of the Space Shuttle to access environmental information needed for the development and successful operation of experiments on the Shuttle. Shuttle environment information is of vital interest to investigators developing payloads or instruments for flight on the Shuttle or one of its carriers (i.e., Spacelab, PAM-D, IUS, etc.). ENVIRONET will contain information describing the environment with respect to the ways it may affect an experiment or experimental equipment. The data will be organized in the following areas:

- Loads and Low Frequency Dynamics
- Vibration and Acoustics
- Electromagnetic Interference
- Thermal and Humidity Environment
- Orbiter Motion (microgravity and pointing)
- Surface Interactions (including glow)
- Molecular Contamination
- Particulate Environment
- Microbial and Toxic Contaminants
- Natural Environment

Environmental measurements taken on previous Shuttle flights will be used to update the STS environment information. The data base will contain bibliographic, numeric, and full text information.

Information for the data base is collected and maintained by three technical panels. One panel represents the community of experimenters, one concentrates on the induced and natural environment in and around the Shuttle, and the third manages the information in the data base. The Experimenters Panel has the responsibility for identifying user requirements and specifying needed environmental data. The Natural and Induced Environment Panel is involved in collecting and organizing appropriate data, making a preliminary assessment of the reliability and traceability of the data, and indicating areas of use to experimenters. The Information Management Panel is researching existing published information and creating a system for information to be compiled, stored, and catalogued.

Access procedures for ENVIRONET were released when the data base was made public in 1985. Once logged onto the system, the user is introduced to extensive "menus" guiding user to the needed information. Further information on ENVIRONET may be obtained from Violeta Vera at NASA Goddard Space Flight Center, (301) 344-7596.

APPENDIX H
LIST OF ATTENDEES

Vincent J. Abreu
U. Mich. Space Phys. Research Lab
2455 Hayward
Ann Arbor, MI 48105
(313) 763-6217

Dr. Marcel Ackerman
Belgium
Aeronomy Institute
Avenue Circulaire 3
B-1180 Brussels, Belgium
2-374-2728   TELEX 21563

Norman Ackerman
Code 732.0
Goddard Space Flight Center
Greenbelt, MD 20771
(301) 344-5115

Wayne Bailey
Teledyne Brown Engineering
300 Sparkman Drive
Huntsville, AL 35807

Gordon Bakken
Wyle Labs
System Engineering
7800 Governors Drive, W.
Huntsville, AL 35807

Dr. Peter Banks
Dept. of Electrical Engineering
Stanford University
Stanford, CA 91109

Mr. Lyle Bareiss
MS-M0487
Martin Marietta Aerospace
Box 179
Denver, CO 80201
(303) 977-8713

Dr. Jack Barengoltz
MS 157/507
Jet Propulsion Laboratory
4800 Oak Grove Drive
Pasadena, CA 91109
(818) 354-2516

Don Bartleson
Lockheed Space Operations
Code LSO-246
Kennedy Space Center, FL 32899

Joseph G. Bastow
MS 144-218
4800 Oak Grove Drive
Pasadena, CA 91109
(818) 354-2500

Ben Bier
Martin Marietta Aerospace
185 S. Douglas St.
El Segundo, CA 90245
(213) 640-0301

Robert Brown
MS 150-100
Jet Propulsion Laboratory
4800 Oak Grove Drive
Pasadena, CA 91109
(818) 354-2611

Mr. Robert Blount
Code WA3
NASA/Johnson Space Center
Houston, TX 77058
(713) 483-2708

Dr. Georgette Burgess
Room KB-229
Massachusetts Inst. of Technology
PO Box 73
Lexington, MA 02173-0073
(617) 863-5500 ext 2870

William J. Boone, III
Code G 6010
Martin Marietta Aerospace
PO Box 179
Denver, CO 80201
(303) 977-1799

Mr. George Caledonia
Physical Sciences, Inc.
Research Park
PO Box 3100
Andover, MA 01810
(617) 475-9030

E. N. Borson
M2-250
The Aerospace Corporation
PO Box 92957
Los Angeles, CA 90009
(213) 648-6943

Dr. William R. Case
Code 725
NASA/Goddard Space Flight Center
Greenbelt, MD 20771
(301)344-8936

Stephen Brodeur
Code 731.1
NASA/Goddard Space Flight Center
Greenbelt, MD 20771
(301) 344-0252

John Cavanaugh
Code 615
NASA/Goddard Space Flight Center
Greenbelt, MD 20771

Appendix H   435

Dr. Robert Chapman
Code PD311
NASA/Johnson Space Center
Houston, TX 77058
(713) 483-4095

Harold A. Comerer, Director
345 East 47th Street
New York, NY 10017
(212) 705-7835

Roger P. Chassay
Mail Code JA62
NASA/Marshall Space Flight Center
Marshall Space Flight Center, AL
(205) 453-1870

Marion Coody
Code MP3
NASA/Johnson Space Center
Houston, TX 77058
(713) 483-5381

Jim Clawson
Rockwell International
1840 NASA Road 1
Houston, TX 77058
(713) 333-0720

Burton G. Cour-Palais
Code SN3
NASA/Johnson Space Center
Houston, TX 77058
(713) 483-5171

Thomas D. Clem
Code 674
NASA/Goddard Space Flight Center
Greenbelt, MD 20771
(301) 344-6319

Michael J. Coyle
Code 732.3
NASA/Goddard Space Flight Center
Greenbelt, MD 20771
(301) 344-5792

Michael A. Comberiate
Code 407.0
NASA/Goddard Space Flight Center
Greenbelt, MD 20771
(301) 344-9074

William D. Cutler
MS-568
Aerospace Corporation
PO Box 92957
Los Angeles, CA 90009
(213) 647-1558

Bonnie P. Dalton
Code LP 240A-3
NASA/Ames Research Center
Moffet Field, CA 94035

Professor Gordon P. Fisher
Cornell University
309 Hollister Hall
Ithaca, NY 14853

Kevin C. Daly
CS Draper Lab
555 Technology Square
Cambridge, MA 02139
(617) 258-2573

Michael C. Fong
Dept. 62-19, Bldg. 104
Lockheed Missiles and Space Co.
1111 Lockheed Way
Sunnyvale, CA 94086
(408) 742-6060

Mr. Maurice Dubin
Code 616
NASA/Goddard Space Flight Center
Greenbelt, MD 20771
(301) 344-5475

Dwight Fortna
Code 302.0
NASA/Goddard Space Flight Center
Greenbelt, MD 20771
(301) 344-5298

Giovanni Fazio
Harvard Smithsonian Center
Astrophysics
60 Garden Street
Cambridge, MA 02138
(617) 495-7458

James H. Fu
MS 144-218
Jet Propulsion Laboratory
4800 Oak Grove Drive
Pasadena, CA 91109
(818) 354-7138

Paul Feldman
Dept. of Physics and Astronomy
Johns Hopkins University
Homewood Campus
Baltimore, MD 21218
(301) 238-7339

Dr. James L. Gallagher
Dept. of Physics
University of Alabama, Huntsville
Science Building
Huntsville, AL 35899
(205) 453-0505

John Garba
MS 157-316
Jet Propulsion Laboratory
4800 Oak Grove Drive
Pasadena, CA 91109
(818) 354-2085

Dr. Henry Berry Garrett
MS 144-218
Jet Propulsion Laboratory
4800 Oak Grove Drive
Pasadena, CA 91109
(818) 354-2644

Mr. Isaac Gillam
Code M, Office of Space Flight
NASA Headquarters
Washington, DC 20546

Robert E. Gold
Applied Physics Lab
Johns Hopkins University
Laurel, MD 20707
(301) 953-5412

Gerald Golub
Mail Code PRC-2621
Planning Research Corp.
Kennedy Space Center, FL 32899
(305) 867-2158

Byron David Green
Physical Sciences, Inc.
Research Park
PO Box 3100
Andover, MA 01810
(617) 475-9030

Dr. John Gregory
University of Alabama
Huntsville, AL 35899
U. AL (205) 895-6028

Dr. Alan J. Grobecker
25542 Orchard Rim Lane
El Toro, CA 92630
(213) 825-7407

Richard C. Hahn
Rensselaer Polytechnic Institute
Materials Research Center
Troy, NY 12181
(518) 266-6012

David F. Hall
M2/271
The Aerospace Corporation
PO Box 92957
Los Angeles, CA 90009
(213) 648-5896

David Hamilton
NASA/Johnson Space Center
Code E542
Houston, TX 77058
(713) 483-4391

Raymond K. Hinkle
Code 731.1
NASA/Goddard Space Flight Center
Greenbelt, MD 20771
(301) 344-0255

Dr. Alan Hedin
Code 614
NASA/Goddard Space Flight Center
Greenbelt, MD 20771
(301) 344-8393

James E. Kahelin
Code D/786, AD35
Rockwell International
12214 Lakewood Blvd.
Downy, CA 90241
(213) 922-5722

Michael Heinemann
AFGL/PHK
Hanscom Air Force Base
Bedford, MA 01731
(617) 861-3176

Dr. Ira Katz
S-Cubed
PO Box 1620
La Jolla, CA 92038
(619) 453-0060

William Henricks
Lockheed Missiles and Space Co.
1075 Highlands Circle
Los Altos, CA 94022

Dennis Kern
MS 144-218
Jet Propulsion Lab
4800 Oak Grove Drive
Pasadena, CA 91109
(818) 354-3158

Appendix H

Dr. Karl Knott
Postbus 299
2200 AG Noordwijk Zh
The Netherlands
Tel 01719-86555
Cables: Spaceurop, Noordwijk

Mr. Raymond Kruger
Code 302.0
NASA/Goddard Space Flight Center
Greenbelt, MD 20771
(301) 344-7668

Dr. Michael Lauriente
Code 420
NASA/Goddard Space Flight Center
Greenbelt, MD 20771
(301) 344-7497

Dr. Y. Albert Lee
Lockheed Missiles and Space Co.
Bldg. 104, Org. 62-18
1111 Lockheed Way
Sunnyvale, CA 94086
(408) 742-5982

Lubert J. Leger, PhD.
Code ES53
NASA/Johnson Space Center
Houston, TX 77058
(713) 483-2059

Mr. Jules Lehmann
Code EM
NASA Headquarters
Washington, DC 20546
(202) 453-1562

Dietrich Lemke
Max Planck Institut fur Astronomie
Konigstohl
D-6900 Heidelberg 1
WEST GERMANY

Byron Lichtenberg
48 Leighton Road
Payload Systems, Inc.
Wellesley, MA 02181

Professor John Lockwood
UNH Space Science Center
DeMeritt Hall
Durham, NH 03824

Melvin H. Lucy
MS 433
NASA/Langley Research Center
Hampton, VA 23665
(804) 865-4624

Carl Maag
MS 157/507
Jet Propulsion Laboratory
4800 Oak Grove Drive
Pasadena, CA 91109

Dan McKeown
Faraday Laboratories
PO Box 2308
La Jolla, CA 92038
(619) 459-2412

Dr. Yves Malmejac
European Space Agency
8-10 Rue Mario Nikis
75738 Paris CEDEX 15
FRANCE
33-1-273-7311 (Dr. Shapland)

Professor Peter McNulty
Physics Department
Clarkson University
Potsdam, NY 13676
(315) 268-2344

Mr. Edward Marian
MS 144-218
Jet Propulsion Laboratory
4800 Oak Grove Drive
Pasadena, CA 91109

Dr. Leonard Melfi
MS 366
NASA/Langley Research Center
Hampton, VA 23655
(804) 865-3718

Dr. David L. Matthews
Inst. for Physical Sci. and Tech.
University of Maryland
College Park, MD 20742
(301) 454-3966

Steve Mende
Lockheed Palo Alto
3251 Hanover Street
Palo Alto, CA 94304
(415) 858-4082

Dr. Billy M. McCormac
Lockheed R DD
Code D91-30/B202
3251 Hanover
Palo Alto, CA 94304
(415) 424-2816

Dr. Ed Mercanti
Code 402.0
NASA/Goddard Space Flight Center
Greenbelt, MD 20771
(301) 344-8108

Appendix H

Mr. Edgar Miller
Code ES61
NASA/Marshall Space Flight Center
Marshall Space Flight Center, AL
(205) 453-5130
(205) 881-4086

C. R. Mullen
MS 8C-05
22210 SE 38th St.
Issaquah, WA 98027
(206) 778-8907

Dr. Tripti Mookherji
Teledyne Brown Engineering
300 Sparkman Dr.
Huntsville, AL 35807

Edmond Murad
AFGL/PHK
Hanscom Air Force Base
Bedford, MA 01731
(617) 861-3046

R. Gilbert Moore
Morton Thickol, Inc.
PO Box 524
Brigham City, UT 84302

Duane Nelson
The Aerospace Corporation
PO Box 92957
Los Angeles, CA 90009

Emily Morey
Code LR
NASA/Ames Research Center
Moffet Field, CA 94035
(415) 694-5471

Dr. Werner Neupert
Code 680.0
NASA/Goddard Space Flight Center
Greenbelt, MD 20771
(301) 344-5523

Jerry L. Moyer
The Bionetics Corporation
Mail Code B10-1
Kennedy Space Center, FL 32899

Dr. Thong Van Nguyen
MS 8C-05
Boeing Aerospace
PO Box 3999
Seattle, WA 98124
(206) 773-8907

Frank On
Code 731.1
NASA/Goddard Space Flight Center
Greenbelt, MD 20771
(301) 344-5398

Richard Payton
Code M0487
Martin Marietta, Inc.
Box 179
Denver, CO 80201
(303) 977-8672

Richard A. Osiecki
Lockheed Palo Alto Research Lab
O/9240 B/205
3251 Hanover Street
Palo Alto, CA 94304
(415) 424-2389

Dennis Peterson
SP-SEO
NASA/Kennedy Space Center
Kennedy Space Center, FL 32899
(305) 867-4925

Lt. Scott E. Palmer
8 Alvin Circle
Salinas, CA 93906

Dr. Duane Pierson
Code SD4
NASA/Johnson Space Center
Houston, TX 77058

Dr. John J. Park
Code 313.2
NASA/Goddard Space Flight Center
Greenbelt, MD 20771
(301) 344-6368

James R. Pignataro
US Army Ballistic Missile Defense
Advanced Technology Center
PO Box 1500
Huntsville, AL 35807-3801
(205) 895-4634

Joseph G. Parravano
University of Toronto
Institute for Aerospace Studies
4925 Dufferin St.
Downsview, Ontario M3H5T6
(416) 667-7701

Charles P. Pike
AFGL/PHK
Hanscom Air Force Base
Bedford, MA 01731
(617) 861-3177

Professor Robert Pond
Dept. of Materials Sci. Engineer
Johns Hopkins University
34th and Charles Sts.
Baltimore, MD 21218

Nancy Pugel
Code 732.2
NASA/Goddard Space Flight Center
Greenbelt, MD 20771
(301) 344-7295

Ken H. Pratt
Rockwell International
Space Transporation Systems
12241 Lakewood Blvd.
Downey, CA 90241

J. Rudy Puleo
The Bionetics Corporation
Mail Code B10-1
Kennedy Space Center, Fl 32899

Mr. George Proca
ESRIN
Via Galileo Galilei
Casella Postale 64
00044 Frascati, ITALY
39-69-4011

Dr. James Ragusa
Code CS-SED
NASA/John F. Kennedy Space Center
Kennedy Space Center, FL 32899
(305) 867-4670

Dr. Nathan Promisel
12519 Davan Drive
Silver Spring, MD 20904
(301) 622-3426

Professor W. John Raitt
Utah State University
Atmospheric Research Center
UMC-34
Logan, UT 84322
(801) 750-2983

Mr. Ian Pryke
European Space Agency
955 L'Enfant Plaza, SW
Suite 1404
Washington, DC 20024
(202) 488-4158

Dr. Edmond Reeves
Code EM
NASA Headquarters
Washington, DC 20546
(202) 453-1562

Arthur Reubens
Code LN
NASA/Johnson Space Cente
Houston, TX 77058
(713) 483-3441

Winfield G. Reynolds
MS/571
The Aerospace Corporation
PO Box 92957
Los Angeles, CA 90009
(213) 647-1516

Mr. Russel Rhodes
Code SP-SEO
NASA/Kennedy Space Center
Kennedy Space Center, FL 32899
(305) 867-7416

Jane T. Riddle
Code 252.0
NASA/Goddard Space Flight Center
Greenbelt, MD 20771
(301) 344-6152

George Roach
Code 402.0
NASA/Goddard Space Flight Center
Greenbelt, MD 20771
(301) 344-0223

Dr. Paul A. Robinson, Jr.
MS 144-218
Jet Propulsion Laboratory
4800 Oak Grove Drive
Pasadena, CA 91109
(818) 354-3882

Donna Romano
345 East 47th Street
New York, NY 10017
(212) 705-7835
TELEX 126022

Christopher Rosander
Code A3-204/13-13
McDonnell Douglas Astronautics Co.
5301 Bolsa Avenue
Huntington Beach, CA 92647
(714) 896-4945

Dr. Allen Rubin
AFGL/PHK
Hanscom Air Force Base
Bedford, MA 01731
(607) 861-3240

Nicolas A. Saflekos
South West Research Institute
PO Drawer 285
6220 Culebra Rd.
San Antonio, TX 78284
(512) 684-5111 ext. 2205

Appendix H   445

Mr. Michael Sander
Code EM
NASA Headquarters
Washington, DC 20546
(202) 453-1560

Dr. Gerald W. Sharp
Eyring Research Institute, Inc.
1455 West 820 North
Provo, UT 84601
(801) 375-2434

James A. Sanders
Martin Marietta Corporation
MSG 6496
PO Box 179
Denver, CO 80201

Dr. Stanley Shawhan
Code EE
NASA Headquarters
Washington, DC 20546
(202) 453-1560

Dr. Erwin R. Schmerling
Code 930.0
NASA/Goddard Space Flight Center
Greenbelt, MD 20771
(301) 344-6989

James W. Smith
MS 144-218
Jet Propulsion Laboratory
4800 Oak Grove Drive
Pasadena, CA 91109
(818) 354-2799

Dr. John J. Scialdone
Code 725.1
NASA/Goddard Space Flight Center
Greenbelt, MD 20771
(301) 344-8547

Kevin A. Smith
Mail Code 01/2210
Bldg 1, Rm 2210
One Space Park
Redondo Beach, CA 90278
(213) 535-2706

Dr. Dai Shapland
8-10 Rue Mario Nikis
75738 Paris CEDEX 15
FRANCE
33-1-273-7311
33-1-273-7654

1st Lt. Michael R. Smith
6595 STG/SH (USAF)
Vandenberg Air Force Base, CA
93437
(805) 866-2367

Dr. Epaminonda Stassinopoulos
Code 633
NASA/Goddard Space Flight Center
Greenbelt, MD 20771
(301) 344-8067

Mr. Robert F. Theis
Code 614
NASA/Goddard Space Flight Center
Greenbelt, MD 20771
(301) 344-7596

Dr. Robert Stencel
Code EZ
NASA Headquarters
Washington, DC 20546
(202) 453-1560

Dr. Alan Thirkettle
ESA-ERNO
Spacelab European Resident Team
Kennedy Space Center, FL 32899

Dr. John Stevens
Hughes Aircraft
Space and Communications Group
P.O. Box 92919
Los Angeles, CA 90009
(213) 647-4975

David A. Tipton
MD-MED-B
NASA/Kennedy Space Center
Kennedy Space Center, FL 32899
(305) 867-3165

Dr. William Swider
AFGL/PHK
Hanscom Air Force Base
Bedford, MA 01731
(617) 861-3891

Dr. Marsha R. Torr
Center for Atmospheric Space Scien
Utah State University
UMC-41
Logon, UT 84322
(801) 750-2779

E. Ray Tanner
Code NA01
NASA/Marshall Space Flight Center
Marshall Space Flight Center, AL
(205) 453-3151

Jack Triolo
Code 732.2
NASA/Goddard Space Flight Center
Greenbelt, MD 20771
(301) 344-8651

Philip Tulkoff
Code 732.3
NASA/Goddard Space Flight Center
Greenbelt, MD 20771
(301) 344-5235

Violeta Vera
Code 614
NASA/Goddard Space Flight Center
Greenbelt, MD 20771
(301) 344-7596

O. Manuel Uy
Space Reliability Group
Applied Physics Laboratory
Johns Hopkins Road
Laurel, MD 20707
(301) 953-4029

Robert G. Wagner
M5/568
The Aerospace Corporation
PO Box 92957
Los Angeles, CA 90009
(213) 647-1560

Dr. A. L. Vampola
M2/259
The Aerospace Corporation
PO Box 92957
Los Angeles, CA 90009
(213) 648-6078

Donald A. Wallace
Telonic Berkeley
PO Box 277
Laguna Beach, CA 92652

Richard J. Vandervoort
Honeywell, Inc.
13350 US Highway 19 S
Clearwater, Fl 33546

Dr. Stephan Walther
ERNO Raumfahrttechnik Gmbtt
Postfach 105909
D-2800
BREMEN1, WEST GERMANY
(421) 539-4138

Dr. William Vaughan
Code ED41
NASA/Marshall Space Flight Center
Marshall Space Flight Center, AL
(205) 453-3100

Jerry L. Weinberg
University of Florida
Space Astronomy Lab.
1810 NW 6th St.
Gainesville, FL 32609

Thomas Wilkerson
Inst. of Phys. Science and Tech.
University of Maryland
College Park, MD 20742
(301) 454-5401

Arthur Winckel
Mail Code SP-AF-2
NASA/Kennedy Space Center
Kennedy Space Center, FL 32899

John M. Winter, Jr.
Dept. of Materials Sci Engineering
Johns Hopkins University
34th and Charles Sts.
Baltimore, MD 21218
(301) 338-7152

John O. Wise
AFGL/PHK
Hanscom Air Force Base
Bedford, MA 01731
(617) 861-3202

Fred C. Witteborn
SSA 245-6
NASA/Ames Research Center
Moffett Field, CA 94035
(415) 965-5520

Romeo Wlochowicz
Canada Centre for Space Science
National Research Council Canada
100 Sussex Drive
Ottawa, CANADA KIA OR6
(613) 992-0872

Donald M. Wong
M5/568
The Aerospace Corporation
PO Box 92957
Los Angeles, CA 90009
(213) 647-1558

Professor Laurence Young
Building 37/219
Massachusetts Inst. of Technology
Cambridge, MA 02139
(617) 253-7759

JPL
Jet Propulsion Laboratory
4800 Oak Grove Drive
Pasadena, VA 91109
FTS 8-792-5359
(818) 354-5359

KSC
John F. Kennedy Space Center
NASA
Kennedy Space Center, Fl 32899
FTS 8-823-7110
(305) 867-7110

MSFC
Marshall Space Flight Center   NASA
Marshall Space Flight Center, AL
FTS 8-872-2121
(205) 453-2121

JSC
Lyndon B. Johnson Space Center
NASA
Houston, TX   77058
FTS 8-525-0123
(713) 483-0123

Marshall Communications Center
(Central Time)
FTS 8-872-4100
FTS 8-872-2145

APPENDIX I
INDEX

American Institute of Aeronautics and Astronautics 139
ARIANE 41
AT&T 45
Abbott, Charles G. 60
accelerometer measurements 73, 219
    low frequency 74
    high frequency 73
Acoustics
    liftoff 215, 217, 218
    transonic 215, 217, 218
active cavity radiometer 56
Active Mesospheric Particle Tracer Experiment 40
airborne bacteria 237
airborne fungi 238
anti-motion sickness drugs 108
Application disciplines 10
Astro Mission 59, 62
    Hopkins Ultraviolet Telescope 62
    Ultraviolet Imaging Telescope 62
    Wisconsin Ultraviolet Photometer Polarimeter Experiment 62
astronomy 65, 106
Astrotech International 46
Atlas 39, 40
Atlas Centaur 39, 40, 42, 46
Atlas Centaur class payloads 15
atmospheric contaminants, manned spacecraft 230, 232-234
atmospheric physics 99, 142
    Absorption Spectrometer 101
    Imaging Spectral Observatory 101, 102, 103
    Imaging Emission Spectrometer 101

back-up equipment 66
Beggs, James A. 28
Broad-Band X-Ray Telescope 61
Centaur 42, 43, 45, 221
charging, internal 149-153
communications technology, advanced 66
Congress 26
Contamination Environment and Effects Handbook 126-136
contamination, payload-to-payload 70
Cosmic Ray Nuclei Experiment 61
cosmic rays 61, 143
customer-friendly environment 30
Cypress Corporation 46
data-base, Shuttle environment 3, 129, 162, 180-184
degradation
       optical 206
       thermal blanket 211
Delta Vehicles 39, 47
Department of Defense 3, 15, 26, 39
Developmental Flight Instrumentation 17, 177, 190
Diffuse X-ray Spectrometer 61
discharge 104
discipline laboratory 57, 59
disturbances, local 155
Dynamic, Acoustic and Thermal Environment Program 177
dynamic load data 177
earth observations 65, 109
Earth Observations Mission 56, 57
Earthnet Program Office 109
electron density near Shuttle 104
electron gun 104
electromagnetic field 155, 156
electromagnetic interference 161-164
electromagnetic levels, minimum 162
electromagnetic measurements 71

environment
    acoustic   216, 217
    atmosphere   141
    auroral   200, 202-203, 204
    cabin   190
    contamination   69, 70
    dynamic   165
    electromagetic   71, 84, 161-164
    electron   198-199, 200
    low-gravity   61, 166, 167-170
    magnetic field   207, 208
    microgravity   109
    micrometeor   143
    natural   143, 152
    particulate   171-174
    payload   79-80
    plasma   204
    pre-launch   173
    radiation, and effects   145-148
    random vibration   81
    thermal and humidity   189-195
    vibroacoustic   215-222
environmental factors   155
environmental requirements   156
Environmental Observation Mission   56, 57, 59
European Space Research Organization (ESRO)   5
European Space Agency   3, 5, 10, 11, 65
    Databases   111-123
    Statistical Software Package   121
European Launcher Development Organization (ELDO)   5
Expendable Launch Vehicles   39
Extravehicular Activity   52
Far Ultraviolet Space Telescope   106
Filament Wound Case   20, 22
Fleetsat Com (FSC)   40
Galileo   15, 17, 20, 42, 148, 150, 193, 207

gaseous background 223-226
General Dynamics 46
Geostationary Operational Environment Satellites (GOES) 40
Geostationary weather satellites 47
German Infrared Laboratory (GIRL) 219, 222
Get Away Special (GAS) 30, 33, 47
Glow 69, 101, 130, 155, 197, 207, 210
Henniker workshop 69
Hitchhiker 33, 37, 47
Hughes Aircraft 45
induced currents 197, 206, 208
Induced Environment Contamination Monitor 69, 172, 223, 224
Inertial Upper Stage 7, 15, 17, 43, 45, 217
Instrument Pointing System 6
integrated circuit response 151
interactive decision making 67
Interim Operational Contamination Monitor 173, 225
Interim Upper Stage (see Inertial Upper Stage)
International Solar Polar Mission (see Ulysses)
International Microgravity Laboratory 10, 56, 57, 59
Internal/External Contamination Monitor (IECM) 89
John Deere 47
Kennedy Space Center (KSC) 42, 49
Large Area Modular Array of Reflectors (LAMAR) 61
Large Format Camera 63
Large-volume payload effects 75
Leasecraft 48
life sciences 56, 65, 108
loads and low frequency dynamics 73, 75, 128, 175-186, 193
    data acquisition 184, 185
    data analysis 185
    data utilization 178
    modelling 181
local electric fields 104
Long Duration Exposure Facility (LDEF) 211
low-gravity research 56, 165, 167

Appendix I

Lyndon B. Johnson Space Center (JSC) 49
magnetosphere 198
Main Engine Program 23
man-tended experiments 8, 9
Manned maneuvering unit 17
material, carbon-phenolic 45
material, carbon-carbon 45
Material Science Laboratory 56, 57, 59
material science 56, 61, 65
McDonnell-Douglas 44, 45
Measurement of Air Pollution from Satellites (MAPS) 60
Metric Camera 108
microbial and toxic contaminants 227-246
Microgravity Research Incorporated 47
micrometeoroid 85, 141
microorganisms
    from STS crews 239, 240
    from spacecraft surfaces 235-236
microphones 73, 219
microwave remote sensing experiment 108, 109
mobile launch platform 19
Module 8, 9
molecular contamination 226
NASA-RECON 120
National Oceanic and Atmospheric Administration (NOAA) 40
non-human crew microbiology 238-241
non-volatile residue 219
orbiter motion 161-164
Optical Effects Module 220
OSS-1 55
OSTA-1 60
OSTA-1 55
OSTA-2 55
Office of Space Flight 13-54
    Customer Services Division 28, 29
Operational Flight Measurements 91

Orbital Transfer Vehicle 52
Orbital Maneuvering Vehicle 52
Orbital Sciences Corporation 45
Orbiter refueling demonstration 52
Pallet 8,9
Pallet acoustics 95-97
particle contamination 125, 126
particles, charged 143-153
Payload Specialists 65-67
Payloads, attached 3
       detached 3
Payload Assist Modules 7, 15, 44, 221
Payload bay vent noise 75
Payload of Opportunity Carrier 59, 60
Plasma Diagnostics Package 161
plasma physics 104
President, The 46
pressure in cargo bay 83
pricing policy 32
QUESTINDEX 120
radiation belts 143
radiation effects 146
      logic upset, single event 146, 148-150
      internal charging 146, 151-153
radiation, energy spectrum of 144
Radioactive Thermal Generators 145
refueling, in-orbit 3, 10
Remote Manipulator System 17, 71
requirements, design 86-87
Research Animal Holding Facility 243-246
response amplitude 73
retrieval 111
SATELDATA-2 Databank 117-118
SPACECOMPS-22 Databank 112-115
STS Schedule 42
STS environmental measurements 69

STS-10 45
Satellite Business Systems 45
Science Magazine 99
scientific disciplines 10
Scout 39
Shuttle 11, 28, 50
Shuttle Atlantis 20, 25
Shuttle Avionics integration Laboratory (SAIL) 71, 161
Shuttle Challenger 20, 25
Shuttle Columbia 17, 25
Shuttle Discovery 20, 25
Shuttle electrical potential 105-106
Shuttle Environment Impact on science 99-110
Shuttle Experimnent and Environment Workshop 11
Shuttle External Tank 7, 20, 24
Shuttle Flights, KSC 31
Shuttle flights, Vandenburg AFB 31
Shuttle High-Energy Astrophysics Laboratory 59
Shuttle Infrared Telescope Facility 57
Shuttle Main Engine 19
Shuttle Orbiter 7, 11, 17, 21
Shuttle Orbiter/Cargo Standard Interface Document 157
Shuttle Payload Bay Cleanliness Ground Testprogram 173, 223
Shuttle Program 28
Shuttle Radar Laboratory 56, 57, 60
Shuttle Upper Stages 41
Shuttle/Spacelab Contamination Environment
   and Effects Handbook 127, 128
        benefits 129,
        data acquisition/retrieval system 131, 133, 134
        development 130, 136
        objectives 128
single event upset 146-148
solar flares 63, 143
solar physics 65
Solar Maximum Mission 17, 50, 60

Solar Optical Telescope 59, 62
Solid Rocket Boosters 7, 20, 22, 24
Space Life Sciences Laboratory 59, 61
Space Plasma Laboratory 59, 61
space plasma physics 65
Space Program 3-4
Space Station 11, 37, 38, 49, 52, 56, 63
Space Station Contamination Workshop 56
Space Transportation Corporation 46
Space Transportation Program 28
Space Transportation System 5-12, 13, 14, 47, 155
Space Transportation System subsystems 20
Spacecraft design 51
Spacecraft use 51
Spacelab 4, 7, 10, 13, 14, 27, 37, 38, 66, 69, 155
Spacelab Payload Accomodation Handbook 77-98, 177
Spacelab Payload Engineering Office 63
Spacelab Payload Program 8, 55, 77
Spacelab video system 66
Spacelab-1 55, 63, 65
Spacelab-1 accelerometers 89, 93
Spacelab-1 contamination and radiation 89
Spacelab-1 Fluid Physics Module 108
Spacelab-1 Results 88, 97, 99
Spacelab-1 Scientific Disciplines 100
Spacelab-1 temperature sensors 89, 92
Spacelab-1 X-ray background 107
Spacelab-2 161
Spacelab-6 (see Space Plasma Program) 61
Student Shuttle Involvement Program 30
Sun 62, 63
Sunlab 63
Supervisory control 67
surface charge 104, 197, 210
surface erosion 197, 207, 210, 211
surface interactions 125, 126, 197-212

Synthetic Aperture Radar (SAR-A)   60, 108, 109
telescience   66
Telstar   45
temperature regulation   67
temperature of cargo bay   82, 83
Tethered Satellite System   48
thermal control, passive   190
thermal environmental measurements   69
Titan   46
toxic effects by chemical compounds   228-229
Tracking Stations   7
Tracking and Data Relay Satellites   6, 7, 43
Transfer Orbit Stage   15, 45, 48
Transpace Carriers Incorporated   47
UV telescope   106
Ulysses   15, 19, 20, 42
user concerns   155
VFI Flight Measurements   90
Vandenburg AFB   19, 49
Venus Radar Mapper (VRM)   43
vibration and acoustics   73, 75, 76, 128
vibrations, structure-borne   215, 217
Vibroacoustic Payload Environment Prediction System   219, 222
wake/ram plasma variations   197, 200, 201, 204
water vapor   224, 225
X-ray sources   61, 106